The Chemical Sensory Informatics of Food: Measurement, Analysis, Integration

ACS SYMPOSIUM SERIES **1191**

The Chemical Sensory Informatics of Food: Measurement, Analysis, Integration

Brian Guthrie, Editor
Cargill, Wayzata, Minnesota

Jonathan Beauchamp, Editor
Fraunhofer IVV, Freising, Germany

Andrea Buettner, Editor
Friedrich-Alexander-Universität Erlangen-Nürnberg, Erlangen & Fraunhofer IVV, Freising, Germany

Barry K. Lavine, Editor
Oklahoma State University, Stillwater, Oklahoma

Sponsored by the
ACS Division of Agricultural Food and Chemistry, Inc.

American Chemical Society, Washington, DC

Distributed in print by Oxford University Press

Library of Congress Cataloging-in-Publication Data

The chemical sensory informatics of food : measurement, analysis, integration / Brian Guthrie, editor, Cargill, Wayzata, Minnesota [and three others] ; sponsored by the ACS Division of Agricultural Food and Chemistry, Inc.
pages cm. -- (ACS symposium series ; 1191)
Includes bibliographical references and index.
ISBN 978-0-8412-3069-9 (alk. paper) -- ISBN 978-0-8412-3070-5 (alk. paper) 1. Food--Sensory evaluation. 2. Chemical senses. I. Guthrie, Brian (Chemist), editor. II. American Chemical Society. Division of Agricultural and Food Chemistry.
TX546.C43 2015
664'.072--dc23

2015020315

The paper used in this publication meets the minimum requirements of American National Standard for Information Sciences—Permanence of Paper for Printed Library Materials, ANSI Z39.48n1984.

Distributed in print by Oxford University Press

Foreword

The ACS Symposium Series was first published in 1974 to provide a mechanism for publishing symposia quickly in book form. The purpose of the series is to publish timely, comprehensive books developed from the ACS sponsored symposia based on current scientific research. Occasionally, books are developed from symposia sponsored by other organizations when the topic is of keen interest to the chemistry audience.

Before agreeing to publish a book, the proposed table of contents is reviewed for appropriate and comprehensive coverage and for interest to the audience. Some papers may be excluded to better focus the book; others may be added to provide comprehensiveness. When appropriate, overview or introductory chapters are added. Drafts of chapters are peer-reviewed prior to final acceptance or rejection, and manuscripts are prepared in camera-ready format.

As a rule, only original research papers and original review papers are included in the volumes. Verbatim reproductions of previous published papers are not accepted.

ACS Books Department

Contents

Indexes

Preface

The Chemical Sensory Informatics of Food: Measurement, Analysis, Integration

Food chemicals provide a variety of information. They inform us of food safety, quality, authenticity, and origin, with direct links to our emotional responses in many cases. This information is key to our survival, whether to avoid disease or to find nutrients, and our enjoyment. Those involved with food production, processing and testing strive to better understand how food chemistry and oral processing provide information about food. Ideally, chemical analytical instrumentation and sensors could be developed to measure, analyze, and predict the chemical sensory information of food. While many research groups endeavor to develop such systems, recent research confirms that the information obtained by humans during food interaction and eating involve extremely complex interactions between the sensory stimuli and the information processes they invoke. Simple chemical analysis of the content of selected stimulants in food will likely not allow the prediction of the total information content that is desired. There is a longstanding need to better understand the generation of complex human sensations produced by food during eating and how they are integrated and translated into perceptions of food quality and safety.

During food consumption, our sensations are triggered by a range of stimuli. Multiple sensory modalities such as taste, aroma, texture, temperature and others have been shown to produce significant interactions and the emergence of "flavor". Their integration into perceptual concepts, such as "fresh" or "fruity", is accordingly, complex. The effective study of sensory and cognitive integration involves a range of scientific disciplines in order to understand the triggers that drive us to eat, and to select specific foods. Physiological processes such as oral and nasal chemistry and biochemistry play important roles in our chemical senses. Behavioral aspects, for example, with regard to oral processing need to be considered. Sensory and cognitive processing and integration, with the underlying processes of neurobiological processing, have also recently provided an increased understanding of the origins of food sensation and on cognition.

The goal of this book is to compile recent advances, research findings and approaches, and current knowledge across the different aligned areas of research, where experts from chemistry, instrumentation, data analytics and physiology, as well as behavioral and sensory sciences focused on these topics. To this end, a series of symposia were organized around three central themes: emerging methods and learnings concerning multi-sensory and cross-modal sensory measurement and integration, advances in chemical sensors, and data analytics of potential use in revealing the information content contained in the chemical signals. This book is a compilation of selected talks from the following ACS symposia:

- *Advances in the Generation and Integration of Food Sensation and Cognition (245th ACS National Meeting & Expo; New Orleans, LA; April 7-11, 2013):*
    - Chemosensory stimuli in foods (taste, aroma, trigeminal, and others)
    - Textural sensations and their characterization
    - Other sensory modalities such as temperature, acoustics, etc. and their characterization with regard to food consumption
    - Behavioral aspects in relation to food consumption and sensation
    - Physiological, biochemical and chemical aspects of oral and nasal processes
    - Concepts and methods for characterization of multi-modal sensory phenomena
    - Sensory rating and hedonics in relation to food consumption
    - Sensory integration with special focus of multi-modal stimuli
    - Underlying neurobiological principles and higher processing
- *Sensor Applications in Food and Agriculture: Identity, Quality, and Safety (247th ACS National Meeting & Expo; Dallas, TX; March 16-20, 2014)*
    - E-nose, e-tongue developments
    - Biosensors for feed food and agriculture
    - Rapid compositional analysis
    - Rapid grading systems
    - Identity tagging and tracing
    - Sensors and array developments
- *Applied Food Chemometrics, Sensometrics, and Qualimetrics (247th ACS National Meeting & Expo; Dallas, TX; March 16-20, 2014)*
    - Advances in, and applications of, instrumental calibration, calibration transfer, discrimination, classification for grading and the determination of composition or contaminants.
    - Advances in, and applications of, data exploration and predictive model development, both linear and non-linear
    - Advances in, and applications of, multivariate analytics for sensory-to-instrument relationships
    - Process analytical technology (PAT) applications in food, feed, bioprocessing and agriculture.

Advances in the Generation and Integration of Food Sensation and Cognition

Physiological processes such as oral food processing and nasal chemistry, and biochemistry, play important roles in the events that connect our chemical senses at the periphery of our nervous system, with information processes in our central nervous system. Important facets of this fascinating research area relate to methods used to understand these sensory and cognitive events and their integration, and the underlying neurobiological processes. This symposium aimed at facilitating the scientific exchange between these different areas of expertise and providing a platform for the interaction between experts from chemistry, physiology, as well as behavioral and sensory sciences.

The chapters contained in this book from this symposium focus on three key topics. The first area involves current knowledge and new advances in methods and learnings in cross-modal and multi-modal sensory perception. In Chapter 1, and as a general overarching introduction to the area, cross-modal interactions between odor and vision are reported. Chapter 2 contains discussion of the cross-modal interaction between aroma and taste, the two base events that produce the emergence of flavor. Similarly, cross-modal odor-taste interactions for salt and sound-chemosensory congruency are reported in Chapters 3 and 4. Chapter 5 demonstrates the effect of food processing on the generation of taste and mouthfeel stimulants. Enhancement of fruity aroma in wine has reported in Chapter 6

Several chapters focus on methods to better understand the emergence of flavor as the second key topic of the book. Two interesting approaches have shown great utility in unraveling the role of individual stimulants in food flavor. In omission studies, individual stimulants are alternatively removed from recombined stimulants blends and submitted to sensory testing in order to determine their role in the overall product flavor. Chapter 7 describes the use of this method to understand how individual volatiles contribute to the perception of strawberry flavor and makes the case for the utility of this technique for commercial applications. Chapter 8 shows the utilization of this approach to understand the volatile origins of fruity flavors in wine. An analogous approach, called serial dilution sensory analysis, is reported in Chapter 9. This new, conceptually exciting approach was used to determine the major sensory contributors to complex flavors of canned coffee and soy sauce.

The third topic involves the role of human physiology in food perception. Starting with the central nervous system, advances in brain science involving the combination and processing of sensory inputs with other information are poised to make a major contribution to our understanding of how food chemistry is converted into information. In Chapter 10 the role of hunger and craving in relation to reward processes in the brain are discussed. Also, methods to understand the collection and integration of sensory stimuli during mastication have contributed to our understanding of sensory dynamics. Chapter 11 shows the use of the Temporal Dominance of Sensation method that captures how attention can control the strength of sensory signals. Research on the sensory dynamics has placed increased focus on understanding the physiological control of these dynamic signals. Chapter 12 provides an example of the role of oro-nasal physiology in the delivery of stimulants to the periphery of the sensory system. Chapter 13 explores the role of food-use and desensitization in taste sensitivity. Also, genetic differences have long been noted in association with the chemical senses. Chapter 14 reports work to explain the role of salivary proteins in bitterness perception and how it could explain differences that cannot be explained by genetics alone. And finally, chapter 15 reports a structure-taste relationship for a natural sweetener for sweetness and bitterness based on the activation of bitter and sweet receptors.

Sensor Applications in Food and Agriculture: Identity, Quality, and Safety

Rapid measurement, whether for composition, identity, quality or contamination is essential in food and agriculture. Sensor technology is rapidly advancing with significant new developments in sensor arrays, e-nose, e-tongue, nano-enabled sensors, labs-on-chips, biosensors, mass spectrometers and others. These new developments have the potential to change current practices spanning from positive impacts on food supply to changes in regulatory data requirements. This symposium aimed at facilitating the scientific exchange between researchers involved with all aspects of sensors and their applications to the measurement challenges faced in food and agriculture.

Three chapters from this symposium are examples of work in sensor technology, real-time measurement and novel probes for food structure that controls much of the dynamic sensory properties of food. Chapter 16 explores the use of nanotechnology in the construction of a sensor to determine antioxidant capacity. Chapter 17 reports on advancements in real-time measurements of food volatiles. And finally, Chapter 18 describes research on probes used to determine the physical state of food that is a major controlling element in sensory dynamics.

Applied Food Chemometrics, Sensometrics, and Qualimetrics

Food materials, and ingredients, are amongst the most complex materials that exist and range from plant and animal tissues, to complex structured food solids, colloids, liquids, glasses, foams and others. As scientists that study food materials, we strive to build analytical methods to characterize the composition and structure of food. Also, we strive to determine the "higher- information" content of foods that stimulate our senses and our cognitive processes and help us determine food characteristics such as "quality", "freshness", "acceptability" and others. The prediction of the composition, structure and the human response to food from instrumental measurements involves the application of data exploration methods to build links and the calibration of models for the prediction of these properties.

Different research communities exist today that focus on the varied aspects in the development of data analytics methods to convert instrumental and/or human data into information. Generally, chemometricians work with data from chemical analytical instruments. Sensometricians work with data from human tasters and consumers. Qualimetricians work with data in production, control and design. The data analytics from each group are important to the delivery of safe and acceptable food. The aim of this symposium was to bring together subject matter experts from these different fields to share their learnings, approaches and latest advancements.

Chapters in this book provide several examples of work in this area. Three chapters report the use of "flavoromics", where chemometric datasets of peak areas from chromatography are used to predict sensometric datasets from human panelists. This approach was used to find the chemical origins of "freshness" in orange products in Chapter 19 and strawberry juices in Chapter 21. And in Chapter 20 a chemometrics approach is applied to LCMS data to determine the origin and cultivars of Chinese wolfberries.

Chapter 22 reports a new approach called path modelling to follow the evolution of sensory attributes during mastication and to link these to physical phenomenon and instrumental measurements. And finally, Chapter 23 reports the use of cheminformatics approach to explore odor structure relationships of odorant groups, with a focus on musks.

This ACS Symposium Series book covers the current state-of-knowledge and the most recent advancements in endeavors to understand the chemical sensory informatics of food. The thematically diverse chapter contributions authored by subject matter experts from across the globe are a testament to the complex and challenging, yet exciting nature of this area of science. Pushing the boundaries of research in this field paves the way forward in our further understanding of food and flavor, from their chemical signatures and ultimately their perception. This book offers us a glimpse of where we are today.

Brian Guthrie
Cargill, Wayzata, Minnesota

Jonathan Beauchamp
Fraunhofer IVV, Freising, Germany

Andrea Buettner
Friedrich-Alexander-Universität Erlangen-Nürnberg
Erlangen & Fraunhofer IVV, Freising, Germany

Barry K. Lavine
Oklahoma State University, Stillwater, Oklahoma

Chapter 1

Painting Flavor

Terry E. Acree*[,1] and Anne J. Kurtz[2]

[1]Cornell University, Geneva New York 14456
[2]Unilever, Trumbull, Connecticut 06611
***E-mail: terry.acree@cornell.edu.**

The chemistry of food includes substances that activate chemosensory, somatosensory, and visual receptors located at the periphery of the nervous system, e.g. in odor, taste, touch, and light sensitive cells. These cells originate signals that travel to different parts of the brain creating recognizably different sensations. Furthermore, these sensations combine to create judgments of a food's identity, valance (pleasantness), and hedonics (liking) shaping a consumer's expectations and attitudes toward the food (*1*, *2*). Flavor is the sum total of these experiences. Although the rules that govern how the sensory modes are summed to express flavor remain a mystery, there are indications that the rules are complex and profound. For example, several sensory studies have shown that the odor descriptors used for white wines are replaced by those used to describe red wines when subjects taste white wine colored red. This paper will review our present knowledge of cross-modal interactions between odor and vision and describe results from studies of the effects of odor-vision congruency on the detection of pure odorants.

Perception

Our total sensory experience of food combines aroma, taste, texture, temperature, spiciness, appearance, etc. with our memory of past meals. That is, our perceptions are multimodal involving many signals from the periphery modifying our memories. Combining sensations with memories in our brain creates perceptions, for example, the smell of Gorgonzola cheese hidden in a box can evoke the aroma of *pasta quarto fromage*. This would be especially true if you were looking at an image of pasta at the same time. However, if you were looking at an image of a toddler in diapers your interpretation of the smell could be fecal instead of food. The odor coming from the box is ambiguous. Since we cannot see it we experience the cheese smell differently in response to different visual cues. This demonstrates two important features of food perception 1) our perception of flavor, appearance and texture are dynamic - they change continuously as we analyze and reinterpret our experiences while the food melts, disintegrates, dissolves and is swallowed; 2) part of our perception comes from our memory of the past. This process of perceiving takes sensory information and past memories yielding an experience that can change instantly or be stable for a long time. These perceptions combine to create judgments of a food's identity, valance (pleasantness), and hedonics (liking) shaping a consumer's expectations and attitudes toward the food (*1*, *2*).

The mind is not taking snapshots as much as it is building a perception over time. You understand this when you look at a picture that is ambiguous. The image in Figure 1 published 1892 can be experienced as either a picture of a rabbit or a duck.

Figure 1. The rabbit-duck ambiguous figure from Fliegende Blätter 1892.

It is visually ambiguous. It was redrawn by Jastrow (*3*) and used to argue that perception was a product of mental activity initiated by the stimulus (Figure 1).

That you can change the picture from a rabbit to a duck and back again, just by how you look at it, demonstrates the dynamic nature of image perception. Similarly, the cheese example demonstrated the same process when smelling hidden cheese but there the ambiguity was resolved by visual information. The power of sensory perception is that the mind will build its perceptions using any information that makes sense out of the world.

The thoughts associated with our perceptions, called schema, are continually modified by sensory input and the schema modify what sensory inputs we attend to. This dynamic process can cycle almost instantaneously allowing us to come to a "split-second" understanding of something in our environment or it can evolve over time as we analyze our sensations. Perceptions are just our experiences of the schema as it changes. Ulrich Neisser put these ideas into a diagram, Figure 2, which describes perception in terms of a "perceptual cycle" (*4*).

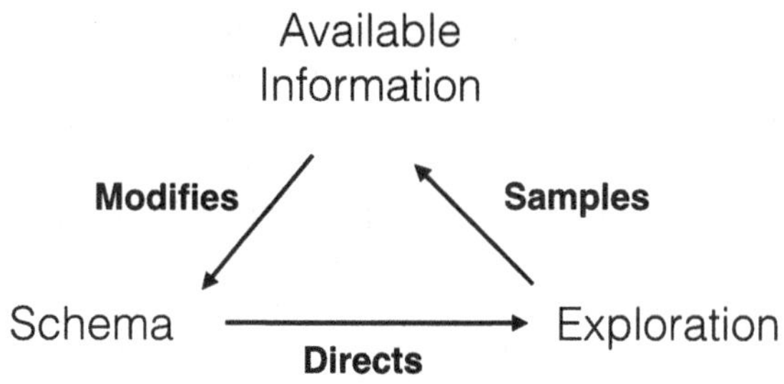

Figure 2. The process of perceptual cycling (4) can explain the multisensory nature of flavor perception as our experience of the schema are instantly modified by available information and directing the search for new information just to modify the schema again.

Neisser's figure implies that perceptions are not still photos of an experience but a moving representation of a continuously streaming input of sensations resulting from active explorations and passive memories. Among the compelling questions that have yet to be answered are: 1) what role does consciousness play in this process? and 2) can we predict the flavor an individual or a group will experience from the composition of the food they eat?

Smell and Vision

Although sight, smell, sound, taste, and touch all contribute to the flavor, consumers experience sight and smell without direct contact with the product. It is usually their first sensation of a product. Indeed, consumers often identify a product just by its odor and appearance and this in turn defines a consumer's attitude (like/dislike) that affects their choices (*1*, *2*, *5–7*). Understanding how sight and smell interact could be valuable for predicting consumer choice as well as flavor perception.

Smell begins at the olfactory epithelium, where odorants bind to olfactory receptors (ORs). Output from the activated ORs propagates through the olfactory bulb then to the piriform cortex where odor quality and odor intensity appear to be delineated in the brain. These modulated signals ultimately reach the orbitofrontal cortex, where olfactory signals first encounter visual signals (*8–11*). The visual signals arrive at the orbitofrontal cortex by an entirely different route, starting at the retina where object features are converted into electrical signals transferred to the lateral geniculate nucleus located in the thalamus, then to the visual cortex by two paths each of which processes different pieces of visual information. Both physiological studies of cells in the orbitofrontal cortex and neuroimaging studies of the orbitofrontal cortex as a whole have shown excitation by these discrete pathways (*9*, *12*, *13*). Knowing where the separate sensory signals combine does not tell us how they interact; however, it is likely that the neural architecture of the olfactory system contributes to the psychophysics of multisensory processing in the same way that visual features are predicted by the anatomy, physiology and perception of the visual system (*14*).

Visual cues have a strong influence on the olfactory experience. For example, studies have demonstrated that color and visual appearance affect our ability to properly identify flavor and influences our overall ratings for liking (*15–17*). Others have shown that visual cues modulate detection (*18–20*), sensory intensity (*21*, *22*), perceived pleasantness (*20*, *23*), preference (*24*), and perceived quality (*8*). For example, it has been observed that adding a color to a clear odorant solution increased the rate of false alarms in an odor detection task (*25*). Demattè (*26*) tested the effects of congruency of visual distractors on odor discrimination and found that presenting a colored figure simultaneously with an odor reduced the accuracy and speed of odor discrimination when the color was incongruent with the odor. However, in the study below, incongruent shape showed no significant effect on odor discrimination.

Color Congruency and Odor Sensitivity

In a study reported by Kurtz (*27*) outlines of fruit were paired with odor stimulant at concentrations above but near their threshold. The odors were "banana" and "cherry" and the outlines were easily identifiable as the corresponding fruit. Whether the shapes were congruent or incongruent had no significant effect on correct responses to odor in a Three Alternetive Forced Choice (3AFC) test. However, the experiments summarized below show that the effect of color on the detection of suprathreshold stimulation was quite different (*27*). Here the color was added to the masking noise and not confined to the shape of the fruit as in Demattè (*26*). Thus, the effect of color could be separated from the effect of shape with greater significance.

Materials and Methods

Odorants

All odorants were diluted in polyethylene glycol (PEG) 400 Lot J33647 (J.T. Baker, Mallinckrodt Baker Inc., Phillipsburg NJ) yielding 2.94 mM benzaldehyde (≥99.5%, Sigma Aldrich, St. Louis MO), 26.91 µM iso-amyl acetate (≥97%, SAFC (Sigma Aldrich), St. Louis MO), 772.6 µM methyl anthranilate (≥99%, SAFC (Sigma Aldrich), St. Louis MO), and 263.5 µM octanal (≥99%, Sigma Aldrich, St. Louis MO) used to determine the congruency of figures and odorants. Benzaldehyde PEG solutions (92 mM to 368 mM) and iso-amyl acetate PEG solutions (1.24 mM to 11.24 mM) were used to determine the individual thresholds for the perithreshold experiments. A suprathreshold level was used to determine benzaldehyde-PEG (736 mM) solutions and iso-amyl acetate-PEG (11.2 mM) solutions were used in the suprathreshold experiments.

Panelists

Eleven female and four male individuals with no reported olfactory or visual impairment whose mean age was 33±11 years, participated in the congruency test. The University Committee of Human Subjects of Cornell University institutional review board (CU-IRB) approved all protocols. All testing took place in an "olfactorium" an isolated room pressurized with deodorized air (*18*, *28*, *29*).

Congruency Test

In order to locate two pairs of congruent figures and odorants, four candidate pairs were tested for congruency using a method adapted from Gilbert et al (*30*). Visual stimuli were presented on a single sheet of paper containing black and white outlines of all four visual stimuli: a single banana, a pair of cherries, a cluster of grapes, and a single lemon located at the four corners. Panelists used five tile pieces (black bottle caps) distributed on the fruit images corresponding to their experience of olfactory stimuli. All five bottle tops could be placed on a single fruit outline or distributed amongst any of the four outlines. Olfactory stimuli consisted of four Teflon squeeze bottles each retrofitted with a Teflon ball for nasal comfort. Each bottle contained a single odorant. Following each trial, the experimenter recorded the placement of the five game pieces. After a 45-second break the subjects evaluated a different odor repeating the same process as before. Participants evaluated each odorant three times. Testing took approximately 15 minutes. The order of odorant presentation was randomized.

Sniff Olfactometer (SO)

A Sniff-Olfactometer was constructed in order to regulate the delivery of olfactory stimuli (300 ms puff delivered in 10 ml air) and combined with the general purpose psychological experimenting system, PsychoPy (*31*, *32*), a scripting language that times the delivery of both the visual and olfactory stimuli and records subject responses. Figure 3 shows a drawing of the Sniff Olfactometer prototype built by DATU, Inc.

The odorant reservoirs consisted of two 250 ml plastic bottles (b) manufactured from PTFE (poly tetrafluroethelene) to reduce scalping from the sample and absorption from the headspace. Solutions of odorants in polyethylene glycol provided a reservoir with a virtually constant headspace concentration over several puffs. Using 10 ml puff volumes it took 4 puffs to reduce the odorant concentration by 10%. Swapping out the bottles for freshly equilibrated ones after 4 puffs guaranteed a reduction in stimulus concentration of less than 10%. The box was made of 3/8" Melamine coated particle board with the cut edges taped to suppress the odorous gases coming from the composite core and the top was made from 3/8" Corian™ a low odor acrylic polymer containing solid surface material made by DuPont. The chin and headrest were made from components of the Table Model Chin Rest (Richmond Products) and the linear actuators were model LA12 (Linak A/S). Instructions and visual cues were presented on an Apple LCD Cinema Display (2560x1440 resolution) attached to a Mac Mini programed with PsychoPy (Nottingham University), Arduino Software (arduino.cc) and ActivePython (activestate.com). Programs written in Python and Arduino C drove the linear actuators controlled by relay circuits and an Arduino UNO board (adafruit.com) using.

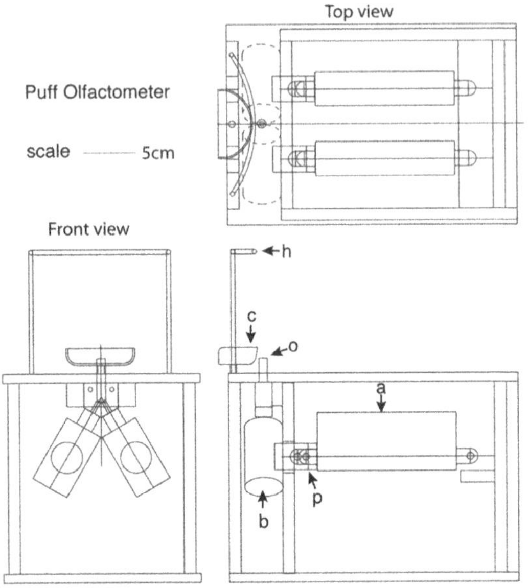

Figure 3. The Sniff Olfactometer (DATU, Inc., Geneva NY) where a. is an actuator, b. is a 250 ml PTFA bottle, p. Plunger, o. odor port, c. chin rest and h. is a headrest.

Thresholds

The individual odor detection thresholds for benzaldehyde and iso-amyl acetate were determined for each of the subjects using a 3AFC test using the SO (*27*). Five concentration levels were used for the two odorants: benzaldehyde: 46 μM, 92 μM, 184 μM, 368 μM, 736 μM and iso-amyl acetate: 0.13 μM, 0.41 μM, 1.24 μM, 3.73 μM, 11.21 μM. The thresholds for the four stimuli, two olfactory and two visual, were determined for each participant. Two bottles were inside of the olfactometer at all times. One bottle contained an odorant, the other a blank, contained only PEG. Subjects were prompted through a series of instructions, indicating when to exhale and inhale, while both visual and auditory prompts (a 440 Hz tone) alerted the subject to the onset of an odor. Five odorant concentrations were tested; a 15-second break took place between each odor trial, with a 1-minute break in between odorant concentration sets. Each odorant concentration was evaluated four times. Two puffs of air were delivered in sequence on each trial. One puff contained an odor the other a blank. The subject used the 'left' and 'right' arrow keys on the keyboard to indicate whether the odorant puff occurred first or second in the sequence. The concentrations of the odorant at 0.75 correct responses were used to determine the threshold for each subject stimulant pair (*33*). The thresholds determined using the SO were within a factor of two of those determined with Teflon hand-squeeze bottles.

Visual thresholds for banana and cherry figures were determined in two separate sessions. Upon entering the testing area, the experimenter adjusted the chair height for each panelist ensuring uniform eye level. All stimuli were presented using a two alternative forced-choice method. The computer program scripted in PsychoPy instructed the subject to evaluate two images side-by-side, one image located on the left side of the screen and one image located on the left side of the screen. Following the first set of instructions were examples of the two reference images: a patch of noise followed by a patch of noise plus either the outline of banana or two cherries on a stem (noise + figure), Figure 4. Six practice trials with feedback followed the presentation of the references. Practice trials, like the test trial presented the subject with two images (noise and noise + figure) presented side-by-side. All images were evaluated using a forced-choice method, where noise and noise+figure were always presented side-by-side on the screen. Subjects were instructed to select the noise+figure image. Presentation location of noise+figure was randomized to appear on either left or right side of the screen. Subjects responded by pressing either the left or right arrow keys on the keyboard, indicating the location of the noise+figure stimulus.

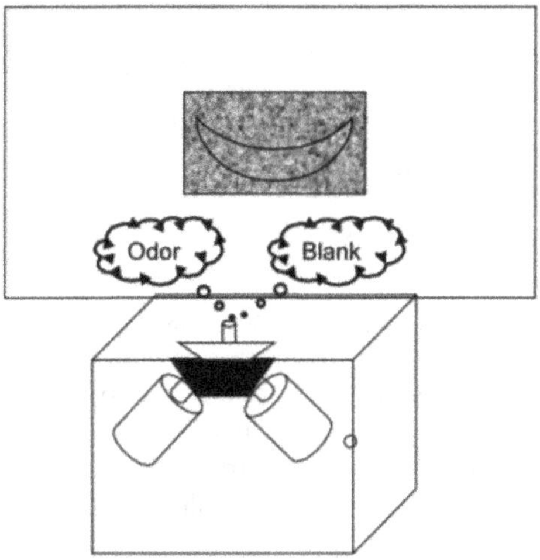

Figure 4. SO setup for the olfactory focus task where the subject is presented with two air puffs, and must decide which puff contains an odorant while a visual patch with or without a figure is presented on a monitor (27).

Subjects were instructed to respond as quickly and accurately as possible. A 440 Hz tone simultaneous with the presentation of a focal crosshair centered in the middle of the screen, alerted the subject to the beginning of each trial. Subjects were told to focus on the crosshair at the start of each trial and to press any key to continue. After the practice trials, subjects were alerted to the start of testing. Each session consisted of 8 test sets, each containing 14 trials. The test sets alternated between increasing and decreasing in visual detection difficulty. Each transparency level was presented two times during each test set. A test set increasing in difficulty would always be followed by a test decreasing in visual detection difficulty. Subjects rested for 1-minute between each test set. There were a total of eight sets of trials. The number of correct responses was tabulated for each mask level presented to each panelist. The mean percent correct response and its Least Significant Interval (LSI) using the pooled error from the analysis of variance (ANOVA)was calculated from the data obtained (*36*). The concentration at the mean correct response of 0.75 was used to define each participant's threshold (*33*).

Experiment

Subjects

Six men and four women completed the cross-modal task for this study, with an average age of 30.5±10 years. Testing consisted of six training sessions and four test sessions. All protocols were approved by the IRB.

Stimulants

Olfactory stimulants were presented at the detection limit for each subject (50% above chance). The visual stimulants were combined with abstract square noise masks between 92 and 94% opacity but well above the detection limit (between 95-100% above chance).

Protocol

A short test in order to assess whether visual stimuli were at the level of recognition was devised. Cherry and banana stimuli with visual masks ranging from 92-94% opacity were presented in a randomized order, where each image was presented a total of 6 times. Panelists had to press 'b' on the keyboard if he or she saw a banana and 'c' if he or she saw a cherry. The accuracy of response indicated the ability to recognize as well as distinguish the two visual stimuli. Stimuli presentation were congruent i.e. benzaldehyde odor was presented with cherry images and iso-amyl acetate was presented with banana images. While the subjects were focusing on the olfactory identification, visual stimuli were presented as congruent i.e. benzaldehyde odor was presented with cherry images or iso-amyl acetate was presented with banana images. Visual stimuli were noise (black), noise (yellow), noise (red), banana + noise (black), banana + noise (yellow), banana + noise (red), cherry + noise (black), cherry + noise (yellow), and cherry +noise (red). There were four practice trials, with direct feedback after each response, followed by two test sets. Each test set contained 12 trials, six control trials and six test trials, a 15-second timed break occurred between each trial presentation, and a 3-minute break separated the set presentations, Figure 5.

These experiments involved measuring perithreshold response to odors in the presence of supra threshold visual stimuli (congruent or incongruent distractors). The experiments were divided into two parts: 1) in the first part visual stimuli included only noise black, yellow (congruent with banana) or red (congruent with cherry) patches. Using the graphic software package GIMP to generate yellow and red patches a color transform was performed on the black noise pattern to create a red noise mask (RGB values were R=255, G=0, B=0 and H=0, S=100, V=100) and yellow noise mask (RGB value of R=255, G=255, B=0, H=60, S=100, V=100). Noise patches black 85%, red 88%, and yellow 93% opaque was superimposed on a plain white patch. There were four practice trials, with direct feedback after each response, followed by two test sets. Each test set contained 6 test trials, a 15-second timed break occurred between each trial presentation, and a 3-minute break separated the sets presented to ten subjects. Thus there were 6-1 or 5 model degrees of freedom and 10 x 2 x 4 x 6 – 6 or 474 pooled error degrees of freedom. 2) In the second part 85%, 88% and 93% opaque black, yellow and red noise masks were superimposed on black fruit outlines to generate three visual patches. The color figure pairing included two congruent combinations and two incongruent combinations. Therefore, there were 5 model degrees of freedom (6 -1) and 474 (10 x 2 x 4 x 6 - 6) pooled error degrees of freedom. Testing lasted approximately 20 minutes. The subjects were asked to choose the puff with an odorant. The mean percent correct response and Least Significant Difference (LSD) were calculated using the pooled error from the ANOVA of the results from this part of the experiment.

Results

Findings from experiments in which the visual images (fruit outlines) were presented well above their threshold on a noise background while performing an olfactory task (detect iso amyl acetate or benzaldehyde) in the presence of visual noise, presented in black, yellow or red showed no significant difference in the detection performance of either iso-amyl acetate or benzaldehyde when the visual noise was black (*27*). As shown in Figure 5, the presence of congruent color in the noise patch during the perithreshold olfactory detection (yellow in the presence of iso-amyl acetate and red in the presence of benzaldehyde respectively) yielded 80% detection. These results are not significantly different than the response to a black noise pattern plus banana or cherry figures.

However, olfactory detection performance decreased to 35% detection for both iso-amyl acetate and benzaldehyde when in the presence of an incongruent noise stimuli, red in the presence of iso-amyl acetate and yellow in the presence of benzaldehyde (Figure 5). That is, the subjects chose the blank puffs as having odor significantly more than the puffs containing the odor when the noise color was incongruent. Thus a yellow noise patch decreases performance in the benzaldehyde detection condition and a red noise patch decreases olfactory detection performance of iso-amyl acetate, $F\ (5, 474) = 18.35, p<0.001$.

Discussion

A simple olfactometer was constructed and combined with PsychoPy, an open source Python scripting program, to produce an integrated technology to study cross-modal interactions between vision and olfaction. The olfactometer, called a Sniff Olfactometer (SO) to distinguish it from the many more complicated machines described in the literature (*34*, *35*), was used to measure in a 2AFC test which of two choices of 10 ml puffs from a bottle had odor. It became clear that the design shown in Figure 3 could be improved by the addition of one more bottle to allow for 3AFC tests and greater experimental design flexibility. However, as described above this system identified congruent odor-figure pairs and produced robust dose-response data. Also, as shown in Figure 5 the protocol yielded LSIs, a graphical measure of significant difference (*36*), near ± 4% on a percent correct scale.

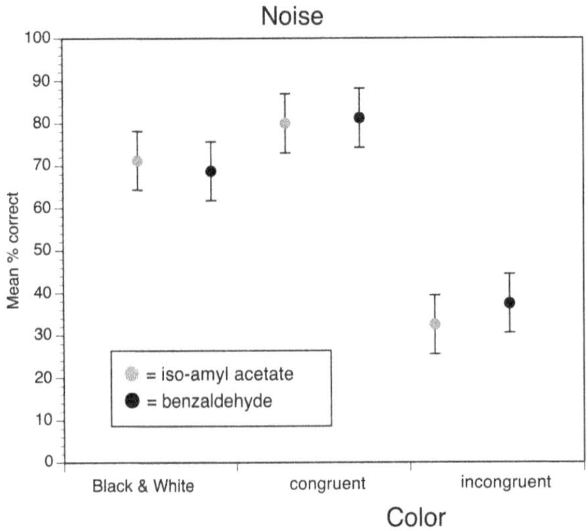

Figure 5. shows the effect of a noise patch on odor detection. Note the strong decrease in correct odor choices caused by noise patches colored incongruently.

To determine effect of a smell on the detection of an image we measure the percent correct response in a forced choice between a black pattern and the same pattern with an outline of either a cherry or a banana superimposed on the pattern. At the same instant the images were presented, the subjects experienced an odor puff of either benzaldehyde (congruent with the cherry outline) or isoamyl acetate (congruent with the banana outline) nears their threshold. The reasoning was that very small cross modal effects might be easily measurable near the threshold of detection. However, this could only be true when a stimulus is very close to the threshold and the stimulus is detected some of the time. To guarantee some detection, thresholds were defined as 75% correct response in the 2AFC threshold measurements determined for each stimulus for each subject. Theoretically, 75% correct response implies that 50% of the responses are due to real detection (*33*) and 25% are due to false positives (random error).

In eight different experiments in which subjects were asked to detect a stimulant when one of the two choices was a blank, all of the scores were significantly above the 50% chance, indicating stimulants were always detected. However, when either a congruent or incongruent odor distractor was simultaneously presented there was no significant difference in their response. For example, there was no significant difference in the percent correct response when the task was to detect a cherry outline in the presence of either benzaldehyde (congruent smell) or isoamyl acetate (incongruent smell). Perhaps due to the low signal strength of the visual information in our experiments the cognitive meaning is dissociated from the image. In order for cognitive interference or enhancement to occur it may be necessary for the visual stimulus to have greater meaning either through signal recognition or another association (*26*).

To address the role of color, another set of 12 experiments were conducted using the same subjects, experimental platform and odorants. The subjects were asked to detect perithreshold levels of odor while viewing either a black, yellow or red noise pattern but no figures. Coloring the pattern congruently showed only a slight but still insignificant increase in odor detection. However, when the distracting pattern was of an incongruent color, the puff with the odor was detected correctly only 32-38% of the time. It seems that experiencing a color incongruent with the odor caused a completely erroneous interpretation of the olfactory experience. Being exposed to congruent colors, however, caused no significant change. If the olfactory detection had only been neutralized by incongruent color, then the correct response should have been near 50% instead of 30 - 40 %. Only incongruent color in the patterns showed strong effects and these were all suppressive. This was again consistent with published results (Demattè, 2009) except that we did not test colored-in outlines or colored shapes just colored noise patterns.

Conclusion

As in a Stroop task, where an incongruent color of a word can overcome the word's meaning, the data presented here indicate that incongruent color of a meaningless pattern can suppress odor detection while an incongruent figure did not. Furthermore, this data shows that incongruent color can alter odor perception, when there is a connection between the olfactory cue and the color. It seems that color-odor congruency may direct consumer expectation more than figure-odor congruency. This further implies that color in advertising, formulation, packaging, and presentation not only alter consumer expectation but color may alter their experience during consumption as well. Doing experiments that are more "ecological", that is, more directly related to the real world will be needed to translate psychophysical data into useful understanding of perception (*7, 37*). As is implied by the title of the book by Spence and Piqueras-Fitzman (*7*), 'The Perfect Meal' is a multisensory experience.

References

1. Deliza, R.; MacFie, H. J. H. The generation of sensory expectation by external cues and its effect on sensory perception and hedonic ratings: A review. *J. Sens. Stud.* **1996**, *11* (2), 103–128.

2. Garber, L. L.; Hyatt, E. M.; Starr, R. G. Placing food color experimentation into a valid consumer context. *J. Food Prod. Mark.* **2001**, *7* (3), 3–24.
3. Jastrow, J. The Mind's Eye. *Popular Sci. Mon.* **1899**, *54*, 299–312.
4. Neisser, U. *Cognition and Reality*; W. H. Freeman and Co.: San Francisco, CA, 1976; p 21.
5. Armand, V. C. Consumer concerns and expectations about novel food processing technologies: Effects on product liking. *Appetite* **2003**, *40* (3), 217–233.
6. Hutchings, J. B. *Expectations and the food industry: The impact of color and appearance*; Kluwer Academic/Plenum Publishers: New York, 2003.
7. Spence, C.; Piqueras-Fizman, B. *The Perfect Meal*; John Wiley & Sons: Blackwell, Chiester, U.K., 2014.
8. Francis, F. J. Quality as influenced by color. *Food Qual. Prefer.* **1995**, *6* (3), 149–155.
9. Gottfried, J. A. Central mechanisms of odour object perception. *Nat. Rev. Neurosci.* **2010**, *11* (9), 628–641.
10. Lundström, J. N.; Boesveldt, S.; Albrecht, J. Central Processing of the Chemical Senses: An Overview. *ACS Chem. Neurosci.* **2011**, *2* (1), 5–16.
11. Xu, W.; Wilson, D. A. Odor-evoked activity in the mouse lateral entorhinal cortex. *Neuroscience* **2012**, *223*, 12–20.
12. Spence, C.; McGlone, F. P.; Kettenmann, B.; Kobal, G. Attention to olfaction: A psychophysical investigation. *Exp. Brain Res.* **2001**, *138* (4), 432–437.
13. Rolls, E. T.; Baylis, L. L. Gustatory, olfactory, and visual convergence within the primate orbitofrontal cortex. *J. Neurosci.* **1994**, *14* (9), 5437–5452.
14. Livingstone, M.; Hubel, D. Segregation of Form, Color, Movement, and Depth: Anatomy, Physiology and Perception. *Science* **1988**, *240* (4853), 740–749.
15. Spence, C.; Levitan, C.; Shankar, M.; Zampini, M. Does food color influence taste and flavor perception in humans? *Chemosens. Percept.* **2010**, *3* (1), 68–84.
16. DuBose, C. N.; Cardello, A. V.; Maller, O. Effects of colorants and flavorants on identification, perceived flavor, intensity, and hedonic quality of fruit-flavored beverages and cake. *J. Food Sci.* **1980**, *45*, 1393–1399.
17. Zampini, M.; Sanabria, D.; Phillips, N.; Spence, C. The multisensory perception of flavor: Assessing the influence of color cues on flavor discrimination responses. *Food Qual. Prefer.* **2007**, *18* (7), 975–984.
18. Engen, T. The Sense of Smell. *Annu. Rev. Psychol.* **1973** (24), 187–206.
19. Gottfried, J. A.; Dolan, R. J. The nose smells what the eye sees: Crossmodal visual facilitation of human olfactory perception. *Neuron* **2003**, *39* (2), 375–386.
20. Osterbauer, R. A.; Matthews, P. M.; Jenkinson, M.; Beckmann, C. F.; Hansen, P. C.; Calvert, G. A. Color of scents: Chromatic stimuli modulate odor responses in the human brain. *J. Neurophysiol.* **2005**, *93* (6), 3434–3441.
21. Zellner, D. A.; Kautz, M. A. Color affects perceived odor intensity. *J. Exp. Psychol.* **1990**, *16* (2), 391–3.

22. Clydesdale, F. M.; Gover, R.; Fugardi, C. The effect of color on thirst quenching, sweetness, acceptability and flavor intensity in fruit punch flavored beverages. *J. Food Qual.* **1992**, *15* (1), 19–38.
23. Zellner, D. A.; Bartoli, A. M.; Eckard, R. Influence of color on odor identification and liking ratings. *Am. J. . Psychol.* **1991**, *104* (4), 547–561.
24. Clydesdale, F. M. Color as a factor in food choice. *Crit. Rev. Food Sci. Nutr.* **1993**, *33* (1), 83–101.
25. Engen, T. The effect of expectation on judgments of odor. *Acta Psychol.* **1972**, *36* (6), 450–458.
26. Demattè, M. L.; Sanabria, D.; Spence, C. When vision matters? *Chem. Sens.* **2009**, *34*, 103–109.
27. Kurtz, A. J. The Role of Color, Congruency, Object Shape and Visual Interactions. Ph.D. Dissertation, Cornell University, 2012.
28. Foster, D.; Scofield, E. H.; Dallenbach, K. M. An Olfactorium. *Am. J. Psychol.* **1950**, *63* (3), 431–440.
29. Stone, H. Techniques for odor measurement: olfactometric vs sniffing. *J. Food Sci.* **1963** (28), 719–725.
30. Gilbert, A. N.; Martin, R.; Kemp, S. E. Correspondence between Vision and Olfaction: The Color of Smell. *Am. J. Psychol.* **1996**, *109* (3), 335–351.
31. Peirce, J. W. PsychoPy—Psychophysics software in python. *J. Neurosci. Methods* **2007**, *162* (1-2), 8–13.
32. Pierce, J. W. Generating stimuli for neuroscience using PsychoPy. *Front. Neuroinformatics* **2009** (2), 10.
33. Lawless, H. T.; Heyman, H. *Sensory Evaluation of Food, Principles and Practices*, 2nd ed.; Springer: New York, 2010; pp 105−108.
34. Dravnieks, A. Instrumental Aspects of Olfactometry. In *Methods in Olfactory Research*; Moulton D. G., Turk A., Jonston J. W., Jr., Eds.; Academic: New York, 1975; pp 1−61.
35. Brattoli, M.; De Gennaro, G.; De Pinto, V.; Demarinis Loiotile, A.; Lovascio, S.; Penza, M. Odour Detection Methods: Olfactometry and Chemical Sensors. *Sensors* **2011**, *11*, 5290–5322.
36. Andrews, A. K.; Snee, R. D.; Sarner, M. H. Graphical Display of Means. *Am. Stat.* **1980**, *34* (4), 196–199.
37. Wansink, B. *Slim by Design*; William Morrow: New York, 2014.

Chapter 2

Influence of Cross-Modal Sensory Interactions on Cheese Flavour Intensity and Character

Jun Niimi,[*,1,2] Amy R. Overington,[3] Patrick Silcock,[1] Phil J. Bremer,[1] and Conor M. Delahunty[2]

[1]Department of Food Science, University of Otago, PO Box 56, Dunedin 9054, New Zealand
[2]CSIRO Animal, Food and Health Sciences, Riverside Corporate Park, 11 Julius Avenue, North Ryde, NSW 2113, Australia
[3]Fonterra Co-operative Group Ltd, Dairy Farm Rd / Private Bag 11029, Palmerston North 4442, New Zealand
[*]E-mail: jun.niimi@otago.ac.nz.

Cross-modal sensory interactions between cheese aroma and taste, and their effect on cheese flavour intensity and difference in flavour character, were determined. NaCl, lactic acid, and aroma were varied at three different levels in combination, according to a 3^3 full factorial design. The change in cheese flavour intensity and difference in flavour character were measured relative to a reference using a panel (n=9). Model solutions were delivered by the simultaneous gustometer olfactometer. Increasing levels of NaCl, lactic acid, or aroma significantly ($p<0.001$) enhanced cheese flavour intensity. A significant interaction between NaCl and lactic acid levels with respect to cheese flavour intensity was detected, indicating that cross-modal sensory interactions were not linear. Flavour character also significantly ($p<0.001$) changed in that raising levels of NaCl decreased the difference in flavour character from the reference, while that of lactic acid or aroma increased the difference.

Introduction

The senses (modalities) that actively take part during the consumption of cheeses are olfaction (smell), gustation (taste), chemesthesis, and texture. Concurrent perception of several of these modalities can influence the perceived intensity of each other through cross-modal sensory interactions (*1*). Interactions can be described by the shift in perceived intensity of one modality due to the perception of a separate modality, causing enhancement or suppression. The perceptual shifts in attribute intensities have been widely tested during the occurrence of cross-modal sensory interactions in complex mixtures of aroma and multiple tastes. In a previous study using the simultaneous gustometer-olfactometer (SGO) (*2*), NaCl, lactic acid, and aroma were important contributors to the perception of cheese flavour intensity (*3*). Little, however, is known regarding effects on flavour character that are concurrent.

Much of the cross-modal sensory interaction research has been focused on changes in perceived flavour intensity – the potential changes in flavour character have been rarely reported. Changes in flavour can be profiled, however, rigorous sensory methodologies such as descriptive analysis can prevent the measurement of cross-modal sensory interactions. This is due to the method requiring extensive partitioning of sensory attributes together with thorough training of assessors, which promotes analytical cognitive processes. Measurements of cross-modal sensory interactions require a more synthetic (holistic) cognitive process (*4*). An alternative to profiling methods that sustains the synthetic cognitive process is to use difference tests from a known reference sample (*5*, *6*). Difference testing has been used for comparisons of model aroma mixtures with the real foods for French fries, boiled beef, coffee, wine, and lavender (*7*, *8*). Using this broad approach allows the measure of intensity changes and to determine the degree of similarity in aroma character. Changes in aroma difference from the original sample have been measured when single compounds or groups thereof were omitted from a mixture.

The objective of this current study was to measure the effect of cross-modal sensory interactions on both cheese flavour intensity and character. Using an experimental design approach, the relative contribution of specific taste characters could be determined. Samples were delivered using the SGO to independently, yet simultaneously, deliver tastes and aroma to the assessors (*2*).

Methods

Experimental Design

A 3^3 full factorial design consisting of three levels of NaCl, lactic acid, and aroma (a total of 27 samples) was used to determine the effect of taste and aroma levels on cheese flavour intensity and difference in cheese flavour character. The concentrations of NaCl and lactic acid were varied by ±0.25 log and ±0.4 log, respectively, from the medium concentration while sucrose, monosodium glutamate (MSG), and caffeine were held constant (Table 1). The medium concentration of tastants was the cheese taste solution developed previously (*9*). Variation in aroma concentrations were the same as that outlined previously (*10*), which was ±0.5 log from the medium concentration (Table 2).

Table 1. Tastant Concentrations Used for the 3^3 Full Factorial Design. NaCl Was Varied by ±0.25 log and Lactic Acid by ±0.4 log, Depicted in Bold Italics.

Tastant	*Tastant level (%, w/w)*		
	Low	*Medium*	*High*
Sucrose*	0.300	0.300	0.300
NaCl**	***0.281***	***0.500***	***0.889***
MSG**§	0.110	0.110	0.110
Lactic acid**	***0.044***	***0.110***	***0.276***
Caffeine**	0.080	0.080	0.080

* Bundaberg Sugar, Spring Hill, QLD, Australia. ** Sigma Aldrich Sydney, NSW, Australia. § Produced using Glutamic acid and NaOH.

Table 2. Aroma Concentrations in Mixtures Used for the 3^3 Full Factorial Design

Compound	*CAS-No*	*Aroma concentration (µg L^{-1})**		
		Low	*Medium*	*High*
2-Butanone#	78-93-3	0.200	0.630	2.000
2-Heptanone#	110-43-0	0.120	0.380	1.200
2-Nonanone#	821-55-6	0.040	0.130	0.400
3-Methylbutanal#	590-86-3	0.020	0.060	0.200
3-Methylbutanoic acid£	503-74-2	0.060	0.190	0.600
Butanoic acid#	107-92-6	0.800	2.530	8.000
Diacetyl#	431-03-8	0.200	0.630	2.000
Ethyl butanoate#	105-54-4	0.120	0.380	1.200
Ethyl hexanoate£	123-66-0	0.040	0.130	0.400
Methional#	3268-49-3	0.020	0.060	0.200

* Adjacent difference in concentration was 0.5 log for each aroma compound. # Givaudan Australia Pty Ltd, Sydney, NSW, Australia. £ Sigma Aldrich Sydney, NSW, Australia.

Samples were presented as an incomplete block design; there were three blocks presented across duplicates equating to 18 samples in each block. In block one, 18 samples from replicate 1 were presented; in block two, nine samples each from replicate 1 and 2 were presented; and in block three 18 samples from replicate 2 were presented. An incomplete design was used to reduce the number of samples for evaluation in each session and hence minimise sample carry-over and sensory fatigue as much as possible. Within block two, identical samples that overlapped across the replicates were not presented. Sample presentation was randomised for each block.

Sensory and SGO Procedures

Assessors experienced in sensory evaluation (n=9) participated in the study and they had been screened in accordance with ISO standards (8586-2) (*11*). These assessors had been previously trained in the use of scales and basic taste discrimination on a wide range of food products, and had participated in descriptive analysis studies. They also had previous experience in the use of the SGO, including a previous study on cheese flavour perception (*3*). The assessors took part in one familiarisation session on the use of scales prior to evaluation. Assessors were instructed on the evaluation procedures for the use of the SGO and rating methodology. The samples were presented to the assessors using the SGO and evaluated with the relative-to-reference method (*3*), where the reference sample was the medium levels of all tastes and aroma compounds (Tables 1 and 2).

The instructions provided were the following:

> *"The SGO instrument presents you with aroma and taste together as a mixture to make flavour. In this study, you will be presented with aroma and taste to make cheese flavour. We wish to measure the overall intensity of this cheese flavour. Throughout the evaluation we also wish to measure the difference in cheese flavour character. The flavour character refers to the overall cheesiness impression that you get from the sample, as opposed to an intensity measure. You will first be presented with the reference. The reference will always be the same. However, it is necessary for you to taste this accurately each time so as your memory for it is consistent. After tasting the reference cleanse your palate with water and wipe the mouth piece. You will then be presented with the test sample. Please rate the* ***Intensity of cheese flavour*** *relative to the reference. Please indicate the* ***Difference in cheese flavour character*** *from the reference."*

The assessors evaluated two attributes using the relative-to-reference method, for both cheese flavour intensity and difference in cheese flavour character. The *cheese flavour intensity* was considered as a combined percept of both taste and aroma together. The scale used for the cheese flavour intensity was the same as that previously reported (*3*). *Difference in cheese flavour character* was measured on a 100 point scale that was anchored at 0 and 100 points with no difference and extreme difference, respectively. The reference used throughout the current study was the sample with medium levels in the experimental design of taste and aroma (Tables 1 and 2). After tasting the sample, the assessors cleansed their palate with water from cups. To prevent any sample carry-over, the assessors wiped the mouthpiece with a paper towel between each reference-sample and sample-reference tasting sequences.

The SGO delivered tastes and aromas through pumps and mass flow controls, respectively, as described previously (*3*). Taste solutions were delivered at 45 mL min^{-1} for 10 s (7.5 mL) and aroma was delivered orthonasally in humidified air at 2 L min^{-1}. The sequence of events used to deliver the taste/aroma were described previously (*3*). Interstimulus breaks of 60 s were provided between the reference and test samples, and between test samples and the reference.

Data Analysis

To determine the effect of NaCl, lactic acid, and aroma concentrations on the cheese flavour intensity and the difference in cheese flavour character, data were analysed using a univariate analysis of variance (ANOVA) with SPSS statistics Ver. 17 (SPSS Inc., Chicago, IL, U.S.A.). Cheese flavour intensity and the difference in cheese flavour character ratings were analysed as the dependent variable. The NaCl, lactic acid, aroma, and replicates were taken as the independent variables and analysed as fixed factors with assessors as a random variable. Significant differences in mean cheese flavour intensity and mean difference in cheese flavour character by the NaCl, lactic acid, or aroma levels underwent post hoc testing using Fisher's least significant difference (LSD). All two-way and three-way interactions between the main variables NaCl, lactic acid, and aroma were analysed. Two-way interaction between assessor and replicates were analysed for the stability of cross-modal sensory interaction effects. Significance testing was performed at an alpha level of 5%.

Results and Discussion

The cheese flavour intensities and difference in cheese flavour character were measured, as NaCl, lactic acid, and aroma levels were varied. The cheese flavour intensity was significantly ($p<0.001$) enhanced by increasing the concentrations of NaCl, lactic acid, and aroma (Figure 1). For each variable, the mean cheese flavour intensity was significantly different across the different levels of that variable. That is, increases in concentrations of NaCl overall, significantly enhanced cheese flavour intensity, as did increases in concentrations of lactic acid and aroma. The enhancement of cheese flavour intensity by NaCl, lactic acid, and aroma was consistent with previously reported results that also determined interaction effects through sample delivery using the SGO (*3*). The similar nature and magnitude of the flavour enhancements demonstrated the stable and reproducible interaction effects between taste and aroma. A significant replicate effect was not detected. Furthermore, a significant replicate × assessor interaction was not detected. This finding confirmed the robustness of cross-modal interactions using experienced sensory assessors (*10*).

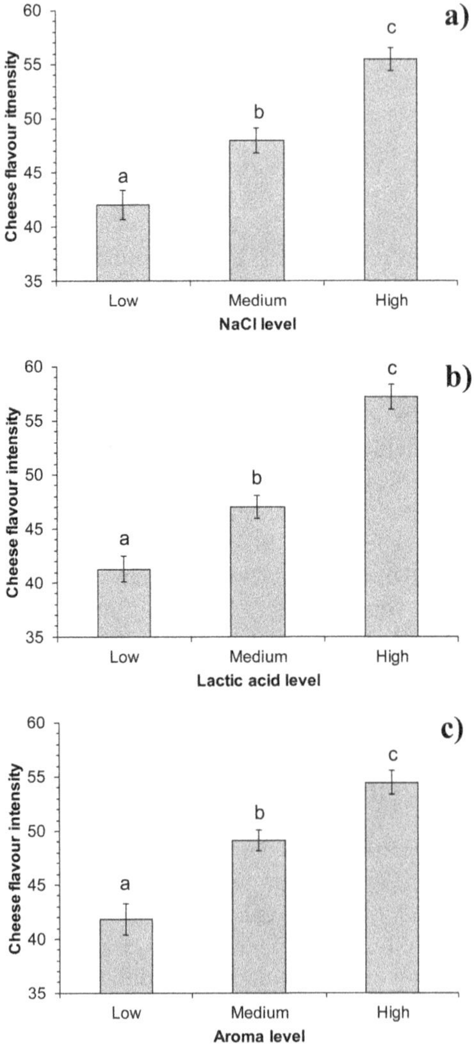

Figure 1. The mean cheese flavour intensities (± standard error) by variation of levels of main effects (a) NaCl, (b) lactic acid, and (c) aroma. Means with the same superscripts are not significantly different according to Fisher's LSD.

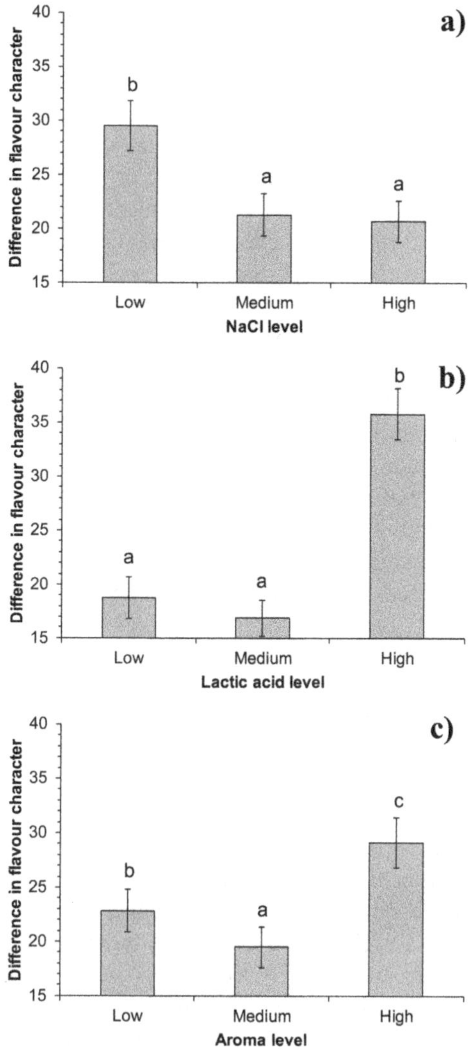

Figure 2. The mean difference in cheese flavour character (± standard error) by variation of levels of main effects (a) NaCl, (b) lactic acid, and (c) aroma. Means with the same superscripts are not significantly different according to Fisher's LSD.

Varying the NaCl, lactic acid, and aroma concentrations significantly changed the magnitude of the difference in cheese flavour character measured from the reference. All differences in flavour character were measured as the difference in cheese flavour character from the medium level of taste and aroma (Figure 2). The difference in cheese flavour character significantly (p<0.001) increased with a reduction in NaCl from a medium to a low level (Figure 2a). However, the cheese flavour character did not significantly change upon an increase in NaCl level from a medium to a high level. The cheese flavour character significantly ($p<0.001$) differed when lactic acid was increased from a medium to a high level, but did not significantly differ when lactic acid was decreased from a medium to a low level (Figure 2b). The difference in cheese flavour character significantly ($p<0.001$) increased when aroma level was either decreased or increased from the medium level (Figure 2c). Increasing the aroma from a medium to a high level resulted in a significantly larger difference in cheese flavour character than decreasing the aroma from a medium to a low level. Significant replicate or replicate × assessor effects were not detected.

There was a significant ($p<0.05$) NaCl × lactic acid interaction on cheese flavour intensity. Increasing lactic acid levels resulted in a greater magnitude of enhancement of cheese flavour intensity at low NaCl concentrations, compared with high NaCl concentrations (Figure 3a). The NaCl × lactic acid interaction showed that the magnitude of enhancement was concentration dependent, where beyond some critical NaCl level no further enhancement of cheese flavour intensity by lactic acid may be observed. This is consistent with literature where magnitudes of enhancement in flavour intensity become smaller, as the taste and aroma components are combined at high concentrations as opposed to low concentrations (*12*, *13*). Large enhancements occur when stimuli are combined at relatively low concentration (*3*). Interestingly, no significant NaCl × lactic acid interaction for difference in cheese flavour character was observed (Figure 3b). Further two-way and three-way interactions of all taste/aroma combinations on cheese flavour intensity or difference in cheese flavour character were not significant.

The results of the effects taken together showed that NaCl was an important variable for the enhancement of flavour intensity and maintenance of flavour character, while lactic acid and aroma, also important for enhancement, have the potential to change flavour character. It appears that NaCl can aid as a compensatory variable for cheese flavour intensity and maintenance of flavour character if other tastes/aroma were reduced in level. Although lactic acid or aroma appeared to be able to enhance the intensity of cheese flavour and potentially compensate for low levels of tastants/aroma, the flavour character would change. Thus optimum ranges of NaCl and lactic acid levels are required in order to maintain an unchanged cheese flavour character. NaCl could only be varied between medium and high levels, lactic acid could only vary from low to medium levels, but aroma could not be altered. However it is likely that aroma level could be altered, on the proviso that the mixture of aroma compounds is adjusted to maintain the same character as the reference. This is a subject for further research.

The difference in cheese flavour character was measured holistically to avoid extensive dissection of attributes that could potentially interfere with the perceptual integration of taste and aroma as a flavour percept (*14*). Indeed, this approach appeared to maintain the cross-modal sensory interaction effects, evident in the enhancement of cheese flavour intensity. While this approach gave a measure of the magnitude of the change in the flavour character from the reference, the measure could not indicate how the cheese flavour changed and identify the attributes responsible for the difference. An observation was made with the rating of the mean difference in cheese flavour character for the blind reference sample. The blind reference sample, being identical to the "reference" taste prior, was expected to be rated with 0 points. However, the sample was rated as a mean of 15.3 points. In sensory evaluations such as descriptive analysis, it is common for assessors to avoid rating attribute intensities of samples as 0 points on a line scale. There is also a possibility that the difference in flavour character of the blind reference was due to gradual adaptation to certain components of the taste/aroma mixture from continual tasting of reference followed by the sample. The study showed that changes in flavour character require consideration when cross-modal sensory interactions on flavour perception in multicomponent mixtures of taste and aroma are studied. To determine the flavour attributes that cause the underlying differences in measured flavour character, a different evaluation approach is required that will provide further detail. This will be subject of further research.

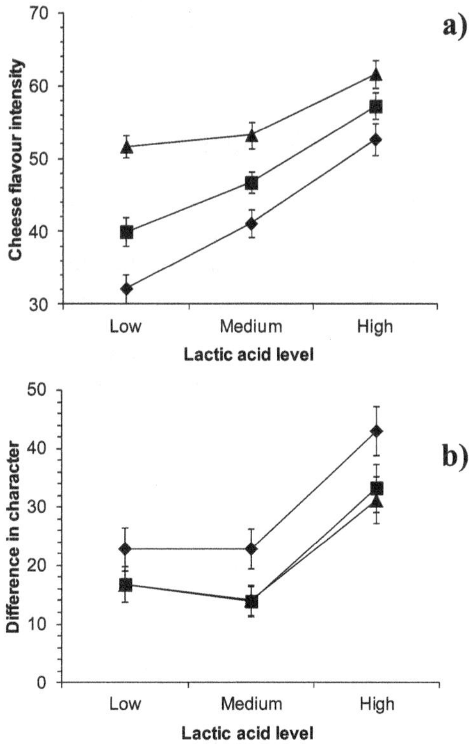

Figure 3. Mean (a) cheese flavour intensity (± standard error) and (b) the difference in flavour character as a function of NaCl and lactic acid levels: ▴, high NaCl; ▪, medium NaCl; ♦, low NaCl.

The differences in flavour character upon changes in tastant levels may be due to taste-taste interactions. A lower NaCl level would not only have decreased saltiness, it could lead to the perceptual change of other interacting characters (*15*). The impact of reduced NaCl on flavour in water soluble extracts of cheese and model cheese taste solution from a taste-taste perspective is the emergence of bitterness (*9*, *16*, *17*). Salt can serve to suppress bitterness perception and without it, bitterness will emerge. The same effect would occur with sour tastes evoked by acids, due to the ability of sourness to also suppress bitterness (*15*). Applying this knowledge to cheese gives an insight to the complexity of factors that contribute to flavour perception in cheese, as tastes such as bitterness are not only due to taste-taste interactions, but are also dependent on other factors including proteolysis during ripening and the effect of salt on the reactions thereof (*18*). How NaCl may have influenced aroma in the current study is unknown. Differences in flavour character due to the increase in lactic acid are most likely caused by the emergence of a dominant sour flavour character. This suggested that lactic acid at the medium concentration blended well to give a flavour character of cheese as a single flavour percept. With regards to aroma, it is difficult to speculate on the changes in flavour character, because of the propensity for aroma character to change, both as a function of aroma compound concentration (*19*) and in mixtures with other compounds (*20*, *21*). This, and the fact that the aroma used in the current study was a mixture of ten compounds, makes the prediction of the changes in aroma character challenging.

Conclusions

Cheese flavour intensity was enhanced by increases in concentrations of NaCl, lactic acid, and aroma levels. The magnitude of the enhancement was dependent on the relative levels of NaCl and lactic acid within the cheese flavour mixture. Cross-modal interactions were thus non-linear. The consequence of enhancement of the flavour intensity by lactic acid and aroma, however, was a change in cheese flavour character compared with the reference. Further investigation is required, however, as the measurement did not reveal how the flavour character changed or which sensory attributes were responsible for the measured differences.

References

1. Delwiche, J. The impact of perceptual interactions on perceived flavor. *Food Qual. Pref.* **2004**, *15*, 137–146.
2. Eddy, A. I.; Delahunty, C. M. A new networked gustometer and olfactometer: applications in cross-modal sensory research, *8th Pangborn Sensory Science Symposium*, Florence, Italy, July 26–30, 2009.
3. Niimi, J.; Eddy, A. I.; Overington, A. R.; Silcock, P.; Bremer, P. J.; Delahunty, C. M. Cross-modal interaction between cheese taste and aroma. *Int. Dairy J.* **2014**, *39*, 222–228.
4. Prescott, J.; Johnstone, V.; Francis, J. Odor-taste interactions: Effects of attentional strategies during exposure. *Chem. Senses* **2004**, *29*, 331–340.

5. Charles, M.; Poinot, P.; Texier, F.; Arvisenet, G.; Vigneau, E.; Mehinagic, E.; Prost, C. The 'Mouth to Nose Merging System': A novel approach to study the impact of odour on other sensory perceptions. *Food Qual. Pref.* **2013**, *28*, 264–270.
6. Teillet, E.; Schlich, P.; Urbano, C.; Cordelle, S.; Guichard, E. Sensory methodologies and the taste of water. *Food Qual. Pref.* **2010**, *21*, 967–976.
7. Grosch, W. Evaluation of the key odorants of foods by dilution experiments, aroma models and omission. *Chem. Senses* **2001**, *26*, 533–545.
8. Johnson, A. J.; Hirson, G. D.; Ebeler, S. E. Perceptual characterization and analysis of aroma mixtures using gas chromatography recomposition-olfactometry. *PLoS One* **2012**, *7* DOI:10.1371/journal.pone.0042693.
9. Niimi, J.; Eddy, A. I.; Overington, A. R.; Heenan, S. P.; Silcock, P.; Bremer, P. J.; Delahunty, C. M. Cheddar cheese taste can be reconstructed in solution using basic tastes. *Int. Dairy J.* **2014**, *34*, 116–124.
10. Niimi, J.; Eddy, A. I.; Overington, A. R.; Heenan, S. P.; Silcock, P.; Bremer, P. J.; Delahunty, C. M. Aroma–taste interactions between a model cheese aroma and five basic tastes in solution. *Food Qual. Pref.* **2014**, *31*, 1–9.
11. ISO. Sensory analysis - general guidance for the selection, training and monitoring of assessors (ISO 8586-2:1994). In *Standardization*; Geneva, Switzerland, 1994.
12. McBride, R. L. Integration psychophysics: the use of functional measurment in the study of mixtures. *Chem. Senses* **1993**, *18*, 83–92.
13. McBride, R. L.; Finlay, D. C. Perceptual integration of tertiary taste mixtures. *Percept. Psychophys.* **1990**, *48*, 326–330.
14. Le Berre, E.; Thomas-Danguin, T.; Béno, N.; Coureaud, G.; Etiévant, P.; Prescott, J. Perceptual processing strategy and exposure influence the perception of odor mixtures. *Chem. Senses* **2008**, *33*, 193–199.
15. Keast, R. S. J.; Breslin, P. A. S. An overview of binary taste-taste interactions. *Food Qual. Pref.* **2003**, *14*, 111–124.
16. Andersen, L. T.; Ardö, Y.; Bredie, W. L. P. Study of taste-active compounds in the water-soluble extract of mature Cheddar cheese. *Int. Dairy J.* **2010**, *20*, 528–536.
17. Toelstede, S.; Hofmann, T. Quantitative Studies and Taste Re-Engineering Experiments toward the Decoding of the Nonvolatile Sensometabolome of Gouda Cheese. *J. Agric. Food. Chem.* **2008**, *56*, 5299–5307.
18. Habibi-Najafi, M. B.; Lee, B. H. Bitterness in cheese: A review. *Crit. Rev. Food Sci. Nutr.* **1996**, *36*, 397–411.
19. Gross-Isseroff, R.; Lancet, D. Concentration-dependent Changes of Perceived Odor Quality. *Chem. Senses* **1988**, *13*, 191–204.
20. Laing, D. G.; Eddy, A.; Best, J. D. Perceptual characteristics of binary, trinary, and quaternary odor mixtures consisting of unpleasant constituents. *Physiol. Behav.* **1994**, *56*, 81–93.
21. Laing, D. G.; Francis, G. W. The capacity of humans to identify odors in mixtures. *Physiol. Behav.* **1989**, *46*, 809–814.

Chapter 3

Strategies To Enhance Saltiness in Food Involving Cross Modal Interactions

T. Thomas-Danguin, G. Lawrence, M. Emorine, N. Nasri, L. Boisard, E. Guichard, and C. Salles*

Centre des Sciences du Goût et de l'Alimentation, UMR1324 INRA, UMR6265 CNRS, Université de Bourgogne, F-21000 Dijon, France
***E-mail: salles@dijon.inra.fr.**

A series of results is reported on cross modal odour-taste interactions as a mean to enhance salty taste in food. Salt-related odours can enhance salty taste in water solutions containing a low level of sodium chloride through odour-induced changes in taste perception. Odour-induced saltiness perception enhancement (OISE) depends on salt concentration (intensity). OISE was also found effective in low-salt content solid model cheese, but texture dependant. A significant saltiness perception enhancement induced by Comté cheese and sardine odours was observed for softer textures only. In ternary odour-sour-salty solutions, sourness enhances saltiness perception additively with salt-related odours. Finally, in cream-based food systems, a strategy combining OISE and heterogeneous distribution of stimuli was found to compensate for over 35% decrease in salt-content without significant loss of acceptability. However, variation of composition of the food matrix influences aroma and saltiness perception. Therefore, it could modulate the overall saltiness perception.

Introduction

Excessive intake of sodium has undesirable effects on health such as hypertension and may contribute to other diseases such as cancer and osteoporosis (*1*). In the developed countries, approximately 75% of the daily salt (NaCl) intake comes from processed foods. Consequently, the World Health Organization and other national health agencies recommend decreasing by 25% NaCl content in targeted foods among which breads, meat products, cheeses, soups and ready to eat meals (*2*, *3*). Decreasing the sodium content in food products has thus become a major issue for the processed-food industry.

However, sodium chloride is a multifunctional ingredient in food. It not only guarantees a sufficient microbiological safety (*4*) but also influences the mechanical and organoleptic properties of food as well as its structure (*5*). In the case of food products which are submitted to fermentation or maturation process during their preparation, the amount of NaCl can influence the development of micro-organisms and thus the overall quality of the final product (*5*). Indeed, salt plays an important role in the regulation of microbial growth and on biochemical activities due to the reduction of water activity. As an example, for cheese making, NaCl acts on cheese draining and allows the formation of the crust; it also influences mineral equilibrium and buffering power, then it has an effect on the organoleptic characteristics of cheeses (*6*, *7*). Moreover, a modification of salt content modifies the structure of food matrix which in turn influences the release kinetics of flavour compounds and their perception (*8*).

Thus, lowering the salt content in food can lead to a significant loss of the overall liking and acceptability of the food product by the consumer, with a negative economic impact. To overcome these difficulties, different strategies were explored, such as the partial substitution of sodium by potassium (*9*, *10*), the progressive decrease of NaCl content until a targeted content was reached (*11*) and the increase of heterogeneity in the tastant distribution in various food products (*12*, *13*). Other strategies have proposed to compensate the loss of salty perception in low-salt foods by the means of cross-modal interactions because food flavor is a multimodal percept involving not only the perception of tastants but also of odourants, thus leading to an interplay between taste and smell percepts (*14*, *15*). The cognitive origin of odour–taste interactions is now generally admitted (*16*) since it is driven by the association between both stimuli (*17*). Neural mechanisms have been highlighted that support the cognitive basis of such interactions (*18*).

The objective of this presentation is to give an overview of a series of results we obtained on cross-modal odour-taste interactions as a means to enhance salty taste in food. Results are reported for saltiness enhancement by aroma obtained in water medium and in model cheeses, and for combined strategies such as taste-taste-smell cross modal interactions in water medium and heterogeneity in stimuli distribution in a cream-based food product. The influence of matrix composition on flavour release and perception is also discussed with regards to these results.

Enhancement of Saltiness Perception by Aroma in Water Solution

We evaluated smell–taste interactions in a simple aqueous media in order to test the hypothesis that aromas could enhance the salty taste (*19*). The approach consisted in first the selection of odours evoking a salty taste, then testing the impact of these odours on saltiness enhancement and last testing the effect of stimuli concentration on saltiness enhancement.

The first step relied on a screening of a large number of odours on the basis of their semantic attribute, taking advantage of the odour–taste cognitive association (*19*). Indeed, food odours may evoke the salty taste only through mental imagery, here induced by the name of this food. Thus, we selected odours that evoked saltiness on the basis of the name of their most representative food source. Eighty-one French panellists (aged 19–73 years, 56 women) participated in a 1-h session. They did not receive any information about the aim of the experiment. Eighty-six food names were used in this experiment. Most of them were pre-selected on the basis of their association with salty food and a few control names were not associated with salty food, such as lemon (for sourness), vanilla, strawberry or milk chocolate (for sweetness). For each food name, panellists were asked to estimate taste intensity (bitterness, sourness, saltiness, and sweetness) of food products as evoked by verbal items (written food names) on four linear scales from 0 to 10 (0: none and 10: extremely strong) or to indicate "not known" if the food name was unknown to them.

The results evidenced first a great consensus between subjects for expected tastes and second significant differences in saltiness between food names. Food names such as anchovy, bacon, smoked salmon, dry sausage, peanuts, "bouillon cube" and sardine were evaluated as the most salty whereas vanilla, orange, fig, strawberry, milk chocolate, cinnamon and lemon were evaluated as the least salty. Anchovy and bacon items were especially considered as the most salt-associated food names. Interestingly, the expected saltiness of the proposed food names was found to be highly correlated with the actual sodium content of the food products. This first experiment allowed selecting food odours associated with the salty taste and control food odours non-associated with the salty taste.

In the second experiment, saltiness of water solutions containing the selected odours, with (0.02 M) and without sodium chloride, were submitted to sensory evaluation in order to validate the effectiveness of odour-induced changes in saltiness (*19*). Fifty-nine panellists (aged 21–67 years, 42 women) participated in two 1-h sessions. For each sample, panellists were asked to rate odour intensity and taste intensity (sourness, bitterness, saltiness, and sweetness) on dedicated linear scales from 0 to 10. Aroma concentrations were chosen according to their intensity and acceptability in a preliminary test; nevertheless odour intensity rating was recorded and included in the data analysis as the aroma concentrations were not fully adjusted for iso-intensity.

The results clearly showed that some odourants have a salty dimension while other odorants do not. Moreover, these "salty aroma" have the ability to significantly enhance the salty taste of solutions containing a low quantity of NaCl: this phenomenon is called Odour-Induced Saltiness Enhancement (OISE) (Figure 1). As an example, anchovy and bacon, both items that were previously found to evoke the highest expected saltiness, were also found to be the two aromas that elicited the highest OISE. For tomato and carrot, even though their odour intensity was quite high, the OISE remained very low or even negative; this observation suggested that these aromas were not associated with saltiness. Carrot and tomato aromas used in this experiment were indeed found to evoke mainly sweetness. Perceptual interactions between tastes usually showed sweetness to reduce saltiness. Thus it is possible that, in our experimental conditions, induced sweetness counteracted actual saltiness of NaCl which could explain why tomato and carrot odours, despite their medium odour intensity, were found not to enhance saltiness. Overall, our results suggest that, to induce taste enhancement, an odour should be associated with that specific taste.

Odour intensity was also found to influence saltiness enhancement but this influence may differ according to the odour quality (*19*). For instance in our experiment (*19*) soy sauce odour did not show any significant OISE (Figure 1). However, in another study, Djordjevic et al (*20*) found a significant OISE using a commercially available soy sauce. Beyond the likely differences in odour quality between the two soy sauce aromas, it is worth noticing that the intensity of our soy sauce odour was low (the lowest of our tested odours), and may be to low to induce a signifiant OISE. In contrast, carrot odour, which was as intense as Comté but not associated with salty taste, did not enhance saltiness (*19*). In order to clarify the effect of aroma intensity on saltiness enhancement, we performed a dedicated experiment in which sixty one consumers tasted 3 concentrations of sardine aroma (0.5, 0.5 and 1 g/L) in salted water solutions (0.02 M NaCl). Additionally, a reference solution with 25% more sodium chloride (0.025 M in Evian water) was used (*21*). The results showed no significant influence of aroma intensity on saltiness enhancement. Therefore, it is suggested that as soon as odour quality is associated to salty taste, OISE can occur, regardeless of the perceived intensity of the aroma.

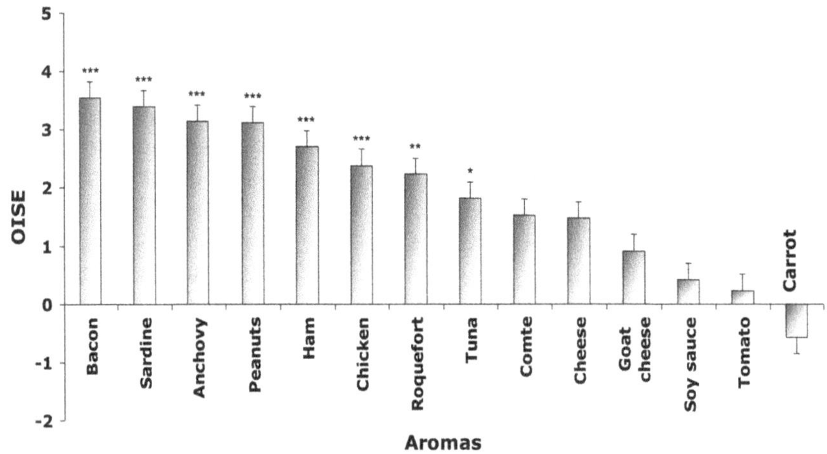

Figure 1. Enhancement of salty taste (OISE) for each aroma solution presented in an aqueous solution containing 0.02 M salt. Adapted from (19).Copyright (2008) with permission from Elsevier.

Saltiness enhancement by addition of congruent aroma also varies according to NaCl concentration (*22*). This is the main result of a study in which sixty four untrained panellists had to evaluate the saltiness of water solutions including NaCl and a tasteless odorant. Following a full factorial design, three concentrations of sodium chloride were used (0.01, 0.02, and 0.04 M) in conjunction with two aroma conditions: sardine aroma was chosen for its saltiness enhancement properties and carrot aroma as a control (*22*). We observed that saltiness of a low concentration of sodium chloride in water (0 – 0.02 M) was enhanced when subjects perceived simultaneously the congruent sardine aroma. However, when NaCl concentration increased to 0.04 M, thus eliciting a high salty taste intensity, no more significant OISE occured. We suggested that this result may come from the negatively accelerating form of the psychophysical functions (*23*). Indeed, when saltiness fall in the upper part of the stimulus–response function, it may be difficult to observe an increase in saltinessy perception. Attentional processes could also account for the modulation of OISE as a function of salt concentration. When taste intensity increased, the taste dimension could have caught subjects' attention so that they were much more focused on taste and somehow precludes attending to the odour, thus preventing OISE.

Enhancement of Saltiness Perception by Aroma in Solid Food

We assessed the efficiency of OISE as a strategy to compensate for NaCl reduction in solid food, especially taking into account texture variation and the influence of cross-modal aroma-texture-taste interactions on saltiness perception (*20*).

To that goal, we performed an experiment using four model cheeses lipoproteic matrices (LPM) with two dry-matter levels (370 and 440 g/kg) and two fat-contents (20 and 40% of dry matter) at a NaCl level of 0.5%. A total of 16 LPMs: 12 flavoured LPMs (Three commercial aromas) and four unflavoured LPMs, were thus prepared for this experiment (*24*). One aroma (Comté cheese) was selected to be congruent with salty food and cheese product, another one (sardine) was chosen to be congruent with salty food but not with cheesy products, and the last one (carrot) was selected to be neither congruent with salty food nor cheesy products. The study was carried out with 27 consumers who rated odour and taste intensities, texture attributes of the model cheeses and their liking for these products in a separate session.

On the basis of intensity data, OISE has been calculated for each LPM (Figure 2). In average, sardine and Comté-cheese odours were found to induce a highly significant saltiness enhancement, which demonstrated that salt-associated odours can enhance saltiness in complex solid-food matrices containing a low amount of sodium chloride. Conversely, carrot odour did not produce a significant OISE.

Nevertheless, the observed enhancement effects were dependant on the composition of the LPMs and their texture characteristics. Sardine and Comté-cheese odours were found to induce a significant saltiness enhancement in LPM with the highest fat level (DxF2, Figure 2). Such an enhancement did not occur in LPM with the low fat level, except for sardine odour in LPM with low fat and low dry matter level (D1F1). No OISE was observed for the firmest LPM (D2F1).

Our data suggest that congruence between odour and food product is not a critical factor for OISE. Indeed, sardine odour, which was less congruent to the LPM food product than Comté-cheese odour, tended to induce more saltiness enhancement. No significant difference in liking was observed between sardine and Comté cheese-flavoured products, but both were significantly more appreciated than carrot-flavoured LPM.

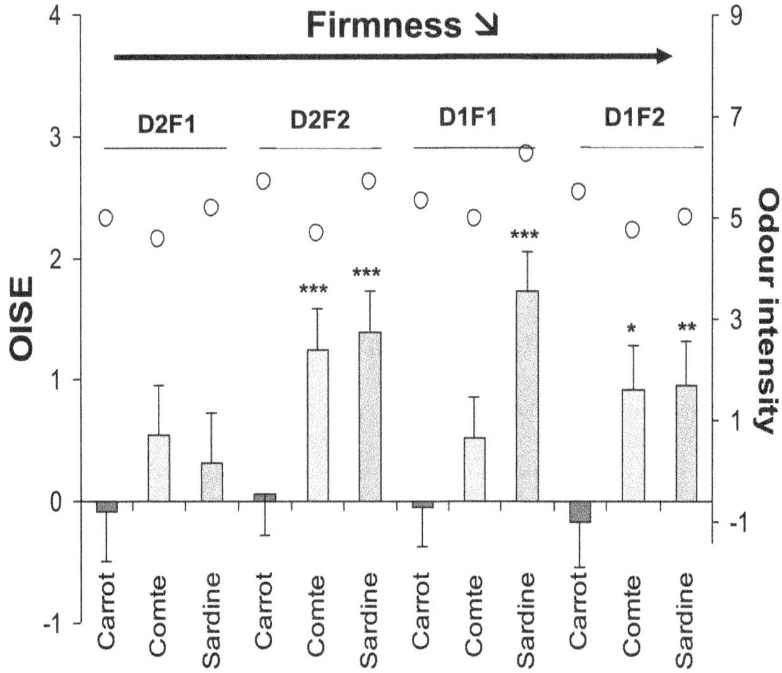

Figure 2. Enhancement of salty taste for "Sardine", "Comté" and "Carotte" aromas in solids lipoprotein matrices. D: Dry matter; F: Fat/D ratio; 1: low level; 2: high level. The circles represent the mean aroma intensity for each product. Adapted from (24). Copyright (2010) with permission from Elsevier.

To sum up, odours can be used to enhance saltiness in water solutions and complex foods containing a small amount of NaCl. Nevertheless, in food products, OISE is dependent on food composition. OISE is no longer observed for high NaCl content and firm textures. NaCl content is a key driving factor of OISE potency. The results of our studies have shown that OISE may compensate for up to 20% decrease in salt content of food.

Combination of OISE with Other Strategies

The objective of the studies reported below was to evaluate if other strategies could be combined to OISE to enhance saltiness perception and therefore increase the compensation level.

Ternary Cross-Modal Interactions: Saltiness-Sourness-Aroma

During consumption, food usually elicits several tastes and taste-taste perceptual interactions have also been reported to impact saltiness (*25*). Salty and sour tastes were found to enhance each other at low intensity levels: thus citric acid has an enhancing effect on saltiness for a low level of NaCl (*26*). In this study we explored the enhancement of salty taste of sodium and potassium chloride salts (NaCl and KCl), induced by sourness in combination with OISE (*21*).

Thirteen water solutions were prepared corresponding to a full factorial design with aroma (no aroma or sardine aroma) and taste (no tastant, 3 single tastants and 2 mixtures) as factors; a reference solution with 25% more sodium chloride (25 mM in Evian water) was also used. The sardine aroma was used at a concentration of 0.5 g/L. The taste factor in the factorial design relied on Evian water (no tastant), sodium chloride (20 mM in Evian water), citric acid (2.5 mM), potassium chloride (40 mM) and the following two mixtures: citric acid + sodium chloride and potassium chloride + sodium chloride. Sixty one panellists tested the samples for their taste attributes and odour intensities (*21*). The analysis of perceived saltiness for the different combinations of salty and aroma solutions indicated that, as expected, the solution containing only citric acid was not salty. The sample containing 40 mM of KCl was less salty compared to the 20 mM NaCl solution; however, the difference between these two means did not reach statistical significance ($p > 0.05$). Solutions containing a mixture of NaCl and KCl were perceived as the saltiest, with no influence of the addition of the sardine aroma. Conversely, mixing NaCl (20 mM) and citric acid did not modify saltiness compared to the sample containing only NaCl, but adding the sardine aroma led to a significant increase in the saltiness of the acid + salt mixture only. In all of the other solutions, with the exception of the one containing 20 mM NaCl, adding the sardine aroma led to a significant increase in perceived saltiness. Thus, complex perceptual interactions can take place in an aroma-salty-sour ternary mixture contributing to an efficient enhancement of saltiness perception, which allows a compensation for more than a 25% reduction of salt content in model solutions.

Heterogeneity in Stimuli Distribution

New approaches reported that stimuli contrasts can enhance sweetness (*27*, *28*), saltiness (*12*, *29*), fat perception (*30*) and aroma perception (*31*, *32*). In this study, we evaluated whether saltiness enhancement induced by stimuli contrast could be combined to OISE. This strategy was based on modifications of the quality and the quantity of stimuli or on the modification of the structure and the composition of the food matrix.

We first investigated whether a heterogeneous distribution of NaCl in a four-layer cream-based hot-served model food could enhance salty perception and consumer liking (*33*). The main ingredients of this cream-based food were: whipping cream, pasteurised eggs, Emmental cheese, modified food starch, wheat flour, xanthan gum, table salt, mineral water and food grade aroma. In these food products the overall added NaCl was 0.8% (w/w), but salt distribution varied throughout the four layers, according to 4 modalities: homogeneous distribution, and 3 different levels of heterogeneity of NaCl distribution within the four layers. Products with a low level of heterogeneity were composed of two layers at 0.4% added NaCl and two layers at 1.2%, products with a medium level of heterogeneity were composed of two salt-free layers and two layers at 1.6% added NaCl, and products with a high level of heterogeneity were composed of three salt-free layers and one salty layer at 3.2% added NaCl. Two control samples with a homogeneous NaCl distribution were also presented, one containing the same amount of salt as the other products and another with 25% more salt. The saltiness intensity evaluation was performed by a panel of 102 consumers (70 women and 32 men, aged 18–65 years) in a single session and the liking evaluation was performed by a panel of 80 consumers (53 women and 27 men, aged 20–66 years) in another session. The results are reported on Figure 3.

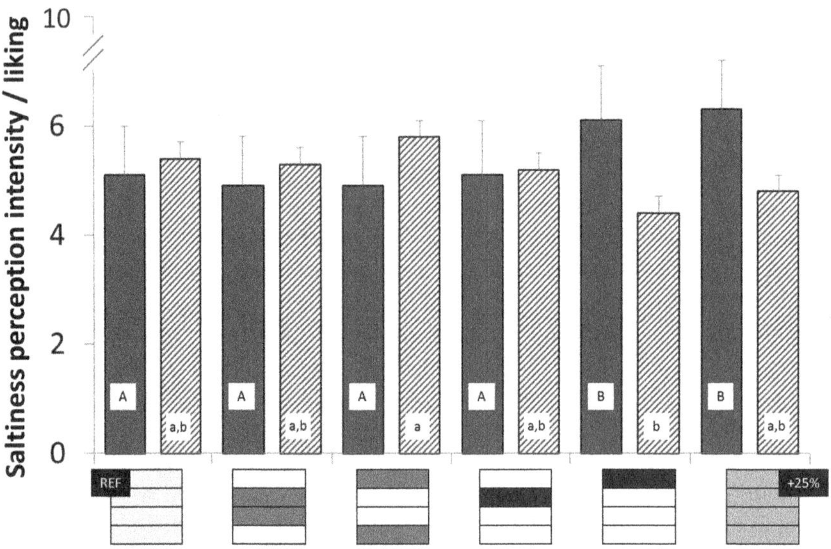

Figure 3. Intensity of salty taste (grey bars) and liking level (hatched bars) according to the heterogeneity of the distribution of salt in a cream-based food matrix made of four layers (schematized in abscise where the intensities of grey represents the salt concentration in the layers). The product REF has a homogeneous distribution of salt and the product +25% has a homogeneous distribution of salt but with an overall content of 25% more compared to the other products. The letters represent the significant differences between the means. Two different letters indicate a significant difference (p<0.05). The error bars represent the standard deviation of the mean.

Among the tested combinations, those with the highest heterogeneity in NaCl distribution, and especially the one with NaCl concentrated in a single external layer, were perceived as saltier than the other combinations with the same overall NaCl content. The saltiness intensity of this combination was equivalent to the saltiness intensity of the homogeneous control product containing 25% more NaCl. Thus, high heterogeneity of NaCl distribution leads to saltiness enhancement as compared to homogeneous distribution. This result suggests that the NaCl concentration in the salted layer is a critical factor that likely drives the saltiness enhancement.

In another experiment, we examined whether the combination of this strategy of heterogeneous spatial distribution of NaCl with heterogeneous spatial distribution of a salt-associated aroma (ham aroma) could improve the saltiness enhancement (*34*). We used food products and a sensory procedure similar to those described in the above experiment. Ten four-layer products (FLPs) containing the same total amount of NaCl and ham aroma, 0.5% (w/w) and 0.05% (w/w), respectively, were produced. The spatial distribution of NaCl and aroma was different from one layer to another. Moreover, a FLP without added ham aroma was used as an unflavoured reference and an unflavoured FLP containing 20% more NaCl was used as a saltier reference.

The results revealed a significant OISE by the ham aroma whatever the NaCl distribution. The heterogeneity of salt distribution also induced saltiness enhancement and confirmed the results obtained in the other experiement (*33*).

The combination of the two strategies allowed obtaining an enhancement of saltiness perception able to compensate a 35% reduction of NaCl content for this type of food, without loss of acceptability. Indeed, the products in which the stimuli distribution was heterogeneous were well accepted by consumers though products perceived as saltier seemed less liked. This appreciation seemed to be increased by the presence of salt-associated aroma.

Influence of Food Matrix Composition on Flavour Release and Perception

The release of NaCl and aroma compounds during the in–mouth process is influenced by food composition such as lipids, protein, water contents (*8*, *35*). These ingredients can physically interact with NaCl and aroma compounds and influence flavour release and perception. Consequently, potential taste – aroma cross modal perceptual interactions can be affected. Moreover, the modulation of the release of flavor compounds during eating can be achieved by changing food composition. This modulation can be combined with the effects described above to increase perception of a low-salt food. The objective of this part is not to describe the effect of changes in food composition on cross modal perceptual interaction but only to present a case study on model cheeses showing in what extent changes in food composition can influence taste and aroma compound release and perception (*36*, *37*).

Model cheeses with analogue cheese technology were developed according to Boisard et al (*37*, *38*). They were made of rennet casein, milk fat, water, melting salt, minerals and acids varying in 3 lipid protein ratios (L28/P20, L24/P24, L20/P28), with (0.1 g/kg) and without added NaCl (0.067 g/kg originated from melting salt). The 6 model cheeses were flavoured with 7 aroma compounds of different chemical classes.

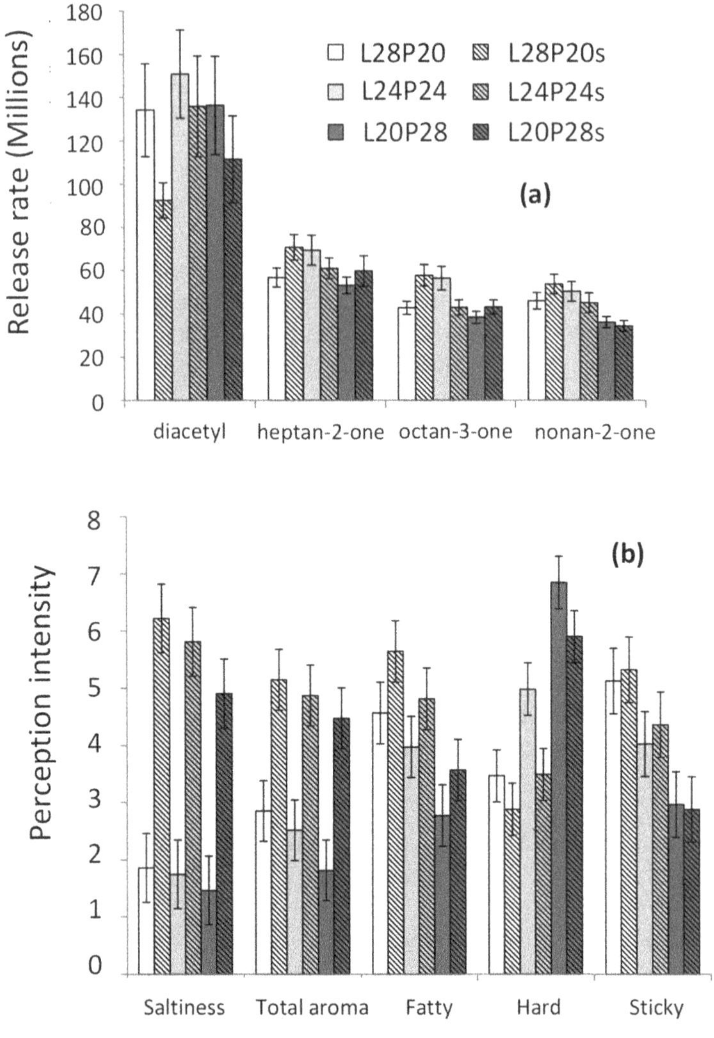

Figure 4. (a) Initial release rate of taste and aroma stimuli from the model cheeses. (b) Sensory perception of the model cheeses: taste and aroma attributes. L/P ratio corresponding to the lipid/protein ratio and calculated from the dry matter content of milk fat and caseins. s for formulations with added salt.

The temporal release parameters of aroma compounds measured by following the *in vivo* release of aroma compounds by atmospheric pressure chemical ionisation mass spectrometry (*37*) were affected by the L/P ratio and NaCl content. As an example, the initial release rate values fell when the L/P ratio significantly decreased from L24/P24 to L20/P28, except for diacetyl which is the more polar aroma compound (Figure 4a). This result suggests that the presence of proteins might reduce the initial release rate for hydrophobic aroma compounds due to hydrophobic effects or because of the increase in product firmness which would limit the diffusivity of aroma compounds. For model cheeses with the higher L/P ratio (L28/P20), the addition of NaCl led to a significant increase in initial release rate for all aroma compounds except diacetyl. This result could be explained by the salting out effect on the release of hydrophobic aroma compounds. In addition, the initial release rate decreased when the hydrophobicity of the compounds increased. Indeed, the initial release rate of diacetyl (log P = 1.34) was much higher than the initial release rate of nonan-2-one (log P = 2.71).

The effect of composition on flavour perception is presented on Figure 4b. Model cheeses with a lower L/P ratio were perceived harder due to the higher amount of proteins. For salted products, the lower L/P ratio was perceived more salty due to differences in composition. This can be explained by perceptual interactions between fat and saltiness perception as the fattier models (L28P20) were perceived more salty while for each L/P ratio the more salted (s) were perceived more fatty. However, this observation could be also explained by differences in NaCl diffusion.

For aroma perception, when the NaCl content increased, the overall aroma perception intensity and the specific aroma notes increased. When L/P ratio decreased, the overall aroma perception intensity and the specific aroma qualities decreased.

Conclusion

OISE is a very interesting strategy to enhance saltiness perception in liquid and in solid foods with reduced NaCl content. However, it is dependent on the texture and NaCl content of the food. The heterogeneity of salt distribution with an important contrast can be efficient to significantly enhance saltiness perception. These strategies can be combined to improve the enhancement effect and thus compensate for a higher reduction of NaCl content in food while maintaining a good acceptability by the consumers. These strategies can thus contribute to prevent a negative impact on the economics of the food industry when producing foods with low NaCl content, as recommended. More investigations are running to evaluate the enhancing impact of aroma on the saltiness perception for more complex and real food systems. These strategies could be combined to an optimisation of the food matrix composition allowing optimizing flavour compounds release and maximizing saltiness perception.

References

1. Strazzullo, P.; D'Elia, L.; Kandala, N.-B.; Cappuccio, F. P. Salt intake, stroke, and cardiovascular disease: meta-analysis of prospective studies. *Brit. Med. J.* **2009**, *339*, b4567–b4576.
2. Dietary Guidelines Advisory Committee. *Report of the Dietary Guidelines Advisory Committee on the Dietary Guidelines for Americans to the Secretary of Agriculture and the secretary of Health and Human Services*; U.S. Department of Agriculture, Agricultural Research Service: Washington DC, 2010.

3. World Health Organization. *Reducing Salt Intake in Populations: Report of a WHO Forum and Technical Meeting*; Geneva, CH, 2007.
4. Taormina, P. J. Implications of Salt and Sodium Reduction on Microbial Food Safety. *Crit. Rev. Food Sci.* **2010**, *50*, 209–227; Erratum **2011**, *51*, 447.
5. Kilcast, D.; Angus, F. In *Reducing salt in foods: Pratical strategies*; Woodhead Publishing Limited: Cambrigde, U.K., 2007.
6. Guinee, T. P.; O'Kennedy, B. T. Reducing Salt in Cheese and Dairy Spreads. In *Reducing Salt in Foods*; Kilcast, D., Angus, F., Eds.; Woodhead Publishing Limited: Cambrigde, U.K., 2007; pp 316–357.
7. Guinee, T. P.; Fox, P. F. Salt in cheese: physical, chemical and biological aspects. In *Cheese: Chemistry, Physics and Microbiology. General Aspects*; Fox, P. F., McSweeney, P. L. H., Cogan, T. M., Guinee, T. P., Eds.; Elsevier Acadelic Press: Amsterdam, Netherlands; Vol. 2, pp 207–259.
8. Guichard, E. Interactions between flavor compounds and food ingredients and their influence on flavor perception. *Food Rev. Int.* **2002**, *18*, 49–70.
9. Toldra, F.; Barat, J. M. Recent patents for sodium reduction in foods. *Recent Pat. Food Nutr. Agric.* **2009**, *1*, 80–6.
10. Toldra, F.; Barat, J. M. Strategies for salt reduction in foods. *Recent Pat. Food Nutr. Agric.* **2012**, *4*, 19–25.
11. Girgis, S.; Neal, B.; Prescott, J.; Prendergast, J.; Dumbrell, S.; Turner, C.; Woodward, M. A one-quarter reduction in the salt content of bread can be made without detection. *Eur. J. Clin. Nutr.* **2003**, *57*, 616–620.
12. Busch, J.; Tournier, C.; Knoop, J. E.; Kooyman, G.; Smit, G. Temporal Contrast of Salt Delivery in Mouth Increases Salt Perception. *Chem. Senses* **2009**, *34*, 341–348.
13. Mosca, A. C.; Bult, J. H. F.; Stieger, M. Effect of spatial distribution of tastants on taste intensity, fluctuation of taste intensity and consumer preference of (semi-)solid food products. *Food Qual. Prefer.* **2013**, *28*, 182–187.
14. Small, D. M.; Prescott, J. Odor/taste integration and the perception of flavor. *Exp. Brain Res.* **2005**, *166*, 345–357.
15. Thomas-Danguin, T. Flavor. In *Encyclopedia of Neuroscience*; Binder, M. D., Hirokawa, N., Windhorst, U., Eds.; Springer-Verlag GmbH: Berlin Heidelberg, 2009.
16. Prescott, J. Psychological processes in flavour perception. In *Flavour Perception*; Taylor, A. J., Roberts D. D., Eds.; Blackwell Publishing: Oxford, U.K., 2004, pp 256–277.
17. Valentin, D.; Chrea, C.; Nguyen, D. H. Taste-odour interaction in sweet taste perception. In *Optimising Sweet Taste in Foods*; Spillane, W. J., Ed.; Woodhead Publishing Limited: Cambridge, U.K., 2006; pp 66–84.
18. Verhagen, J. V.; Engelen, L. The neurocognitive bases of human multimodal food perception: Sensory integration. *Neurosci. Biobehav. Rev.* **2006**, *30*, 613–650.
19. Lawrence, G.; Salles, C.; Septier, C.; Busch, J.; Thomas-Danguin, T. Odour-taste interactions: A way to enhance saltiness in low-salt content solutions. *Food Qual. Prefer.* **2009**, *20*, 241–248.

20. Djordjevic, J.; Zatorre, R. J.; Jones-Gotman, M. Odor-induced changes in taste perception. *Exp. Brain Res.* **2004**, *159*, 405–8.
21. Nasri, N.; Septier, C.; Béno, N.; Salles, C.; Thomas-Danguin, T. Enhancing salty taste through odour-taste-taste interactions: Influence of odour intensity and salty tastants' nature. *Food Qual. Prefer.* **2013**, *28*, 134–140.
22. Nasri, N.; Béno, N.; Septier, C.; Salles, C.; Thomas-Danguin, T. Cross-modal interactions between taste and smell: Odour-induced saltiness enhancement depends on salt level. *Food Qual. Prefer.* **2011**, *22*, 678–682.
23. Murphy, C.; Cain, W. S.; Bartoshuk, L. M. Mutual action of taste and olfaction. *Sens. Proc.* **1977**, 204–211.
24. Lawrence, G.; Salles, C.; Palicki, O.; Septier, C.; Busch, J.; Thomas-Danguin, T. Using cross-modal interactions to counterbalance salt reduction in solid foods. *Int. Dairy J.* **2011**, *21*, 103–110.
25. Keast, R. S. J.; Breslin, P. A. S. An overview of binary taste-taste interactions. *Food Qual. Prefer.* **2003**, *14*, 111–124.
26. Breslin, P. A. S. Interaction among salty, sour and bitter compounds. *Food Sci. Technol.* **1996**, *7*, 390–399.
27. Burseg, K. M. M.; Brattinga, C.; de Kok, P. M. T.; Bult, J. H. F. Sweet taste enhancement through pulsatile stimulation depends on pulsation period not on conscious pulse perception. *Physiol. Behav.* **2010**, *100*, 327–331.
28. Mosca, A. C.; van de Velde, F.; Bult, J. H. F.; van Boekel, M. A. J. S.; Stieger, M. Enhancement of sweetness intensity in gels by inhomogeneous distribution of sucrose. *Food Qual. Prefer.* **2010**, *21*, 837–842.
29. Noort, M. W. J.; Bult, J. H. F.; Stieger, M.; Hamer, R. J. Saltiness enhancement in bread by inhomogeneous spatial distribution of sodium chloride. *J. Cereal Sci.* **2010**, *52*, 378–386.
30. Mosca, A. C.; Rocha, J. A.; Sala, G.; van de Velde, F.; Stieger, M. Inhomogeneous distribution of fat enhances the perception of fat-related sensory attributes in gelled foods. *Food Hydrocolloids* **2012**, *27*, 448–455.
31. Nakao, S.; Ishihara, S.; Nakauma, M.; Funami, T. Inhomogeneous Spatial Distribution of Aroma Compounds in Food Gels for Enhancement of Perceived Aroma Intensity and Muscle Activity during Oral Processing. *J. Texture Stud.* **2013**, *44*, 289–300.
32. Nakao, S.; Ishihara, S.; Nakauma, M.; Funami, T. Effects of Inhomogeneous Spatial Distribution of Aroma Compounds on Perceived Aroma Intensity and Human Eating Behavior for Neutral pH Gels. *Food Sci. Technol. Res.* **2013**, *19*, 675–683.
33. Emorine, M.; Septier, C.; Thomas-Danguin, T.; Salles, C. Heterogeneous salt distribution in hot snacks enhances saltiness without loss of acceptability. *Food Res. Intl.* **2013**, *51*, 641–647.
34. Emorine, M.; Septier, C.; Andriot, I.; Martin, C.; Salles, C.; Thomas-Danguin, T. Combined heterogeneous distribution of salt and aroma in food enhances salt perception. *Food Funct.* **2015** DOI:in press10.1039/c4fo01067a.
35. Lawrence, G.; Buchin, S.; Achilleos, C.; Bérodier, F.; Septier, C.; Courcoux, P.; Salles, C. In Vivo Sodium Release and Saltiness Perception in

Solid Lipoprotein Matrices. 1. Effect of Composition and Texture. *J. Agric. Food Chem.* **2012**, *60*, 5287–5298.

36. Boisard, L.; Andriot, I.; Martin, C.; Septier, C.; Boissard, V.; Salles, C.; Guichard, E. The salt and lipid composition of model cheeses modifies in-mouth flavour release and perception related to the free sodium ion content. *Food Chem.* **2014**, *145*, 437–444.
37. Boisard, L.; Tournier, C.; Semon, E.; Noirot, E.; Guichard, E.; Salles, C. Salt and fat contents influence the microstructure of model cheeses, chewing/swallowing and in vivo aroma release. *Flavour Fragrance J.* **2014**, *29*, 95–106.
38. Boisard, L.; Andriot, I.; Arnould, C.; Achilleos, C.; Salles, C.; Guichard, E. Structure and composition of model cheeses influence sodium NMR mobility, kinetics of sodium release and sodium partition coefficients. *Food Chem.* **2013**, *136*, 1070–7.

Chapter 4

Influence of Auditory Cues on Chemosensory Perception

Han-Seok Seo*,[1] and Thomas Hummel[2]

[1]Department of Food Science, University of Arkansas, 2650 N. Young Avenue, Fayetteville, Arkansas 72704, U.S.A
[2]Smell & Taste Clinic, Department of Otorhinolaryngology, TU Dresden, Fetscherstrasse 74, 01307 Dresden, Germany
***E-mail hanseok@uark.edu.**

Previous studies have demonstrated that sounds elicited by mastication or swallowing processes can affect consumers' perception of food texture and quality. However, little is known about the modulatory role of auditory cues with respect to chemosensory perception. People often perceive odors, flavors, and taste substances in the presence of various sounds. This review addresses past and current findings associated with influences of auditory cues on perception of chemosensory cues. It specifically focuses on three main points: 1) cross-modal correspondences between auditory and chemosensory cues, 2) effects of congruent sound on chemosensory perception, and 3) effects of background sound on chemosensory perception. Although interest on this topic has been growing, further studies are necessary to answer many questions, including those regarding the mechanisms underlying cross-modal interaction between auditory and chemosensory cues.

Introduction

It is rare to consume foods or drinks without hearing any sound. Even though people may consume foods in silence, people at least hear the sounds elicited by mastication or swallowing processes. Also, when people have meals with their family, friends, or colleagues, mutual conversation is typically present during meal consumption (*1*, *2*). A recent survey reported that more than half (58.8%) of the 244 North American respondents preferred eating while having conversations with others; only 3.7% preferred eating in silence (*1*). Many people are also used to eating or drinking while listening to background music or noise. These phenomena represent the most substantial impact of auditory cues with respect to food perception and acceptability in everyday life. In other words, it is significant that eating has an intimate connection with a variety of auditory cues.

Past studies have demonstrated that auditory cues play an important role not only in judging textural characteristics such as crispness (*3–5*) and carbonation (*6*), but freshness as well (*5*). For example, if a sound produced by mastication of potato chips is louder, people seem to perceive the chips as being crisper and fresher (*5*). Similarly, people are likely to rate carbonated water to be more carbonated with an increase in overall sound level and/or high frequency components of the sound emitted from the carbonated beverage (*6*).

In previous studies examining the influence of auditory cues on food perception, little attention has been paid to chemosensory aspects, such as the senses of smell, taste, and trigeminal function, of the food or drink samples. Notably, there is growing evidence that chemosensory perception can be altered by auditory cues like background music or noise. This study will review past and current studies of cross-modal interaction between auditory and chemosensory cues. The influence of auditory cues on trigeminal sensation is not included in this review because of the limited number of publications on that topic. Previous research regarding the cross-modal interaction between the auditory and chemosensory cues can be classified into three main streams: 1) cross-modal correspondence between auditory and chemosensory cues, 2) influence of congruency between the bimodal cues, and 3) influence of irrelevant sound, whether background music or noise, on chemosensory perception. This review will thus discuss earlier findings emphasizing these topics.

Cross-Modal Correspondences between Auditory and Chemosensory Cues

When describing a relationship between bimodal cues, earlier studies have often used the term, "cross-modal correspondence". "Cross-modal correspondence" refers to "*a compatibility effect between attributes or dimensions of a stimulus (i.e., an object or event) in different sensory modalities (be they redundant or not)*" (*7*). It has long been known that such cross-modal correspondence is innate (*8*) and universal among individuals (*7*). It is also considered as "weak synaesthesia", defined as "*cross-sensory correspondences expressed through language, perceptual similarity, and perceptual interactions during information processing*" (*9*) or "synaesthetic congruency" (see below) (*7*, *10*).

Cross-Modal Correspondences between Auditory and Olfactory Cues

Many studies have demonstrated cross-modal correspondences between visual and auditory cues (*11–13*). There is growing empirical evidence that cross-modal correspondences occur between auditory and olfactory cues (*14–19*). For example, people could consistently match certain auditory frequencies (e.g., 200 Hz and 1,000 Hz) with specific fragrant odors (*14*). Furthermore, recent studies have shown that certain odors sniffed via the nose could be matched not only with specific frequencies, but also with timbres of musical notes (*17*, *19*). For example, fruit odors could be paired with high-pitched sounds.

A cross-modal correspondence was also found between auditory cues (e.g., pitches and timbres of musical notes) and flavors perceived through the mouth (*15*, *16*, *18*). A cross-modal correspondence between flavors and musical pitches has not been consistently observed, and is likely to be mediated by characteristics of "potency" (e.g., strong *vs.* weak) and "activity" (e.g., active *vs.* passive) of the bi-modal cues (*15*, *18*). On the other hand, previous studies have often reported a cross-modal correspondence between flavors and musical timbres, and this seems to be linked by subjective hedonic valence (e.g., pleasantness *vs.* unpleasantness) (*15*, *18*); i.e., pleasant flavors can be matched with pleasant musical timbres based on individuals' subjective hedonic valence. This is an example of "the indirect hypothesis" that accounts for the cross-modal correspondence between auditory and olfactory cues. Deroy et al. (*20*) proposed that emotional similarity between auditory and olfactory cues can mediate such cross-modal correspondence. To explain the cross-modal correspondence between auditory and olfactory cues, Deroy et al. (*20*) proposed another hypothesis, "the amodal hypothesis", suggesting that perception of auditory or olfactory cues may take place in a common amodal dimension such as space. This hypothesis is, to some extent, supported by recent animal studies demonstrating that auditory and olfactory inputs converge in the mammalian cerebral cortex through the olfactory tubercle and the auditory cortex (*21–25*). Furthermore, Deroy et al. (*20*) explained the cross-modal correspondence using "the transitivity hypothesis." That is, if a certain dimension (X) corresponds to another dimension (Y), and the dimension (Y) also corresponds to a dimension (Z) in a third sensory modality, a cross-modal correspondence will occur between dimensions X and Z (*20*). However, these hypotheses are based on theoretical aspects, and future empirical research would be needed to validate them.

Cross-Modal Correspondences between Auditory and Gustatory Cues

Like olfactory cues, gustatory cues have been shown to exhibit a cross-modal correspondence with not only frequencies (*18*, *26–30*), but also with musical notes (*15*, *18*). Using the implicit association test (IAT, a measure of the strength of individuals' implicit association between concepts or attributes), Crisinel and Spence (*28*) demonstrated that sour and bitter tastes could be associated with high-pitched and low-pitched sounds, respectively. In that study, the researchers presented the names of typically sour- or bitter-tasting food items instead of presenting actual taste solutions or food items. In the follow-up study, participants responded more quickly and accurately when the names of sweet-tasting food items were paired with high-pitched sounds than when they were paired with low-pitched sounds (*29*). Furthermore, using actual food samples (dark chocolate, marzipan-filled chocolate, and milk chocolate), Crisinel and Spence (*18*) demonstrated cross-modal correspondences between sound pitch and taste quality. Participants were asked to match the pitch of four instrumental sounds (piano, strings, woodwinds, and brass), ranging from C2 (64.4 Hz) to C6 (1,046.5 Hz) in intervals of two tones, to each chocolate sample presented in the study, followed by intensity ratings of bitterness, sweetness, and pleasantness of the sample. The study showed that, if a higher-pitched sounds was chosen, chocolate samples were rated sweeter and less bitter. Similarly, in another study Crisinel and Spence (*15*) demonstrated that low-pitched sounds produced by brass instruments were often matched with the bitter taste of a caffeine solution, while high-pitched sound played on a piano were frequently paired with the sweet taste of sucrose solution (Figure 1).

Effects of Congruent Sound on Chemosensory Perception

A large number of previous studies investigating cross-modal interaction involving the sense of smell have highlighted how "congruency" between bimodal cues affects olfactory perception. Schifferstein and Verlegh (*31*) defined congruency as "*the extent to which two stimuli are appropriate for combination in a food product.*" Based on this definition, Seo (*32*) characterized the congruency as "*the extent to which bimodal stimuli are appropriate for combination in everyday life*", extending the concept of congruency to all kinds of cues occurring in everyday life. Spence and colleagues proposed three types of congruency: "synaesthetic congruency", "spatiotemporal congruency", and "semantic congruency" (*7*, *10*). Among these, synaesthetic congruency, characterized as "*the correspondences between more basic stimulus features (e.g., pitch, lightness, brightness, size) in different modalities*" (*7*) appears to be most aligned with the concept of "cross-modal correspondence" addressed above. Spatiotemporal congruency, characterized as "*the proximity between two unisensory events in time and space*" (*10*), has rarely been observed in cross-modal interaction between olfactory and auditory cues. In a related study by La Buissonnière-Ariza et al. (*33*), participants were asked to localize left or right unilateral auditory cues (a high-pitched alert sound for 150 ms) as soon as possible in the presence of one of four chemosensory conditions: odourless air puffs (somatosensory cue), phenylethanol odor (olfactory cue), eucalyptol odor (mixed olfactory and trigeminal cue), and no additional cue. The chemosensory or somatosensory cue was delivered to either the left or right nostril. The researchers tested whether participants' reaction times to the auditory cue might differ as a function of spatial congruency between auditory and chemosensory cues, but no significant effect of spatial congruency was observed in the reaction time to the auditory cue. Finally, semantic congruency is characterized as "*a cross-modal match (vs. mismatch) in terms of the identity or meaning of the unisensory component stimuli*" (*10*). For example, an odor of potato chips is semantically congruent with the sound of biting crispy chips, but not with the sound of drinking a glass of water. In this section, the effect of congruent sounds on olfactory perception based on the semantic congruency among the three types of congruency will be addressed.

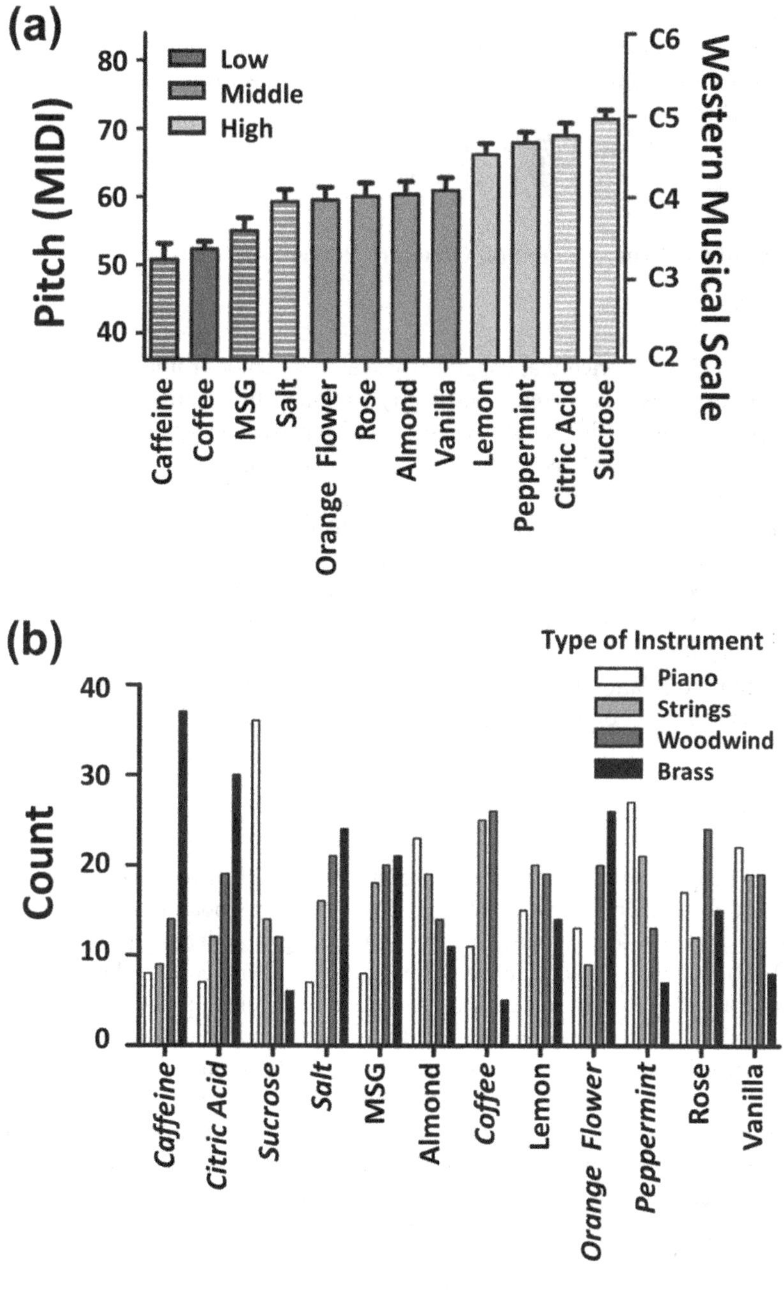

Figure 1. Cross-modal correspondences of auditory and olfactory cues. Figure 1 (a) represents mean pitch matched to each taste or flavor. MIDI (musical instrument digital interface) note numbers (shown on the left-hand y-axis) were used to code the pitch of the chosen notes. The western musical scale is depicted on the right-hand axis. Basic tastes are represented with lined bars. Taste and flavors are grouped as low-pitched (dark gray), middle-pitched (gray), or high-pitched (light gray) groups according to the pitch that they were matched to. Figure 1 (b) represents the number of times each type of instrument matched to each taste or flavor; the maximum count was 68. A taste or flavor showing a significant preference with respect to the choice of instrument is marked in Italics. (Source: Reproduced with permission from reference (15). Copyright 2010 Springer.)

Effects of Congruent Sound on Olfactory Perception

Odor Intensity

It has been reported that congruent visual (e.g., color or image) or gustatory (e.g., taste) cues can increase sensitivity or intensity of olfactory cues. For example, a study by Christensen (*34*) showed that participants perceived processed cheese odors more intensely when the cheese was appropriately colored than when it was inappropriately colored. However, unlike gustory cues, congruent visual cues do not always increase odor intensity (*35*, *36*). Similarly, previous studies have demonstrated that auditory cues do not affect the intensities of their congruent odors (*37*, *38*). For example, as participants rate a pair of auditory and olfactory cues to be more congruent, they appear to rate the olfactory cue as significantly more pleasant. However, previous studies (*38*) reported no significant correlation between individual ratings for the degree of congruency and intensity of an olfactory cue.

Odor Pleasantness

Do people like coffee aroma significantly more while listening to the sound of drinking coffee than while listening to the sound of biting potato chips? To answer this question, Seo and Hummel (*37*) presented congruent, incongruent, or neutral sounds both before and during the presentation of either a coffee odor or a potato chip odor to 22 German participants. As shown in Figure 2, participants liked the coffee odor significantly more when it was presented with a congruent sound (the sound of drinking coffee) than when presented with an incongruent sound (the sound of biting potato chips). Similarly, potato chip odor was rated more pleasant when accompanied with the sound of eating the potato chips than with either the sound of drinking coffee or white noise (*37*). Congruent sound-enhanced odor pleasantness was also obtained in other studies using congruent background sounds (*38*). For example, German participants rated coffee odors more pleasant when presented with the sound of coffee in a popular German advertisement than with a Christmas carol. Can this trend also be observed even when people are unaware that the background sound is coffee advertisement music? The coffee advertisement music may have no effect on pleasantness of coffee aroma if individuals are unaware of the type of advertisement music. Since individuals have no semantic connection between the coffee aroma and the advertisement music, the congruency effect may be absent when that particular pair of auditory and olfactory cues is presented.

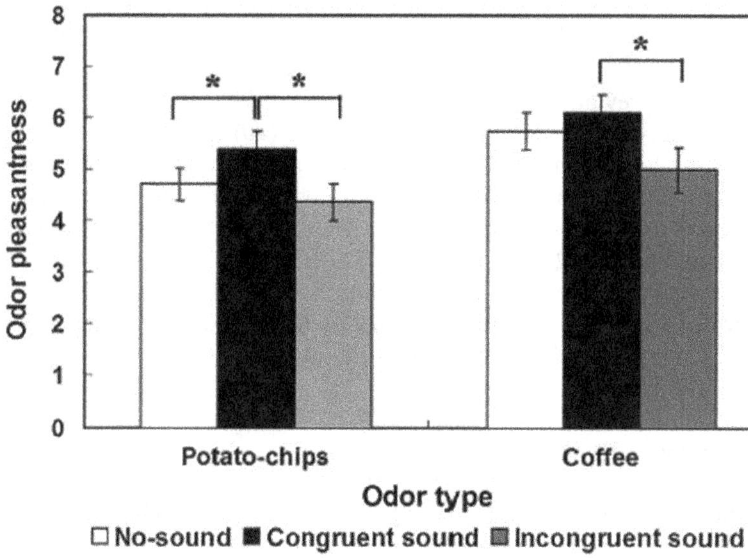

Figure 2. Mean ratings of odor pleasantness in relation to three different auditory cues: no-sound (white noise), congruent sound, and incongruent sound. White noise was expressed as "no-sound" meaning no distinctive sound. Odor pleasantness in the presence of auditory cue was rated on a visual analogue scale ranging from 0 (extremely unpleasant) to 10 (extremely pleasant). The pleasantness ratings of odors were significantly higher when an odor was presented with a congruent sound than when presented with an incongruent sound. An asterisk represents a significant difference at $P < 0.05$. The error bars represent the standard errors of the mean. (Source: Reproduced with permission from reference (37). Copyright 2011 Oxford University Press.)

Previous studies have shown that auditory cues facilitate identifying corresponding odors, in turn increasing pleasantness of the odors (*38*). A question may be raised as to why auditory cues affect odor pleasantness, but not odor intensity. A plausible explanation for this is that a pleasantness rating is a "synthetic" task, while an intensity rating is largely an "analytic" task (*31*, *39*). In other words, when participants rate their hedonic response to an odor sample, they tend to consider not only how pleasant the odor itself is, but also how harmonious the accompanying auditory cue is, possibly resulting in an increase in the impact of auditory cues on odor pleasantness ratings. However, when participants rate the intensity of the odor sample, they are more likely to focus on how strong the odor is apart from the presence of auditory cues, which possibly decreases the effect of auditory cues on odor intensity ratings (*38*).

Effects of Congruent Sound on Gustatory Perception

Taste Intensity

Based on previous findings with respect to cross-modal correspondence between auditory and gustatory cues, Crisinel et al. (*40*) conducted an experiment to determine whether synaesthetically congruent music could affect taste intensity. They designed two soundtracks, each mainly composed of either low-pitched or high-pitched sound, and representing either bitter taste or sweet taste, respectively. Twenty participants (12 women, aged 17 – 33 years) received cinder toffee samples in the presence of either the "bitter" or the "sweet" soundtrack. The cinder toffee samples were rated more bitter when tasted while listening to the bitter soundtrack than while listening to the sweet soundtrack.

Taste Pleasantness

Little is known about the effect of congruent sound on taste pleasantness. In the above study (*40*), the participants were also asked to rate pleasantness of taste/flavor of the cinder toffee sample in the presence of either the "sweet" or the "bitter" soundtrack; taste and flavor were not separately asked for in that study. The results demonstrated that liking of the taste/flavor of the cinder toffee was not significantly different for sweet and bitter soundtrack conditions.

Effects of Background Sound on Chemosensory Perception

People frequently experience chemosensory cues, such as aroma, flavor, and taste, in the presence of background sound ("external sound which is unwanted or irrelevant to the chemosensory cue"). For example, when we drink a cup of coffee at a restaurant, we are usually exposed not only to coffee aroma and flavor, but also to background music, conversation, and the sounds of coffee machines. Can background music or noise influence consumers' perception of coffee aroma and flavor?

Sound levels at many restaurants and dining places have become quite loud; noise is identified as the second most common complaint of restaurant-goers (*41*). Furthermore, emotional arousal or unpleasantness caused by background noise may alter perception of foods eaten in a noisy location. In fact, a recent survey reported that only one person (0.4%) among 244 North American respondents (136 women; 207 Caucasians; aged 19 - 78 years) preferred eating at a noisy restaurant. In a modern society, background noise has become common in many public places, including on subways and airplanes. Spence et al. (*42*) pointed out an interesting observation that many people specifically ask air stewards for tomato juice on airplanes. They suggested a plausible reason for this phenomenon; it might be due to the fact that umami taste, which is rich in tomato juice, appears to be relatively unaffected by the loud noise during flight. Keeping this background in mind, this section will review earlier findings of the association between background sound and chemosensory perception.

Effects of Background Sound on Olfactory Perception

Odor Discrimination

Mozart's music has been found to facilitate cognitive performance; this has been referred to as the "Mozart effect" (*43*). Rauscher et al. (*43*) showed that Mozart's music improved participants' performance with respect to standard IQ spatial-reasoning tasks. Since an odor-discrimination task, i.e., picking the odd odorant out from among three odorants (including two identical odorants and one different odorant), seems to demand significant cognitive function (*44*), Seo et al. (*45*) tested whether such a "Mozart effect" is observed in performing an odor discrimination task. More specifically, the researchers asked 36 German participants (27 women; aged 19-36 years) to perform the odor discrimination tasks of the Sniffin' Sticks test, making 16 sets of three alternative forced choices (3-AFC), while listening to either a Mozart composition (sonata for two pianos in D major, K448) or to no added sound (silence). Contrary to the researchers' expectations, the Mozart effect was not present in the experiment; there was no significant difference between participants' performances with respect to the odor discrimination task either with or without Mozart's sonata. It was argued that, since the given task was too easy, there appeared to be a ceiling effect, possibly resulting in a lack of performance improvement.

A significant influence of background sound on performance in an odor discrimination task has also been found. Seo et al. (*45*) asked 38 participants (29 women; aged 19-40 years) to perform the odor discrimination task of the Sniffin' Sticks test under either background noise conditions or silent conditions. Under background noise conditions, both verbal noise (audio book of a comedian's humorous speech) and non-verbal noise (the sounds of a crowded party) were presented over headphones. Participants showed degraded performance on the odor discrimination task when presented with background noise when compared to a silent condition. Furthermore, verbal noise, in comparison to non-verbal noise, yielded significantly higher detrimental effects on performance in the odor discrimination task. In other words, when compared to in a silent condition (baseline), participants' poor performance in discriminating different odors was more pronounced while listening to a comedian's humorous speech than while listening to the sound of a crowded party.

Odor Sensitivity

It is well known that odor sensitivity (or threshold) tasks demand less cognitive load than does the odor discrimination task (*44*). If so, there is a question as to whether background noise would affect performance on an odor sensitivity task. Previous research has demonstrated that background noise, whether verbal or non-verbal, has no effect on participants' performance on odor sensitivity tasks when compared to a silent condition (*46*). However, significant interaction between the type of background noise and an extraversion of personality trait has been demonstrated. More specifically, as shown in Figure 3, introverts demonstrated worse performance on an odor sensitivity task in the presence of verbal noise (audio book of a comedian's humorous speech) than under silent conditions. In contrast, extroverts showed better performance on an odor sensitivity task when presented with verbal noise rather than silence. Similarly, Koelega (*47*) reported that male extroverts, in comparison to male introverts, were more sensitive to odors in the presence of distracting noise (*47*). According to Eysenck's arousal theory of extraversion (*48*), extroverts have higher optimum arousal levels than introverts, so extroverts tend to seek more stimulation to reach their optimum arousal level; introverts are likely to need less stimulation. In this sense, when background noise, especially verbal noise, was presented, extroverts' arousal levels might reach their optimum values, in turn leading to improved performance on odor sensitivity tasks (*46*). Because background sound-induced olfactory performance can vary with respect to individuals' personal traits, sensory professionals, marketers, and business owners should consider personality traits of their main target group when manipulating background sound conditions in their shops or restaurants.

Odor Intensity

It appears that perceived intensity of olfactory cues is little influenced by background sound. When odors (phenylethanol or 1-butanol) were presented in the presence of background sounds varying in hedonic tone (e.g., baby laughing, baby crying, jazz drum, and screaming), there was no significant difference in the odor intensity for any of the background sound conditions (*37*). In another study, Fiegel et al. (*49*) tested whether flavor intensity can be either increased or decreased by particular background music genres. In that study, researchers asked 99 North American participants (53 women; 90 Caucasians; aged 18 - 30 years) to eat milk chocolate or bell peppers in the presence of four different genres of background music ("Air on the G string"): classical, jazz, rock, and hip-hop. In that study, flavor intensity of chocolate or bell peppers did not differ as a function of background music genre. However, since those studies did not compare flavor intensity both with and without background sound, it is inconclusive as to whether either presence or absence of background sound can alter perceived intensity of an olfactory cue. Including the "no sound" condition as a baseline, Woods et al. (*50*) examined changes in flavor intensities of rice cakes under both quiet (45-55 dB) and loud (75-85 dB) white-noise conditions. In that study, by subtracting the baseline (no sound condition) ratings from the quiet sound and loud sound conditions, they found that background noise-induced flavor intensity did not differ under quiet and loud noise conditions.

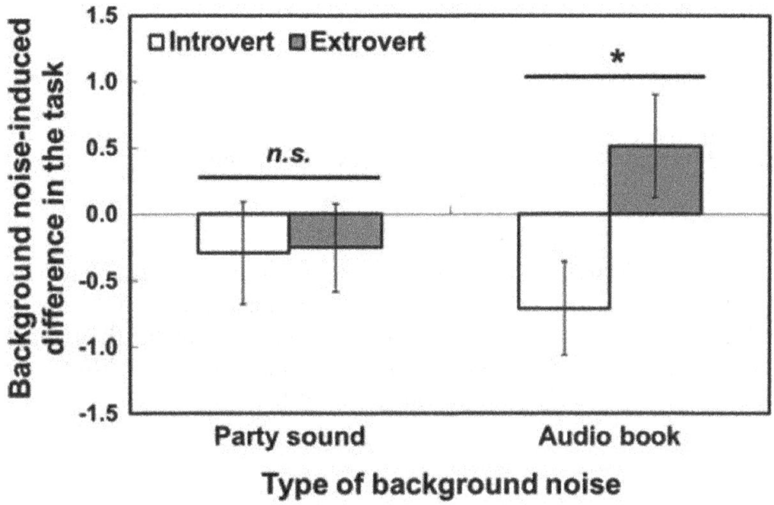

Figure 3. Comparison of background noise-induced difference in the odor sensitivity task between introvert and extrovert groups. The background noise-induced difference indicates the mean score differences between noisy and silent conditions: i.e., "party sound" – "silent condition" or "audio book" – "silent condition". The extrovert group showed better performance on the odor sensitivity task in the presence of audio book sound rather than silence. However, the introvert group showed worse performance on the odor sensitivity task with audio book sound than silence. The n.s. and asterisk represent no significance and significance at P < 0.05, respectively. The error bars represent standard errors of the means. (Source: Reproduced with permission from reference (46). Copyright 2012 Springer.)

Odor Pleasantness

In contrast to odor intensity, odor pleasantness appears to be somewhat influenced by background sound; i.e., hedonic valence of background sound can modulate the hedonic tone of olfactory cues (*37*). For example, olfactory cues, regardless of their hedonic tone, tend to be rated as more pleasant when they are presented with pleasant sounds (e.g., baby laughing and jazz drum) than when presented with unpleasant sounds (e.g., baby crying sound and screaming) (*37*). Furthermore, the more that participants liked a background sound, the more pleasant the subsequent odor became. Similarly, a positive correlation between the pleasantness ratings of food flavors and musical stimuli was also obtained in other studies (*49*). These findings show that a "halo/horns effect" can be present between auditory and olfactory cues. The halo effect, strictly a "horn effect", was observed between the hedonic ratings of white noise combined with odors. In a recent study by Velasco et al. (*51*), the researchers presented either pleasant or unpleasant odors along with three types of "musical" stimuli (pleasant, unpleasant, or white noise). The odors were rated the most unpleasant when accompanied by white noise; the latter was rated as more unpleasant than any of the musical pieces.

Effects of Background Sound on Gustatory Perception

Taste Discrimination

Little research has been performed to examine the effect of background sound on performance of a taste discrimination task. McFadden et al. (*52*) examined the effect of background noise presented over headphones on the discriminatory ability of weak solutions of sucrose or sodium chloride (NaCl) from distilled water. The results demonstrated no significant effect of background noise on the taste discrimination task.

Taste Intensity

Previous studies have demonstrated influences of background sound on taste intensity. In the aforementioned study conducted by Woods et al. (*50*), participants tasted savory and sweet foods under either quiet (45-55 dB) or loud (75-85 dB) white-noise conditions. The foods also were tasted under a silent condition to provide a baseline. In that experiment, sweetness and saltiness of foods were rated to be less intense under loud noise conditions than under relatively quiet noise conditions. However, the opposite findings have also been found. Stafford et al. (*53*) asked participants to taste alcoholic drinks under four noise conditions: music (modern genres), shadow (listening and repeating a news story), shadow and music, and silence. In that study, alcoholic drinks tasted sweeter and less bitter under loud music conditions than under silent conditions. Stafford et al. (*53*) explained that the inconsistent findings of with respect to background noise-induced sweetness might result from differences in the type of background noise (white noise *vs.* music) and the type of test samples (foods *vs.* alcoholic drinks); this suggests that the effect of background sound on taste intensity may vary with experimental context.

Taste Pleasantness

Can background music or noise alter liking of taste? The answer appears to remain unclear. According to Ferber and Cabanac's study (*54*), sweet solutions were more favored in the presence of loud sound (90 dB), whether music or noise, than in the presence of quiet sound (70 dB) or under silent conditions. However, the significant effect of background sound on taste pleasantness was obtained in neither salty solution, nor in a mixture of sweet and salty solutions.

Conclusion

To summarize, auditory cues can highly affect individuals' perceptions of chemosensory cues. First, it was observed that certain auditory cues (pitches or timbres) can be matched with specific odors, flavors, or tastes. Second, congruency between auditory and olfactory cues plays a key role in modulating chemosensory intensity and pleasantness. Finally, background music, noise, or silence affects chemosensory intensity, sensitivity, discrimination, and pleasantness. However, since the effects of auditory cues on chemosensory perception have not been consistently described in earlier studies, more work needs to be conducted before generalizing the findings. In addition, the neural mechanisms underlying cross-modal interaction between the auditory and chemosensory cues still remains unclear.

Finally, further studies should consider many influential factors when designing the tests or interpreting the results. These include stimulus-driven factors (e.g., intensity/amplitude, familiarity, hedonic tone of auditory or chemosensory cues, etc.), individually-driven factors (e.g., age, gender, demographics, personality, etc.), and environmentally-driven factors (e.g., culture, place, season, etc.). Individuals' perceptions and preferences of odors or musical pieces were found to be affected by demographics, experience, personality, mood, and culture (*55–59*); such factors may influence cross-modal interaction between the auditory and chemosensory cues. For example, German people are more likely to match a Christmas carol to a cinnamon odor, while North American people appear to match the carol to a peppermint odor, reflecting their dietary consumption and culture (*38*). Thus, when North American (or German) food companies design their new food products for winter season, they may consider mint (or cinnamon) flavored cookies accompanied with a Christmas carol to boost consumers' acceptability.

References

1. Pellegrino, R.; Luckett, C. R.; Shinn, S. E.; Mayfield, S.; Gude, K.; Rhea, A.; Seo, H. S. Effects of background sound on consumers' sensory discriminatory ability among foods. *Food Qual. Pref.* **2015**, *43*, 71–78.
2. Sobal, J.; Nelson, M. K. Commensal eating patterns: a community study. *Appetite* **2003**, *41*, 181–190.
3. Vickers, Z. M.; Wasserman, S. S. Sensory qualities of foods sounds based on individual perceptions. *J. Texture Stud.* **1980**, *10*, 319–332.
4. Christensen, C. M.; Vickers, Z. M. Relationships of chewing sounds to judgments of food crispness. *J. Food Sci.* **1981**, *46*, 574–578.

5. Zampini, M.; Spence, C. The role of auditory cues in modulating the perceived crispness and staleness of potato chips. *J. Sens. Stud.* **2004**, *19*, 347–363.
6. Zampini, M.; Spence, C. Modifying the multisensory perception of a carbonated beverage using auditory cues. *Food Qual. Prefer.* **2005**, *16*, 632–641.
7. Spence, C. Crossmodal correspondences: a tutorial review. *Atten. Percept. Psychophys.* **2011**, *73*, 971–975.
8. Schifferstein, H. N. J.; Tanudjaja, I. Visualising fragrances through colours: The mediating role of emotions. *Perception* **2004**, *33*, 1249–1266.
9. Martino, G.; Marks, L. E. Synesthesia: Strong and weak. *Curr. Direct. Psychol. Sci.* **2001**, *10*, 61–65.
10. Knöferle, K.; Spence, C. Crossmodal correspondences between sounds and tastes. *Psychon. Bull. Rev.* **2012**, *19*, 992–1006.
11. Marks, L. E. On colored-hearing synesthesia: cross-modal translations of sensory dimensions. *Psychol. Bull.* **1975**, *82*, 303–331.
12. Ward, J.; Huckstep, B.; Tsakanikos, E. Sound-colour synaesthesia: to what extent does it use cross-modal mechanisms common to us all? *Cortex* **2006**, *42*, 264–280.
13. Adeli, M.; Rouat, J.; Molotchnikoff, S. Audiovisual correspondence between musical timbre and visual shapes. *Front. Hum. Neurosci.* **2014**, *8*, 352.
14. Belkin, K.; Martin, R.; Kemp, S.; Gilbert, A. N. Auditory pitch as a perceptual analogue to odor quality. *Psychol. Sci.* **1997**, *8*, 340–342.
15. Crisinel, A. S.; Spence, C. As bitter as a trombone: Synesthetic correspondences in nonsynesthetes between tastes/flavors and musical notes. *Atten. Percept. Psychophys.* **2010**, *72*, 1994–2002.
16. Crisinel, A. S.; Spence, C. Crossmodal associations between flavoured milk solutions and musical notes. *Acta Psychol.* **2011**, *138*, 155–161.
17. Crisinel, A. S.; Spence, C. A fruity note: Crossmodal associations between odors and musical notes. *Chem. Senses* **2012**, *37*, 151–158.
18. Crisinel, A. S.; Spence, C. The impact of pleasantness ratings on crossmodal associations between food samples and musical notes. *Food Qual. Prefer.* **2012**, *24*, 136–140.
19. Crisinel, A. S.; Jacquier, C.; Deroy, O.; Spence, C. Composing with cross-modal correspondences: Music and odors in concert. *Chem. Percept.* **2013**, *6*, 45–52.
20. Deroy, O.; Crisinel, A. S.; Spence, C. Crossmodal correspondences between odors and contingent features: odors, musical notes, and geometrical shapes. *Psychon. Bull. Rev.* **2013**, *20*, 878–896.
21. Budinger, E.; Heil, P.; Hess, A.; Scheich, H. Multisensory processing via early cortical stages: connections of the primary auditory cortical field with other sensory systems. *Neuroscience* **2006**, *143*, 1065–1083.
22. Budinger, E.; Scheich, H. Anatomical connections suitable for the direct processing of neuronal information of different modalities via the rodent primary auditory cortex. *Hear. Res.* **2009**, *258*, 16–27.
23. Wesson, D. W.; Wilson, D. A. Smelling sounds: olfactory-auditory sensory convergence in the olfactory tubercle. *J. Neurosci.* **2010**, *30*, 3013–3021.

24. Cohen, L.; Rothschild, G.; Mizrahi, A. Multisensory integration of natural odors and sounds in the auditory cortex. *Neuron* **2011**, *72*, 357–369.
25. Varga, A. G.; Wesson, D. W. Distributed auditory sensory input within the mouse olfactory cortex. *Eur. J. Neurosci.* **2013**, *37*, 564–571.
26. Holt-Hansen, K. Taste and pitch. *Percept. Mot. Skills* **1968**, *27*, 59–68.
27. Holt-Hansen, K. Extraordinary experiences during cross-modal perception. *Percept. Mot. Skills* **1976**, *43*, 1023–1027.
28. Crisinel, A. S.; Spence, C. Implicit association between basic tastes and pitch. *Neurosci. Lett.* **2009**, *464*, 39–42.
29. Crisinel, A. S.; Spence, C. A sweet sound? Food names reveal implicit associations between taste and pitch. *Perception* **2010**, *39*, 417–425.
30. Simner, J.; Cuskley, C.; Kirby, S. What sound does that taste? Cross-modal mappings across gustation and audition. *Perception* **2010**, *39*, 553–569.
31. Schifferstein, H. N. J.; Verlegh, P. W. J. The role of congruency and pleasantness in odor-induced taste enhancement. *Acta. Psychol.* **1996**, *94*, 87–105.
32. Seo, H. S. Multimodal Integration in Smell and Taste Perception. Doctoral dissertation, Technical University of Dresden, Dresden, Germany, 2011.
33. La Buissonnière-Ariza, V.; Frasnelli, J.; Collignon, O.; Lepore, F. Olfactory priming leads to faster sound localization. *Neurosci. Lett.* **2012**, *506*, 188–192.
34. Christensen, C M. Effects of color on aroma, flavor and texture judgments of foods. *J. Food Sci.* **1983**, *48*, 787–790.
35. Zellner, D. A.; Kautz, M. A. Color affects perceived odor intensity. *J. Exp. Psychol. Hum. Percept. Perform.* **1990**, *16*, 391–397.
36. Zellner, D. A.; Whitten, L. A. The effect of color intensity and appropriateness on color-induced odor enhancement. *Am. J. Psychol.* **1999**, *112*, 585–604.
37. Seo, H. S.; Hummel, T. Auditory-olfactory integration: Congruent or pleasant sounds amplify odor pleasantness. *Chem. Senses* **2011**, *36*, 301–309.
38. Seo, H. S.; Lohse, F.; Luckett, C. R.; Hummel, T. Congruent sound can modulate odor pleasantness. *Chem. Senses* **2014**, *39*, 215–229.
39. Mattila, A. S.; Wirtz, J. Congruency of scent and music as a driver of in-store evaluations and behavior. *J. Retailing* **2001**, *77*, 273–289.
40. Crisinel, A. S.; Cosser, S.; King, S.; Jones, R.; Petrie, J.; Spence, C. A bittersweet symphony: Systematically modulating the taste of food by changing the sonic properties of the soundtrack playing in the background. *Food Qual. Prefer.* **2012**, *24*, 201–204.
41. Spence, C. Noise and its impact on the perception of food and drink. *Flavour* **2014**, *3*, 9.
42. Spence, C.; Michel, C.; Smith, B. Airplane noise and the taste of umami. *Flavour* **2014**, *3*, 2.
43. Rauscher, F. H.; Shaw, G. L.; Ky, K. N. Music and spatial task performance. *Nature* **1993**, *365*, 611.

44. Hedner, M.; Larsson, M.; Arnold, N.; Zucco, G. M.; Hummel, T. Cognitive factors in odor detection, odor discrimination, and odor identification tasks. *J. Clin. Exp. Neuropsychol.* **2010**, *32*, 1062–1067.
45. Seo, H. S.; Gudziol, V.; Hähner, A.; Hummel, T. Background sound modulates the performance of odor discrimination task. *Exp. Brain Res.* **2011**, *212*, 305–314.
46. Seo, H. S.; Hähner, A.; Gudziol, V.; Scheibe, M.; Hummel, T. Influence of background noise on the performance in the odor sensitivity task: effects of noise type and extraversion. *Exp. Brain Res.* **2012**, *222*, 89–97.
47. Koelega, H. S. Extraversion, sex, arousal and olfactory sensitivity. *Acta Psychol.* **1970**, *34*, 51–66.
48. Eysenck, H. *The biological basis of personality*; C.C. Thomas Publisher: Springfield, MO, 1967.
49. Fiegel, A.; Meullenet, J. -F.; Harrington, R. J.; Humble, R.; Seo, H. S. Background music genre can modulate flavor pleasantness and overall impression of food stimuli. *Appetite* **2014**, *76*, 144–152.
50. Woods, A. T.; Poliakoff, E.; Lloyd, D. M.; Kuenzel, J.; Hodson, R.; Gonda, H.; Batchelor, J.; Dijksterhuis, G. B.; Thomas, A. Effect of background noise on food perception. *Food Qual. Prefer.* **2011**, *22*, 42–47.
51. Velasco, C.; Balboa, D.; Marmolejo-Ramos, F.; Spence, C. Crossmodal effect of music and odor pleasantness on olfactory quality perception. *Front. Psychol.* **2014**, *5*, e1352.
52. McFadden, D.; Barr, E. A.; Young, R. E. Audio analgesia: Lack of a cross-masking effect on taste. *Percept. Psychophys.* **1971**, *10*, 175–179.
53. Stafford, L. D.; Fernandes, M.; Agobiani, E. 2012. Effects of noise and distraction on alcohol perception. *Food Qual. Prefer.* **2012**, *24*, 218–224.
54. Ferber, C.; Cabanac, M. 1987. Influence of noise on gustatory affective ratings and preference for sweet or salt. *Appetite* **1987**, *8*, 229–235.
55. Christenson, P. G.; Peterson, J. B. Genre and gender in the structure of music preference. *Commun. Res.* **1988**, *15*, 282–301.
56. LeBlanc, A.; Sims, W. L.; Siivola, C.; Obert, M. Music style preferences of different age listeners. *J. Res. Music Edu.* **1996**, *44*, 49–59.
57. Chen, D.; Dalton, P. The effect of emotion and personality on olfactory perception. *Chem. Senses* **2005**, *30*, 345–351.
58. Hummel, T.; Kobal, G.; Gudziol, H.; Mackay-Sim, A. Normative data for the "Sniffin' Sticks" including tests of odor identification, odor discrimination, and olfactory thresholds: an upgrade based on a group of more than 3,000 subjects. *Eur. Arch. Otorhinolaryngol.* **2007**, *264*, 237–243.
59. Seo, H. S.; Hummel, T. In *Food Flavors – Chemical, Sensory and Technological Properties*; Jeleń, H., Ed.; Chemical and Functional Properites of Food Components Series; CRC Press: Boca Raton, FL, 2012; pp 35–63.

Chapter 5

Contribution of Phenolic Compounds to Sensory Profiles of Blackcurrant Juices

Oskar Laaksonen and Baoru Yang[*]

Food Chemistry and Food Development, Department of Biochemistry, University of Turku, FI-20014, Turku Finland
[*]E-mail: baoru.yang@utu.fi.

Blackcurrant (*Ribes nigrum*) juice was produced with or without enzymatic assistance in laboratory and industrial scales. Phenolic profiles (proanthocyanidins, anthocyanins, flavonols, hydroxycinnamic acids) and taste (sweetness, sourness, bitterness) and astringent (mouthdrying, puckering) characteristics of the juice were analyzed. The compositional and sensory data were processed with multivariate regression models. Compared with the non-enzymatic process, the enzyme-aided process resulted in higher contents of phenolic compounds along with higher astringencies and bitterness in juices produced at both laboratorial and industrial scales. The mouth-drying astringency of the juices was positively associated with the contents of all subgroups of phenolic compounds and molecular size of proanthocyanidins but negatively with the procyanidin/prodelphinidin ratio. Puckering astringency correlated with sourness and lower juice pH as well as with phenolic variables. High pectin content may have masked the astringency of the non-enzymatic juices. Increased astringency and bitterness as a result of the enzymatic process may affect negatively the consumer acceptance of blackcurrant juices.

Introduction

Berries, fruits and vegetables form an important part of a healthy human diet. They are rich in dietary fiber, micronutrients, and potential bioactive constituents. Dietary patterns rich in these elements may be associated with lower risk of most chronic diseases. However, despite their health benefits, daily intakes of fruits and vegetables remain inadequate. Western-type dietary patterns are characterized by high consumption of meat and products with low content of essential nutrients and high content of salt. Various health authorities around the world recommend increase in consumption of berries, fruits and vegetables. Sensory properties characterized by high intensities of astringency and bitterness are often factors limiting the use of berries and berry products by the consumers.

Phenolic compounds are commonly associated with astringency and bitterness in food. Astringency can be described as drying, puckering and rough sensation in oral cavity (*1–3*). A commonly accepted hypothesis is that polymeric tannins in food bind and precipitate salivary proteins resulting in astringent sensation in the mucous membranes. The structural characteristics of tannin molecules affect the binding to proteins (*4*) and therefore the sensory properties of food. Some phenolic compounds, such as flavonol glycosides, do not bind to salivary proteins, but elicit the astringent sensation by different mechanisms (*5*). Flavonols (quercetin, myricetin and kaempferol) and flavan-3-ols ((+)-catechin, (–)-epicatechin and (–)-epigallocatechin) can activate the human bitter taste receptors (*6, 7*). Additionally, a procyanidin trimer activated some bitter receptors whereas a dimer did not (*7*). The sensory properties of food are influenced by not only the content of individual compounds but also the interactions between different components as well as with food matrixes.

Blackcurrant (*Ribes nigrum*) is the second largest cultivated berries in Europe, just after strawberry. The health benefits of blackcurrant berries are supported by traditional use and modern research. Juice pressing is the most important industrial processing of blackcurrant berries. In this study, we aim to investigate the effects of cultivars and processing technologies on the composition and sensory properties of blackcurrant juices with multivariate statistical models. Special attention was paid to different groups of phenolic compounds and correlation of these compounds with astringencies and bitterness that are often perceived as negative attributes of blackcurrant juices.

Materials and Methods

Samples

The juice samples were produced in laboratory scale from five different Finnish blackcurrant cultivars, four commercial cultivars 'Mortti', 'Mikael', 'Marski', 'Ola' and a new breed, 'Breed15' (*8, 9*). Berries were harvested in 2010 from southern Finland from the test filed of MTT Piikkiö, Agrifood Research Finland. Two juice processing methods were applied for each cultivar. The first process was carried out without enzymes and the second with the aid of a commercial enzyme product, Pectinase 714L (Biocatalysts Ltd, Cardiff, UK). The juice processing in laboratory scale did not include pasteurization. All juice samples were frozen at -20 °C right after the processing until analyses.

For the industry-scale processing (*10*), blackcurrant berries of the cultivar 'Mortti' were harvested from the cultivation field of Saarioinen Oy (Huittinen, Finland) in 2011 and processed with facilities of Saarioinen Oy. Four different juices were produced: two without the aid of enzymes (No enzymes juices 1 and 2) and two with enzymes (Enzyme juices 3 and 4) The juices were pasteurized and bottled in Marjajaloste Meritalo Oy (Salo, Finland). Thereafter the juices were stored in dark at +4 °C for further analyses.

Compositional Analyses

Anthocyanins, flavonol glycosides (and possible flavonol aglycons) and hydroxycinnamic acids in the juice samples were analyzed using methods as described previously (*11*, *12*). For qualitative and quantitative analyses of proanthocyanidins, a method reported by Engström and colleagues was used (*13*). Sugars and acids were analyzed by gas chromatography as trimethylsilyl derivatives in duplicates using the method previously applied in our laboratory (*14*). All results from aforementioned analyses are presented as sums of individual compounds (total contents) of each group or as their ratios.

Sensory Evaluation

Sensory characteristics of the juice samples were evaluated using generic descriptive analysis by two panels (*8*, *10*). The sensory evaluation was focused on taste (sourness, sweetness and bitterness) and two astringency (mouthdrying and puckering) attributes. The intensities of these attributes were rated on a continuous graphical scale from 0 (none) to 10 (very strong) with references. Reference samples were water solutions of citric acid (0.1%) for sourness, fructose (0.07%) for sweetness, caffeine (0.07%) for bitterness, ammonium aluminum sulphate (0.2%) for mouth-drying astringency and aluminum sulphate (0.2%) for puckering astringency. The panelists were trained to focus on the sub-qualities of the astringency references instead of the whole astringent sensation. The data were collected using Compusense-*five* software (Compusense Inc., Guelph, Canada).

Statistical Analyses

Partial least squares regression (PLS) method was applied for standardized data with X-variables (predictors) as chemical compound sums and their ratios and Y-variables (responses) as the sensory properties. The models are shown as correlations loading plots where the outer ellipse indicates 100% of explained variance and the inner 50% and the samples are presented as downweighted variables. Full cross validation was used to estimate the number of factors for a statistically reliable model. Multivariate models were conducted using Unscrambler 10.3 (Camo Process AS, Oslo, Norway).

Results and Discussion

Chemical Profiles of the Juices

Table 1 presents the averaged juice yields and compositional profiles of the blackcurrant juices pressed from berries of different cultivars. The main differences between averaged juices were observed between non-enzyme and enzyme juices rather than between laboratory and industry scales. Juice yields were notably higher in the processes with enzymatic assistance than in those without the use of enzyme. In the laboratory processing (*8*), the yields varied among the five cultivars with the highest yields obtained from berries of the cultivar 'Mikael' in both enzyme-aided and non-enzymatic processes. Viscosity of the juice of 'Mikael' was the lowest, whereas that of 'Mortti' was the highest with the lower yield. In the industrial scale processing (*10*), the yields were calculated based on the amounts of press residues. The juice yield for non-enzymatic pressing was approximated 31 % (No enzyme 1). Part of this juice was further processed by clarification and filtration to produce the second juice of the non-enzymatic pressing (No enzyme 2). Due to the high viscosity, some water was added to the juice during the filtration and clarification. The use of alternative cultivars with lower viscosity may increase the yields of non-enzymatic processing in industrial scale.

Although there was not notable difference in sugar contents between juices from the two processes in the laboratory scale, in industry scale pressing the non-enzymatic process resulted in higher contents of sugars than the enzymatic process. Contrary to this, more variation was found in the acid content among laboratory scale juices. All in all, the ratio between sugars and acids was lower in enzyme-aided juices than in non-enzymatic juices.

The enzyme-aided juices had significantly higher contents of all phenolic compound subclasses than the non-enzymatic juices (Table 1). Cultivars 'Mortti' and 'Ola' had the highest contents of phenolic compounds among the five cultivars in spite of the process. Flavonol aglycons do not typically exist in free form in abundance, but can be released due to hydrolysis caused by different processes. No free flavonol aglycons were detected in the laboratory-scale juices which were not pasteurized. However, in corresponding industry-scale juices, free flavonol aglycones were observed indicating that heat-treatment may have broken the glycosidic bonds.

Mean degree of polymerization (Table 1) indicates the average number of flavanol monomeric units present in the numerous oligomeric and polymeric condensed tannins present in the juice samples. The mDP value of proanthocyanidins was significantly higher in enzyme-aided juices than in the non-enzymatic juices (Table 1). The PC:PD was also higher in enzyme-aided juices, which indicated the higher PD contents compared to PC in the skins compared to the flesh of blackcurrant berries.

Table 1. Chemical Characteristics of Juices Averaged within Processes. Data Are from References (*8–10*).

	Laboratory scale [a]		*Industry scale* [a]	
	No enzymes	*Enzymes*	*No enzymes*	*Enzymes*
Juice yield (%)	62–70	71–77	approx. 31	80–91
pH	2.8	3.0	3.0	3.0
°Brix	16.4	15.0	12.8	13.0
Sugars (SUG)	9950 ± 725	9960 ± 1450	8770 ± 1470	7550 ± 150
Acids (ACID)	2850 ± 290	3530 ± 600	2540 ± 410	2670 ± 80
Phen.compounds (PHE) [b]	68	416	179	411
Anthocyanins (ANT)	44.6 ± 27	275 ± 50	138 ± 9.8	227 ± 2.6
Flavonol glycosides (FG)	2.5 ± 1.5	8.2 ± 2.3	6.5 ± 0.6	8.7 ± 0.5
Free flavonol aglycons (FG)	-	1.1 ± 0.7	0.9 ± 0.2	1.0 ± 0.4
Hydroxycinnamic acids (HCA)	2.5 ± 0.9	5.3 ± 1.3	3.9 ± 0.6	5.5 ± 0.2
Proanthocyanidins (PA) [b]	16.7	126	29.7	169
Procyanidins (PC)	6.9 ± 1.7	17.9 ± 7.7	11.7 ± 0.4	21.9 ± 6.6
Prodelphinidins (PD)	9.8 ± 7.8	108 ± 35	18.0 ± 1.5	146 ± 48
Mean degree of polymerization (mDP)	5.2	13.3	4.0	21.1

[a] Contents presented as mg/100 mL. Averages of five cultivar juices in laboratory scale; averages of two juices in industrial scale [b] Phenolic compounds, PHE, is the sum of phenolic compound classes; proanthocyanidins, PA, is the sum of PC and PD.

Sensory Profiles of the Juices

Taste and astringency profiles of juices from different processes are shown in Figure 1. The results are averages of five cultivars. All juices were significantly sour, and this attribute was not affected by the processes. Sweetness was rated lower in enzyme-aided juices than in the non-enzyme juices, but statistical difference was found only between juices from industry scale processing. Enzyme-aided juices were significantly more bitter, mouthdrying astringent and puckering astringent.

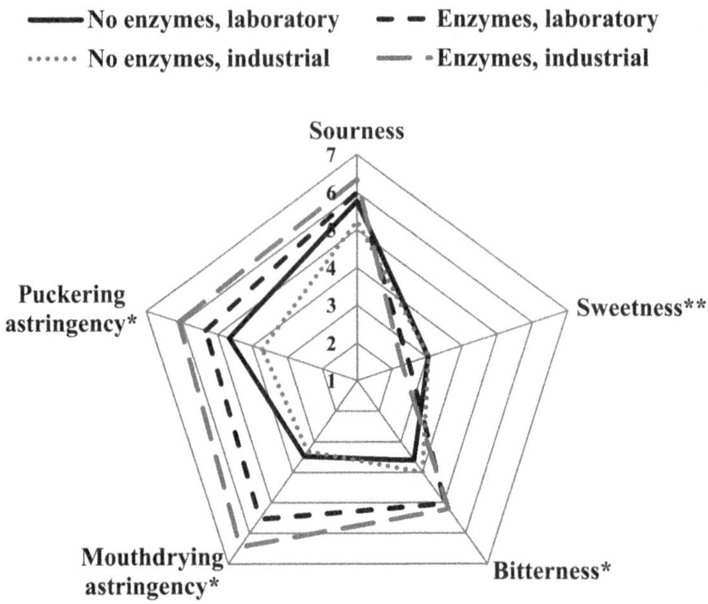

*Figure 1. Sensory profiles of averaged juices (2×5 juices in laboratory process; 2×2 juices in industrial process). Original ratings on the line scale between 0–10. * Significant difference between enzymatic and nonenzymatic processes; ** difference only between juices from industrial scale processing (t-test, p<0.05). Data are from references (8) and (10).*

Interactions between Chemical Composition and Sensory Properties

Proanthocyanidins (condensed tannins) have been reported to be more astringent than ellagitannins (a group of hydrolyzable tannins), the former ones having lower thresholds for detection of astringency (*15*). The small glycosylated phenolic compounds contribute to velvety and mouth-drying sensation without bitterness, whereas proanthocyanidins and phenolic acids may be more puckering as well as bitter (*16*, *17*). Flavonol glycosides and phenolic acid derivatives may have very low sensory threshold for astringency (*16–18*). Additionally, some indole and nitrile compounds in currants may contribute to astringency (*18*, *19*). Organic acids can also contribute to astringency and especially due to their impact on pH (*20*). Higher mDP of proanthocyanidins has been reported to increase perceived astringency (*21*, *22*), although various other structural characteristics may be more important factors (*23*, *24*). Monomeric units of proanthocyanidins have been reported to be more bitter than astringent, while proanthocyanidins of higher molecular weights are generally more astringent than bitter (*25*, *26*). Additionally, monomeric (–)-epicatechin can be more bitter and astringent than (+)-catechin in the equal concentrations (*27*).

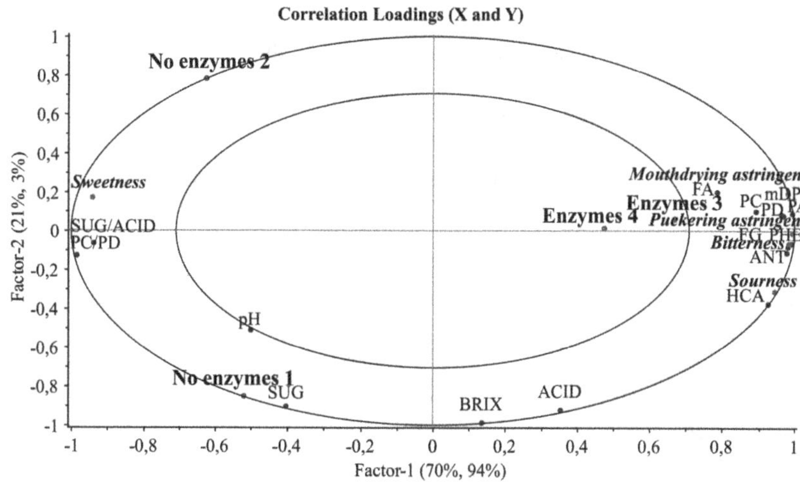

Figure 2. PLS regression correlation loadings plot of the interactions between chemical composition (X; 14 variables) and five sensory attributes (Y) in four juice samples. Samples are presented in the plot as downweighted variables. Enzyme samples 3 and 4 from enzyme-aided process, No enzyme samples 1 and 2 processed without enzymes. Data are from reference (10).

In this study, PLS regression models were created to examine the interactions between chemical (X) and sensory (Y) data. In the first model, 91% of the variation in the X explained 97% of the variation in Y with two factors (Figure 2; the data from reference (*10*)). However, the majority of the variation in Y is explained already on the validated first factor and the second contained only little variation. In order not to overfit the model, only the first factor is taken into account. Enzyme-aided juices (Enzymes 3 and 4) are located on the right together with all the phenolic variables, mDP of proanthocyanidins, bitterness (R^2 0.970; validated R^2 0.823), mouthdrying astringency (R^2 0.963; validated R^2 0.866), and puckering astringency (R^2 0.992; validated R^2 0.875). The contents of procyanidin (PC) and prodelphinidin (PD) are located on the right, whereas the PC/PD ratio is on the left side of the plot highlighting the higher content of proanthocyanindins and lower PC/PD ratio in enzyme-aided juices. Sweetness (R^2 0.889; validated R^2 0.694) is located on opposite side together with sugar/acid ratio. Sugar/acid ratio correlated negatively with sourness (R^2 0.891; validated R^2 0.679), which was weakly explained by the absolute content of acids and pH. No enzyme juices 1 and 2 are located on the left, but separated along the second factor. This was mainly due to filtration and dilution process that was only conducted to the No enzyme juice 2.

In the second model (*8*, *9*), 75% of the variation in chemical variables explained 82% of the variation in the sensory data with two validated factors (Figure 3). Figure 3 highlights again the significant difference between processes as enzyme-aided juices are located on the right and non-enzymatic juices on the left. Having only four samples and one validated factor in the first model may have resulted in less reliable results in comparison to the model with ten samples. In full cross validation, the first model was validated with maximum of three samples and with an assumption that the samples are relatively similar. In the second model, there were more similar samples included for both processes thus confirming the differences between non-enzyme and enzyme-aided processes.

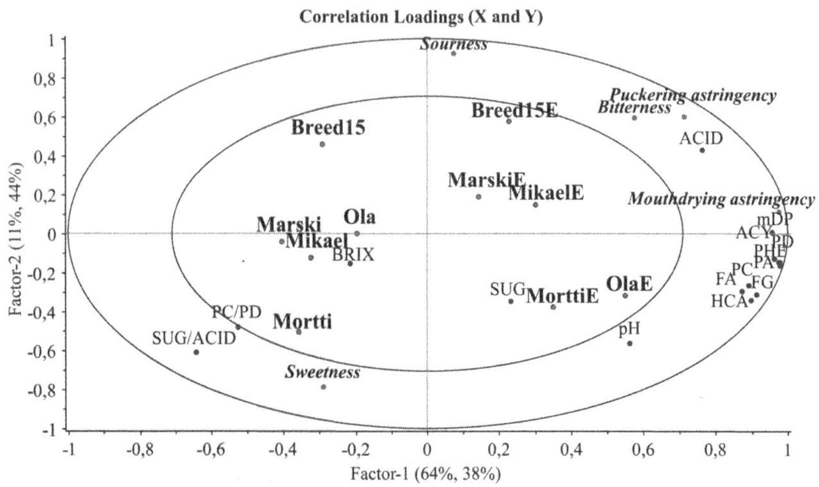

Figure 3. PLS regression correlation loadings plot of the interactions between chemical composition (X; 15 variables) and five sensory attributes (Y) in ten juice samples (E refers to enzyme-aided process). Samples are presented in the plot as downweighted variables. Cultivar names without E represents juices pressed without enzymes. Original data are from references (8) and (9).

Of the five sensory attributes, mouthdrying astringency was the best explained in the second model (Figure 3) with 0.965 R^2-value (validated value 0.897). All subgroups of the phenolic compounds (from Table 1) correlated with mouthdrying astringency indicating their significant role in this sensory attribute. Puckering astringency (R^2, 0.917; validated R^2, 0.679 with two factors) in this model correlates to some extent with sourness and negatively with pH on the second factor, but also with the phenolic variables on the first factor. Sweetness (validated R^2, 0.212 with two factors) and bitterness (validated R^2, 0.125 with two factors) were relatively poorly explained with the chemical variables. However, sugar/acid ratio correlates positively with sweetness and negatively with sourness. Bitterness, on the other hand, is higher in enzyme-aided processing, but not directly explained by any of the chemical variables. The most bitter cultivars ('Breed15' and 'Marski') in the processes did not contain high contents of any of the phenolic variables in comparison to other cultivars.

The enzyme treatment breaks down the pectins and at the same time releases phenolic compounds from the berry skins. Overall, the pectin content is high in the non-enzymatic juices contributing to the high viscosity. Higher viscosity of solutions may result in lower perceived astringency (*28*). This lowering effect may be dependent of polysaccharide type and concentration (*29*).

The second model (Figure 3) also shows the differences between blackcurrant cultivars on factor 2. Cultivar-specific characteristics in the sensory properties remained across the different juice pressing processes. The new cultivar, 'Breed15', was the most sour of the juices in both processes, whereas 'Mortti' and 'Ola' were sweeter and less sour than the rest. The juices of 'Mortti' and 'Ola' contained more phenolic compounds and sugars and less acids than those of 'Breed15'.

Conclusions

Both juice processing technology and cultivar significantly affect the chemical composition and sensory quality of blackcurrant juice. Enzyme-treatment increases the content of phenolic compounds, the average size of proanthocyanidins, and the prodelphinidin/procyanidin ratio resulting in higher astringency and bitterness of the juices. All subgroups of phenolic compounds contribute to mouth-drying and puckering astringent subqualities. However, the latter is also related sourness, pH, and sugar/acid ratio. Bitterness was partly explained with the phenolic contents. It was significantly higher in the enzyme-aided juices, but in the comparison of the juices produced from different cultivars the phenolic variables did not show equal correlation with this sensory attribute.

Non-enzyme pressing provides an alternative processing technology to produce juices and purees of more pleasant sensory profiles mostly due to the lower content of phenolic compounds and higher sugar/acid ratio. The pectin content of non-enzyme juices may be exploited to mask the astringency of phenolic compounds. However due to low juice yields, this process may require innovative strategies also for the press residue.

References

1. Bajec, M. R.; Pickering, G. J. *Crit. Rev. Food Sci. Nutr.* **2008**, *48*, 858–875.
2. Scollary, G. R.; Pásti, G.; Kállay, M.; Blackman, J.; Clark, A. C. *Trends Food Sci. Technol.* **2012**, *27*, 25–36.
3. McRae, J. M.; Kennedy, J. A. *Molecules* **2011**, *16*, 2348–2364.
4. Dobreva, M. A.; Green, R. J.; Mueller-Harvey, I.; Salminen, J.-P.; Howlin, B. J.; Frazier, R. A. *J. Agric. Food Chem.* **2014**, *62*, 9186–9194.
5. Schwarz, B.; Hofmann, T. *Eur. Food Res. Technol.* **2008**, *227*, 1693–1698.
6. Roland, W. S. U.; van Buren, L.; Gruppen, H.; Driesse, M.; Gouka, R. J.; Smit, G.; Vincken, J.-P. *J. Agric. Food Chem.* **2013**, *61*, 10454–10466.
7. Soares, S.; Kohl, S.; Thalmann, S.; Mateus, N.; Meyerhof, W.; De Freitas, V. *J. Agric. Food Chem.* **2013**, *61*, 1525–1533.

8. Laaksonen, O.; Mäkilä, L.; Tahvonen, R.; Kallio, H.; Yang, B. *Food Chem.* **2013**, *138*, 2421–2429.
9. Laaksonen, O.; Salminen, J.-P.; Mäkilä, L.; Kallio, H.; Yang, B. Unpublished results.
10. Laaksonen, O. A.; Mäkilä, L.; Sandell, M. A.; Salminen, J.-P.; Liu, P.; Kallio, H. P.; Yang, B. *Food Bioprocess. Technol.* **2014**, *7*, 2877–2888.
11. Laaksonen, O.; Sandell, M.; Nordlund, E.; Heiniö, R.-L.; Malinen, H.-L.; Jaakkola, M.; Kallio, H. *Food Chem.* **2012**, *130*, 31–41.
12. Zheng, J.; Yang, B.; Ruusunen, V.; Laaksonen, O.; Tahvonen, R.; Hellsten, J.; Kallio, H. *J. Agric. Food Chem.* **2012**, *60*, 6581–6593.
13. Engström, M. T.; Pälijärvi, M.; Fryganas, C.; Grabber, J. H.; Mueller-Harvey, I.; Salminen, J.-P. *J. Agric. Food Chem.* **2014**, *62*, 3390–3399.
14. Zheng, J.; Yang, B.; Tuomasjukka, S.; Ou, S.; Kallio, H. *J. Agric. Food Chem.* **2009**, *57*, 2977–2987.
15. Hofmann, T.; Glabasnia, A.; Schwarz, B.; Wisman, K. N.; Gangwer, K. A.; Hagerman, A. E. *J. Agric. Food Chem.* **2006**, *54*, 9503–9509.
16. Hufnagel, J. C.; Hofmann, T. *J. Agric. Food Chem.* **2008**, *56*, 1376–1386.
17. Hufnagel, J. C.; Hofmann, T. *J. Agric. Food Chem.* **2008**, *56*, 9190–9199.
18. Schwarz, B.; Hofmann, T. *J. Agric. Food Chem.* **2007**, *55*, 1394–1404.
19. Schwarz, B.; Hofmann, T. *J. Agric. Food Chem.* **2007**, *55*, 1405–1410.
20. Lawless, H. T.; Horne, J.; Giasi, P. *Chem. Senses* **1996**, *21*, 397–403.
21. Chira, K.; Pacella, N.; Jourdes, M.; Teissedre, P.-L. *Food Chem.* **2011**, *126*, 1971–1977.
22. Chira, K.; Schmauch, G.; Saucier, C.; Fabre, S.; Teissedre, P.-L. *J. Agric. Food Chem.* **2009**, *57*, 545–553.
23. Quijada-Morín, N.; Regueiro, J.; Simal-Gándara, J.; Tomás, E.; Rivas-Gonzalo, J. C.; Escribano-Bailón, M. T. *J. Agric. Food Chem.* **2012**, *60*, 12355–12361.
24. Wollmann, N.; Hofmann, T. *J. Agric. Food Chem.* **2013**, *61*, 2045–2061.
25. Arnold, R. A.; Noble, A. C.; Singleton, V. L. *J. Agric. Food Chem.* **1980**, *28*, 675–678.
26. Peleg, H.; Gacon, K.; Schlich, P.; Noble, A. C. *J. Sci. Food Agric.* **1999**, *79*, 1123–1128.
27. Kallithraka, S.; Bakker, J.; Clifford, M. N. *J. Sens. Stud.* **1997**, *12*, 25–37.
28. Peleg, H.; Noble, A. C. *Food Qual. Pref.* **1999**, *10*, 343–347.
29. Troszyńska, A.; Narolewska, O.; Robredo, S.; Estrella, I.; Hernández, T.; Lamparski, G.; Amarowicz, R. *Food Qual. Pref.* **2010**, *21*, 463–469.

Chapter 6

Enantiomeric Distribution of Ethyl 2-Hydroxy-4-methylpentanoate in Wine, A Natural Enhancer of Fruity Aroma

Georgia Lytra,[1,2] Sophie Tempere,[1,2] Gilles de Revel,[1,2] and Jean-Christophe Barbe*,[1,2]

[1]Univ. Bordeaux, ISVV, EA 4577 Œnologie, F-33140 Villenave d'Ornon, France
[2]INRA, ISVV, USC 1366 Œnologie, F-33140 Villenave d'Ornon, France
***E-mail jean-christophe.barbe@agro-bordeaux.fr.**

Ethyl 2-hydroxy-4-methylpentanoate enantiomers were assayed in 55 commercial wines using chiral gas chromatography. White wines presented only the R form, whereas red wines contained both enantiomers, in various ratios according to aging (average ratio: 95:5, m/m) with an average total concentration of about 400 μg/L. The olfactory threshold of ethyl (2R)-2-hydroxy-4-methylpentanoate (126 μg/L) was almost twice that of the S- form (55 μg/L). The olfactory threshold of the mixture of ethyl (2R)-2-hydroxy-4-methylpentanoate and ethyl (2S)-2-hydroxy-4-methylpentanoate (95:5, m/m) was 51 μg/L, indicating a synergistic effect. Sensory analysis revealed that fruity character was perceived at concentrations 2.2, 4.5, and 2.5 times lower, when the matrix was supplemented with ethyl (2R)-2-hydroxy-4-methylpentanoate, ethyl (2S)-2-hydroxy-4-methylpentanoate, and the mixture of ethyl (2R)-2-hydroxy-4-methylpentanoate and ethyl (2S)-2-hydroxy-4-methylpentanoate (95:5, m/m), respectively, at their average concentrations in red wines, demonstrating a synergistic effect of this ester on fruity aroma perception. Sensory profiles of aromatic reconstitutions highlighted the contribution of this compound to black-berry and fresh fruit descriptors.

Introduction

Ethyl 2-hydroxy-4-methylpentanoate (**1**) is a compound used in flavor chemistry. **1** has been identified in several distillates, such us grape brandies (*1*), as well as freshly-distilled Calvados and Cognac (*2*).

Concerning wine, **1** was first characterized in dry white wines made from the Romanian cultivar, Feteasca Regala (*3*), and later in dry white Chardonnay wines (*4*). This ethyl ester was also found in aged Madeira wines and some types of sherry (*5*), as well as in dry red wines, at an average concentration of about 400 µg/L (*6*).

Its enhancer effect was reported by Luccarelli, Mookherjee, Wilson, Zampino and Bowen (*7*) who demonstrated that, when **1** was mixed with C4-C10 alkanoic acids, it enhanced natural, ripe, tropical fruit flavors in food.

Although **1** clearly has an asymmetrical carbon atom in position 2 (Figure 1), to our knowledge, no previous work investigated the possibility of two enantiomers.

a b

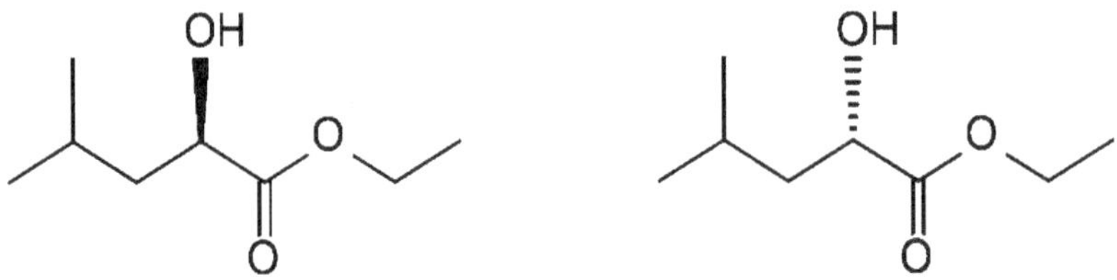

Figure 1. (a): Ethyl (2R)-2-hydroxy-4-methylpentanoate (CAS number =60856-83-9) and (b): ethyl (2S)-2-hydroxy-4-methylpentanoate (CAS number =60856-85-1).

This paper reports the separation, distribution, and concentrations of **1** enantiomers in wines from various vintages and origins and evaluates the organoleptic impact of this compound in red wines, by determining their olfactory thresholds and studying their perceptive interactions.

Materials and Methods

Samples. **1** was assayed in wines from several vintages and origins: 42 red (vintages 1981-2010) and 13 white wines (vintages 1989-2008). Pays d'Oc Merlot was used to evaluate the organoleptic impact of **1** on quantitative odor perception, while Margaux wine (vintage 2000) was used to evaluate the organoleptic impact of **1** on qualitative odor perception. Wine samples from the 2010 vintage were collected and analyzed 3 months after alcoholic fermentation. Dilute alcohol solution was prepared using double-distilled ethanol and microfiltered water (12%, v/v).

Aromatic Reconstitution. Sample preparation was as described by Lytra, Tempere, de Revel and Barbe (*8*) using liquid–phase extraction technique. Reversed-phase (RP) HPLC was performed on this raw extract, under the chromatographic conditions optimized by Pineau, Barbe, Van Leeuwen and Dubourdieu (*9*). The 25 fractions obtained in dilute alcohol solution were then directly evaluated by three trained assessors. For aromatic reconstitutions, fractions were retained and added individually or blended together to reproduce the initial concentrations in the original wines, adding double-distilled ethanol and microfiltered water to obtain an ethanol content of 12% (v/v).

Ethyl 2-Hydroxy-4-methylpentanoate Enantiomer Quantification. Chromatographic conditions and sample preparation were as optimized by Lytra, Tempere, de Revel and Barbe (*8*) using liquid–phase extraction technique. The enantiomers of **1** were separated by chiral gas chromatography on a γ-cyclodextrin phase. Gas chromatography analyses were carried out on an HP 6890 GC system coupled to an HP 5973i quadrupole mass spectrometer. The mass spectrometer was operated in electron ionization mode at 70 eV with selected-ion-monitoring (SIM) mode.

Ester and Acetate Analyses in HPLC Fractions. Chromatographic conditions and sample preparation were as optimized by Antalick, Perello and de Revel (*10*) using the Solid-phase microextraction technique (SPME). Gas chromatography analyses were carried out on an HP 5890 GC system coupled to an HP 5972 quadrupole mass spectrometer. The mass spectrometer was operated in electron ionization mode at 70 eV with selected-ion-monitoring (SIM) mode.

Sensory Analyses

General Conditions. Sensory analyses were performed as described by Martin and de Revel (*11*). Judges were all research laboratory staff at ISVV, Bordeaux University, selected for their experience in assessing fruity aromas in red wines.

Olfactory Thresholds. The olfactory thresholds of ethyl (2R)-2-hydroxy-4-methylpentanoate (**1a**), ethyl (2S)-2-hydroxy-4-methylpentanoate (**1b**), and the mixture of **1a** and **1b** (95:5, m/m) were determined by 15 judges in a three-alternative, forced-choice presentation (3-AFC) (*12*). The impact of a mixture of **1a** and **1b** (95:5, m/m) in dilute alcohol solution was evaluated using an additive model (*13*), as developed by Miyazawa, Gallagher, Preti and Wise (*14*).

Particular "olfactory thresholds" of fruity HPLC fractions (18 to 22), corresponding to an initial wine volume of 0.3, 0.6, 1.3, 2.5, 10, 20, 40, 80, 160 ml, diluted in 50 mL matrix, were determined by 19 judges, using four different matrices (dilute alcohol solution and dilute alcohol solution supplemented with 400 μg/L **1a**, 20 μg/L **1b** or 420 μg/L of the mixture of **1a** and **1b** (95:5, m/m), in a three-alternative, forced-choice presentation (3-AFC) (*12*).

Data Analysis. The results of the three-alternative, forced-choice tests were statistically interpreted and the olfactory threshold value was determined using an adaptation of the ASTM – E1432 method (*15*). Sigma Plot 8 (SYSTAT) software was used for graphic resolution and ANOVA transform for non-linear regression (*16*).

Sensory Profiles for red-berry, black-berry, fresh-, and jammy-fruit aromas were evaluated by 15 judges on a 0-7 point intensity scale. Two samples of aromatic reconstitutions in dilute alcohol solution were presented. The first consisted of HPLC fruity fractions (18 to 22) and the second contained the same HPLC fruity fractions, supplemented with 550 µg/L of the mixture of **1a** and **1b** (95:5, m/m).

Statistical data were analyzed using R analysis of variance (ANOVA) software (R v2.15.0 - R Development Core Team 2009, Vienna, Austria, R Foundation for Statistical Computing): the homogeneity of variance was tested using Levene's and the normality of residuals was tested using Shapiro-Wilk Test. All descriptors are mean-centered per panelist and scaled to unit variance. The statistically significant level was at 5% ($p<0.05$).

Results and Discussion

Ethyl 2-Hydroxy-4-methylpentanoate Enantiomer Distribution and Concentrations

In dry wines of the same age, **1** levels were generally higher in reds than whites (maximum concentration in red wines: 660µg/L, Margaux, 2005). Results concerning **1** levels in red and white wines were in agreement with those reported by Falcao, Lytra, Darriet and Barbe (*6*).

Table 1. Concentrations of Ethyl 2-Hydroxy-4-methylpentanoate Enantiomers (µg/L)

Sample type	*Vintages*	*Average Concentration (µg/L)*			*Average Ratio of R/S*
		1	*1a*	*1b*	
Red Wine	1980 - 1990	**408**±118	**371**±113	**32**±16	**91:9** ± 5
	1991 - 2000	**449**±125	**431**±122	**18**±13	**96:4** ± 3
	2001 - 2009	**361**±119	**354**±114	**7**±6	**98:2** ± 1
	2010	**135**±47	**135**±47	**0**±0	**100:0** ± 0
White Wine	1980 – 2000	**342**±236	**341**±237	**2**±2	**99:1** ± 1
	2001 - 2010	**182**±121	**182**±121	**0**±0	**100:0** ± 0

± Standard Deviation over the Average Concentration; 1, ethyl 2-hydroxy-4-methylpentanoate; 1a, ethyl (2R)-2-hydroxy-4-methylpentanoate; 1b, ethyl (2S)-2-hydroxy-4-methylpentanoate.

Generally, white wines only contained **1a**, while aged red wines presented both enantiomeric forms in varying ratios, according to age. Young red wines (2010) contained only the R- form. Table 1 shows the impact of ageing on the R/S ratio in red wines.

The maximum enantiomeric ratio was 100% for the R form (all 2010 vintage red wines) and 15% for the S form (Margaux, 1990). In red wines, the highest **1b** levels were found in the oldest samples (Table 1). However, two white wines produced surprising results: an exceptionally high **1** level compared to red wines (827µg/L in Pessac-Leognan, 1989) and the presence of the S-form in an enantiomeric ratio R/S of 97:3 (Bordeaux, 1994).

Direct Organoleptic Impact of Ethyl 2-Hydroxy-4-methylpentanoate on Quantitative Odor Perception

The olfactory threshold of **1a** in dilute alcohol solution was 126 µg/L, almost twice that of **1b** (55 µg/L), clearly indicating that the thresholds were strongly dependent on the odorant's stereochemistry. The olfactory threshold of the mixture of **1a** and **1b** (95:5, m/m) was 51 µg/L, indicating that both enantiomeric forms contributed to its perception in wine and confirming the direct impact of this ester on aroma perception. These results are in agreement with those found in the literature, where the olfactory threshold and descriptors for several odoriferous compounds differed according to the stereoisomer (*17*, *18*)

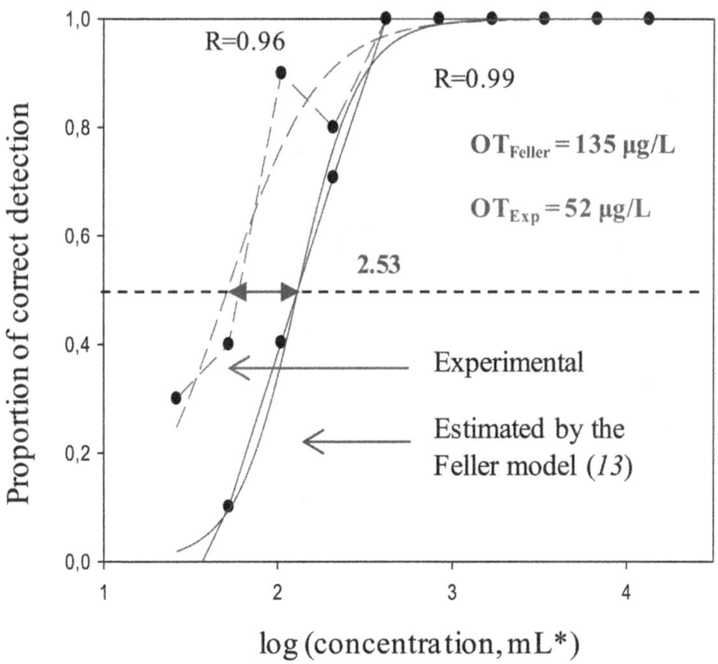

Figure 2. Comparison between probability of experimental detection and estimated by the feller model detection of the mixture of ethyl (2R)-2-hydroxy-4-methylpentanoate and ethyl (2S)-2-hydroxy-4-methylpentanoate (95:5, m/m) in dilute alcohol solution.

Table 2. Distribution of Esters and Acetates with Fruity Notes in HPLC Fractions

Compounds	*F16*	*F17*	*F18*	*F19*	*F20*	*F21*	*F22*
			Ethyl Esters				
Ethyl propanoate	-	-	x	-	-	-	-
Ethyl 2-methyl-propanoate	-	-	x	x	-	-	-
Ethyl butanoate	-	-	x	-	-	-	-
Ethyl 2-methylbutanoate	-	-	-	x	x	-	-
Ethyl pentanoate	-	-	-	-	x	-	-
Ethyl 3-methylbutanoate	-	-	-	x	x	x	-
Methyl hexanoate	-	-	-	-	x	-	-
Ethyl hexanoate	-	-	x	x	x	x	x
Ethyl (E)-hex-2-enoate	-	-	-	-	-	x	-
Ethyl 2-hydroxy-4-methylpentanoate	x	x	-	-	-	-	-
Methyl octanoate	-	-	-	-	-	-	x
Ethyl octanoate	-	-	-	-	-	-	x
Ethyl 2-phenylacetate	-	-	-	x	x	-	-
			Higher Alcohol Acetates				
Propyl acetate	-	-	x	-	-	-	-
2-methylpropyl acetate	-	-	x	-	-	-	-
Butyl acetate	-	-	-	x	x	-	-
3-methylbutyl acetate	-	-	-	x	x	x	-
Hexyl acetate	-	-	-	-	-	-	x
2-phenylethyl acetate	-	-	-	x	x	-	-

Olfactory thresholds results suggest a hyper-additive effect in a binary mixture, which is perceived as more intensely aromatic than the sum of the two compounds. An additive model was used to confirm the synergistic impact of mixing **1a** and **1b** (95:5, m/m). This method was applied to a particular binary model mixture, consisting of both enantiomers. After modeling the detection curve for the S-enantiomeric form, the response probabilities p(S) for the range of concentrations of the mixture used were calculated. The measured probability of detecting the mixture was higher than the calculated probability (Figure 2), revealing a perceptual synergistic effect between the two enantiomeric forms.

Distribution of Aromatic Compounds during HPLC Factionation

Applying HPLC to a wine extract resulted in 25 fractions. By sensory analysis, fractions 17-22 were selected for their intense fruitiness. Table 2 shows the ethyl esters and acetates with fruity notes in fractions 17-22, determined using headspace solid-phase microextraction. **1** was the only ester eluted in fraction 17. Thus, it was easy to obtain a pool of fruity wine esters without **1**, in order to investigate its indirect impact, by partial aromatic reconstitution of HPLC fruity fractions 18-22.

Indirect Organoleptic Impact of Ethyl 2-Hydroxy-4-methylpentanoate on Quantitative Odor Perception

As shown in Figure 3, in dilute alcohol solution, the "olfactory threshold" of fruity HPLC fractions (18-22), a pool of fruity wine esters excluding **1**, was (a) 2.2, (b) 4.5, and (c) 2.5 times higher than that of dilute alcohol solution supplemented with (a) 400 µg/L **1a**, (b) 20 µg/L **1b**, and (c) 420 µg/L of the mixture of **1a** and **1b** (95:5, m/m), respectively. These results, obtained using the average concentrations found in red wines of **1a**, **1b,** and the mixture of **1a** and **1b** (95:5, m/m), demonstrated that **1** had a synergistic effect on the perception of fruity aromas in wine.

These results are in agreement with those presented by Falcao, Lytra, Darriet and Barbe (*6*), where **1** omission was clearly perceived and simultaneous omission of **1** and ethyl butanoate was perceived even more clearly, suggesting perceptive interactions between **1** and another ethyl ester. In addition, evidence in literature established the additive effect of low-impact odorants on fruity wine aroma; Pineau, Barbe, Van Leeuwen and Dubourdieu (*9*) demonstrated that, in some complex mixtures in dearomatized red wine, very small variations in the concentrations of some ethyl esters were perceived, even at concentrations far below their individual olfactory thresholds.

In the literature, the behavior of other compounds was also studied by comparing two detection thresholds. Romano, Perello, Lonvaud-Funel, Sicard and de Revel (*19*), demonstrated that adding isobutyric and isovaleric acids to wine resulted in a remarkable increase in the olfactory threshold for ethylphenols.

Lytra, Tempere, Le Floch, de Revel and Barbe (*20*) reported that the individual presence of ethyl propanoate, ethyl-3-hydroxybutanoate, butyl acetate, or 2-methylpropyl acetate, at subthreshold levels, in dilute alcohol solution, or ethyl (2S)-2-methylbutanoate (*21*) at suprathreshold levels, resulted in a decrease in the olfactory threshold of the fruity pool (constituted by the other red-wine esters), reflecting an individual quantitative contribution of these compounds to overall aroma intensity. More recently, using the same methodology, these authors (*22*) demonstrated the additive effect of a compound that does not present fruity aromas (dimethyl sulfide) on the fruity aroma of a complex mixture (containing 12 red-wine esters), emphasizing the importance of perceptive interactions for the intensity of fruity aromas in red wine.

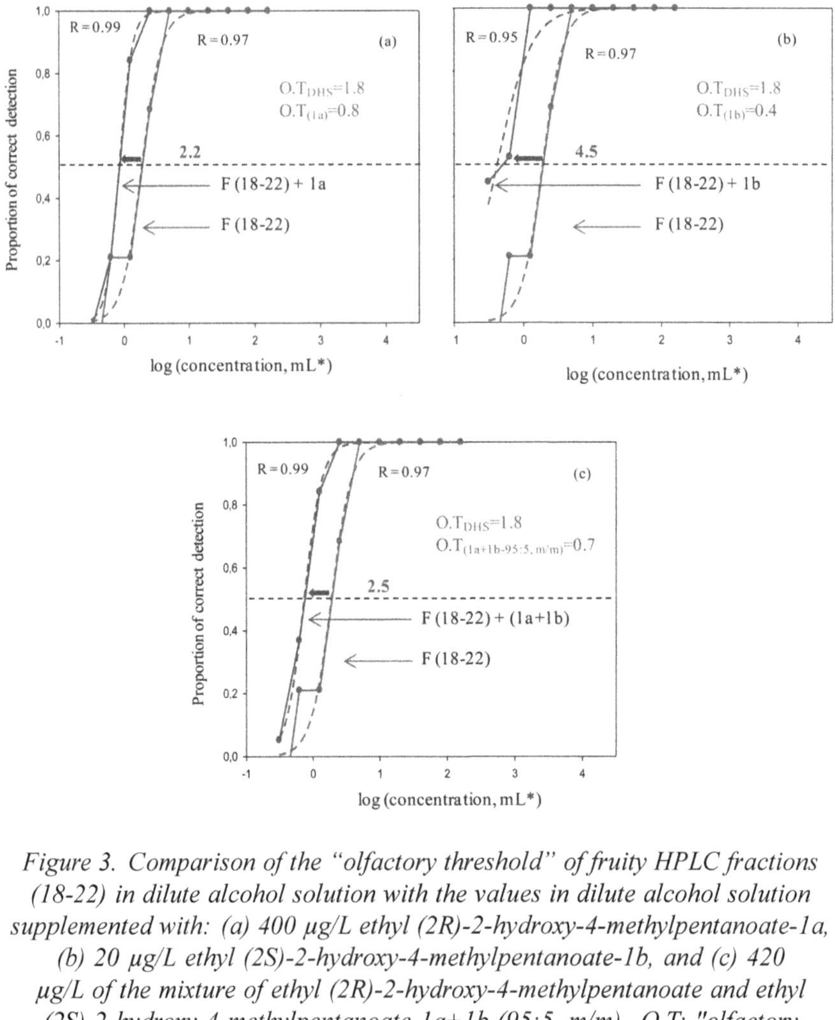

Figure 3. Comparison of the "olfactory threshold" of fruity HPLC fractions (18-22) in dilute alcohol solution with the values in dilute alcohol solution supplemented with: (a) 400 µg/L ethyl (2R)-2-hydroxy-4-methylpentanoate-1a, (b) 20 µg/L ethyl (2S)-2-hydroxy-4-methylpentanoate-1b, and (c) 420 µg/L of the mixture of ethyl (2R)-2-hydroxy-4-methylpentanoate and ethyl (2S)-2-hydroxy-4-methylpentanoate-1a+1b (95:5, m/m). O.T: "olfactory thresholds" of fruity HPLC fractions (18-22) expressed in wine volume (mL) diluted in 50 mL matrix.

Organoleptic Impact of Ethyl 2-Hydroxy-4-methylpentanoate on Qualitative Odor Perception

Significant results for the descriptors evaluated are summarized in Figure 4. The average scores for red-berry fruit intensity were identical after adding the mixture of **1a** and **1b**, whereas jammy fruit intensity was significantly lower. The average scores for black-berry and fresh fruit aromas were significantly higher for the aromatic reconstitution of HPLC fruity fractions (18 to 22) supplemented with the mixture of **1a** and **1b** (95:5, m/m). These results confirmed the sensory importance of this ester, suggesting that it is an active contributor to the black-berry and fresh fruit nuances in the wine studied. These results are in agreement with previous observations of Lytra, Tempere, de Revel and Barbe (*23*) demonstrating that **1** was the only ester eluted from fraction 17 and that, at a concentrations of 550 µg/L, the mixture of **1a** and **1b** (95:5, m/m) (adduced from fractions 17) contributed actively to black-berry and fresh-fruit notes, revealing that **1** alone played the same aromatic role as fraction 17.

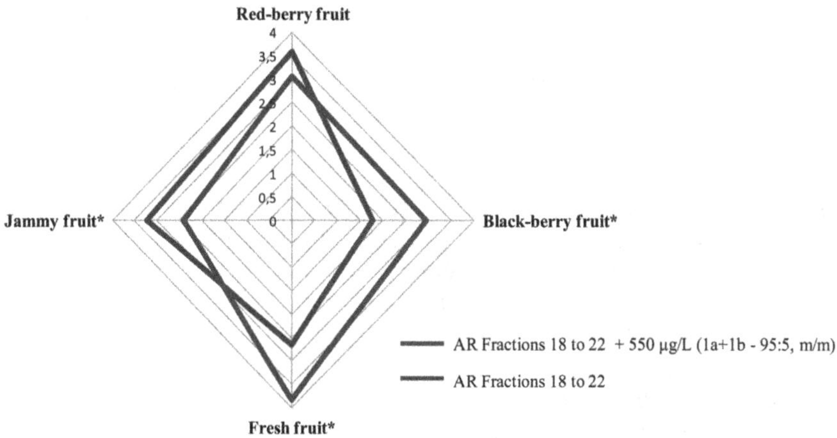

*Figure 4. Aromatic impact of ethyl 2-hydroxy-4-methylpentanoate addition to the complex fruity aromatic reconstitutions made by HPLC Fractions. * $p<0.05$; AR, aromatic reconstitutions; 1a+1b, mixture of ethyl (2R)-2-hydroxy-4-methylpentanoate and ethyl (2S)-2-hydroxy-4-methylpentanoate (95:5, m/m).*

These findings highlighted the indirect role of ethyl 2-hydroxy-4-methylpentanoate in red wine aroma, showing that this ester contributed to a synergistic effect that enhanced the perception of fruity character. Finally, it was clearly demonstrated that this compound acted as a natural enhancer for black-berry and fresh fruit notes in red wine.

Acknowledgments

The authors thank the Bordeaux Wine Council (Conseil Interprofessionnel des Vins de Bordeaux, CIVB) for their financial support. They also thank Château Luchey-Halde and Château Brane-Cantenac for providing wine samples.

References

1. Schreier, P.; Drawert, F.; Winkler, F. *J. Agric. Food Chem.* **1979**, *27* (2), 365–372.
2. Ledauphin, J.; Saint-Clair, J. F.; Lablanquie, O.; Guichard, H.; Founier, N.; Guichard, E.; Barillier, D. *J. Agric. Food Chem.* **2004**, *52*, 5124–5134.
3. Câmpeanu, G.; Burcea, M.; Doneanu, C.; Namolosanu, I.; Visan, L. *Analusis* **1998**, *26*, 93–97.
4. Li, H.; Tao, Y. S.; Wang, H.; Zhang, L. *Eur. Food Res. Technol.* **2008**, *227*, 287–292.
5. Campo, E.; Cacho, J.; Ferreira, V. *J. Chromatogr., A* **2006**, *1137*, 223–230.
6. Falcao, L. D.; Lytra, G.; Darriet, P.; Barbe, J.-C. *Food Chem.* **2012**, *132*, 230–236.
7. Luccarelli, D., Jr.; Mookherjee, B. D.; Wilson, R. A.; Zampino, M. J.; Bowen, D. R. International Flavors and Fragrances Inc., New York, 1984, U.S. Patent 4526798.
8. Lytra, G.; Tempere, S.; de Revel, G.; Barbe, J.-C. *J. Agric. Food Chem.* **2012**, *60*, 1503–1509.
9. Pineau, B.; Barbe, J. C.; Van Leeuwen, C.; Dubourdieu, D. *J. Agric. Food Chem.* **2009**, *57*, 3702–3708.
10. Antalick, G.; Perello, M. C.; de Revel, G. *Food Chem.* **2010**, *121*, 1236–1245.
11. Martin, N.; De Revel, G. *J. Int. Sci. Vigne Vin* **1999**, 81–93.
12. ISO 13301 (2002) Sensory analysis - Methodology - General guidance for measuring odour, flavour and test detection thresholds by a threealternative forced-choice (3-AFC) procedure. In *Analyse Sensorielle*; AFNOR: Paris, France, 2002.
13. Feller, W. *An Introduction to Probability Theory and its Applications.* In Wiley series in Probability and Mathematical Statistics, 3rd Edition.; New York, 1968; Volume 1.
14. Miyazawa, T.; Gallagher, M.; Preti, G.; Wise, P. *Chem. Senses.* **2008**, *33*, 363–369.
15. Cometto-Muñiz, J. E.; Abraham, M. H. *Pharmacol., Biochem. Behav.* **2008**, *89* (3), 279–291.
16. Tempere, S.; Cuzange, E.; Malak, J.; Bougeant, J. C.; De Revel, G.; Sicard, G. *Chemosens. Percept.* **2011**, *4*, 99–115.
17. Kinlin, T. E.; Muralidhara, R.; Pittet, A. O.; Sanderson, A.; Walradt., J. P. *J. Agric. Food Chem.* **1972**, *20*, 1021–1028.
18. Brenna, E.; Fuganti, C.; Serra, S. *Tetrahedron: Asymmetry* **2003**, *14*, 1–42.
19. Romano, A.; Perello, M. C.; Lonvaud-Funel, A.; Sicard, G.; de Revel, G. *Food Chem.* **2009**, *114*, 15–19.
20. Lytra, G.; Tempere, S.; Le Floch, A.; de Revel, G.; Barbe, J.-C. *J. Agric. Food Chem.* **2013**, *61*, 8504–8513.
21. Lytra, G.; Tempere, S.; de Revel, G.; Barbe, J.-C. *J. Agric. Food Chem.* **2014**, *61*, 5005–5010.
22. Lytra, G.; Tempere, S.; Zhang, S.; Marchand, S.; de Revel, G.; Barbe, J.-C. *J. Int. Sci. Vigne Vin* **2014**, *48*, 75–85.
23. Lytra, G.; Tempere, S.; de Revel, G.; Barbe, J. - C. *J. Agric. Food Chem.* **2012**, *60*, 12260–12269.

Chapter 7

Measuring Flavor Interactions Using Fractional Omission Testing

Neil Desforges,*,[1] Kate O'Mahony,[2] Perrine Delime,[2] Joanne Hort,[2] and Andrew J Taylor[1]

[1]WALTHAM ® Centre for Pet Nutrition, Mars Petcare, Waltham-on-the-Wolds, LE14 4RT, United Kingdom
[2]Division of Food Sciences, University of Nottingham, LE12 5RD, United Kingdom
***E-mail: neil.desforges@effem.com.**

Sensory omission is an experimental method used to identify key compounds contributing to a flavor. A novel method for omission testing involving the same-different test and a surety rating have been applied to a strawberry flavor. Results from separate omission experiments determined the orthonasal impact of removing different fractions of each individual volatile and the retronasal impact of different tastants on volatile omission. R-indices were calculated from the surety rating and were used to assess significant differences. All nine volatiles were significant on omission when tested orthonasally. Three volatiles remained significant when half of their concentration was removed or diluted in mineral water. Orthonasal testing was more sensitive than retronasal testing with a higher number of significant observations. This new approach utilizing the R-index indicates the relative contribution of a volatile to the overall perceived flavors.

Introduction

From a commercial perspective, it is important for industry to develop mixtures with the minimum number of volatiles necessary to represent the target flavors or fragrance. In the creation of food flavor models, the challenge is to determine which volatiles are needed to reproduce the flavors, as not all the volatiles in food contribute to sensory perception of flavors.

Sensory omission testing involves omitting one volatile or a group of volatiles from a flavor mixture and comparing it to the original. It has several applications, such as the identification of key odorants in raw materials (*1*), but is often used in the assessment of the contribution of individual volatiles to the overall flavors and the elucidation of interactions between the volatiles (*2*).

Grosch (*3*) reviewed sensory omission work published pre-2001 and indicated that although attribute rating (*4*) and similarity rating (*5*) have occasionally been used, the triangle test (ISO 4120:2004) was the most popular method used in sensory omission experiments (*1, 4–6*). The approach using the triangle test reported in the literature has often been limited statistically with assessor numbers too low (between 5 and 19) to accurately draw any significant conclusions, especially in terms of claims for similarity where power (1-β) and hence number of judges become critical. Furthermore, omission tests tend to focus on identifying 'key' volatiles and have not assessed the 'relative' contribution of volatiles to the overall flavors. Hence there is considerable scope to improve the approach used in such omission studies.

The Same-Different test (ASTM E2139-05) is simple, efficient and intuitive for naïve assessors and has shown high sensitivity due to low memory requirements (*7*). The major limitation of the same-different test is that it can be subject to response bias when samples are similar and the same/different decision lies in the region of uncertainty. This can cause a certain level of cautiousness for the assessors who may place an emphasis on getting the answer correct (*8*). Response bias is a cognitive mechanism which is independent of the assessor's sensitivity to the attributes within the sample; therefore it reduces the discriminatory power of the same-different test (*9*).

To overcome the issue of response bias, a sureness rating can be added to the test so that the assessor can indicate the level of confidence they had in answering the "same or different?" question. The sureness rating can be analysed in conjunction with the "same" or "different" answer, by calculating the R-index statistic (*9*). Therefore, the R-index removes response bias and provides a discrimination index, which in turn provides a metric to compare the 'relative' contribution of volatiles to the overall flavors. This approach has not been used in this context previously.

The aim of this study was to apply this methodology to a commercial strawberry flavor containing nine volatiles. The sensory test measured the relative importance of each individual volatile within the strawberry flavor when one volatile was completely removed and the relative importance of each individual volatile when the strawberry flavor was diluted into water. The strawberry flavor was assessed again when the relative importance of each individual volatile was determined when half the amount of one volatile was removed. The method was also used to investigate cross-modal interactions between volatiles and tastants. The strawberry flavor was diluted into water and was assessed retronasally for the omission of volatiles in the absence and presence of sucrose and citric acid. The results from the orthonasal and retronasal testing were compared and discussed.

Materials and Methods

Materials

A strawberry flavor, based on a commercially available product, and composed of nine volatiles (all Sigma Aldrich, UK) was used throughout this study (Table 1). Propylene glycol (PG) (Sigma Aldrich, UK) was used as a solvent as it is easily miscible with the related volatile compounds and works effectively as a volatile carrier (*10*). Other consumables included Evian™ mineral water (Danone™ Group, France) used as a palate cleanser and as a solvent. Plain, unsalted matzo crackers (Rakusens Limited, UK) were also used for palate cleansing.

Table 1. Strawberry Flavor Model

Volatile	*Aroma*	*Concentration (g/kg)*
2,3-Butandione	buttery	0.00500
Butanoic acid	sweaty, rancid	0.920
Gamma-decalactone	fatty, peach like	1.33
Ethyl butanoate	fruity	5.00
Ethyl hexanoate	green, pineapple	3.36
Methyl dihydrojasmonate	jasmine	0.00300
4-Hydroxy-2,5-dimethyl-3-furanone	caramel	10.7
Methyl(E)-3-phenylprop-2-enoate	fruity	2.70
cis-3-Hexen-1-ol	leaf like	10.8

Preparation of the Original Strawberry Flavors in PG

Samples were prepared by pipetting the volatiles (Table 1) into Duran® GL 45 laboratory glass bottles (SCHOTT, USA) using a calibrated balance (allowing a 5% error). Samples were then diluted in PG and mixed on a roller bed for 30 minutes. Samples were refrigerated at 4°C and used up to 8 days after preparation. All samples were removed from the refrigerator at least one hour prior to testing to ensure samples were at room temperature (20 ° C ± 2°C).

Preparation of the Orthonasal Omission Samples in PG

Omission samples were prepared as described above. For n-1 samples, one volatile was removed from the original flavor model (n) and for the n-0.5 samples, half of the volatile was removed from the original flavor model (n).

Nine omission samples were prepared for each test with each omission sample omitting either one volatile or half the volatile from the original flavor model. For sensory testing, samples were kept at room temperature (20 ± 2°C) and used within 1 week.

Preparation of the Orthonasal Omission Samples in Mineral Water

Omission samples (n-1) were prepared as described above, by omitting one volatile from the original flavors model (n). Nine omission samples were prepared, each omission sample omitting one volatile from the original flavor model. For sensory testing, strawberry flavor and omission samples in PG were diluted at 0.75% w/w with mineral water. Samples were kept at room temperature (20 ± 2°C) and used within 24 hours.

Preparation of the Retronasal Omission Samples in PG

Omission samples (n-1) were prepared as described above, by omitting one volatile from the original flavors model (n). Nine omission samples were prepared, each omission sample omitting one volatile from the original flavor model. For sensory testing, strawberry flavor and omission samples in PG were diluted at 0.75% w/w in mineral water; mineral water alone was used for the blank taste. Sucrose was added at 2% v/v for a sweet taste and citric acid was added at 0.05% v/v for an acidic taste. Samples were kept at room temperature (20 ± 2°C) and used within 24 hours.

Sensory Testing

Subjects

Naïve assessors (80% female, aged between 18 and 25) were recruited from students at the University of Nottingham. Assessors were recruited for the orthonasal and retronasal testing separately. Ethical approval for the use of human subjects were reviewed internally by Division of Food Sciences ethics committee.

Sensory Sessions

Six sensory sessions were carried out in isolated booths. Each session involved nine discrimination tests to compare each one of the nine omission samples (n-x) with the original mixture (n). Sessions one, two and three involved 100 assessors carrying out nine same-different tests on the flavors delivered orthonasally. Session one evaluated n-1 omission samples in PG, session two evaluated n-0.5 omission samples in PG and session three evaluated n-1 omission samples in mineral water. Sessions four, five and six involved 100 assessors carrying out nine same-different tests on the flavors delivered retronasally. Session four contained blank tastants, session five contained a sweet taste and session six contained an acid taste. The order of presentation for the samples was randomized over each session. FIZZ software (Biosystèmes, France) was used to provide a randomized balanced design for the same-different test within each sensory session. Assessors were instructed to fast (except water) at least one hour prior to the sessions. They were instructed to assess samples from left to right and were allowed to re-evaluate the samples if necessary. Within a session, after three and six tests, assessors were allocated a five minute break to limit sensory fatigue and carryover effects.

Orthonasal Delivery

Screw-top 20 mL glass bottles containing 10 mL of sample were presented to the assessors. Assessors were instructed to sniff the samples and replace the lid immediately to prevent the volatiles dispersing throughout the test area.

Retronasal Delivery

Assessors were instructed to sip from a 20 mL sample through the straw of a lidded pot (thus avoiding orthonasal detection). Mineral water and crackers were provided as a palate cleanser between samples to minimise carry over effect.

Same-Different Testing

The protocol used in this study was an extension of the same-different test using a sureness rating (*11*). The same-different test with sureness rating can be regarded as a version of the degree of difference (DOD) test proposed by Aust (*12*). For each same-different test, the assessors assessed the two samples and stated whether they thought they were the same or different. Secondly, the assessors were asked to state the sureness level of their decision, represented by a four point surety scale ('very unsure', 'unsure', 'sure', 'very sure'). A complete randomized balanced design was used for the sample presentation, with half of the assessors presented with a 'same pair' and the other half presented with a 'different pair'.

Data Analysis

The R-index was calculated from the surety data for each omission test (*9*). According to the critical value tables published by Bi and O'Mahony (*13*), where there were fifty presentations of the same samples, an R-index of 59% or higher indicated a significant difference.

Results and Discussion

Orthonasal Testing

The R-index values for the n-1 and n-0.5 orthonasal omission experiments are presented in Table 2. Each line of Table 2 corresponds to an omission test comparing the original flavor model with an omission sample omitting one volatile. The R-index can therefore be used as a measure of the relative importance of each individual volatile within the strawberry flavors. The results for n-1 omission testing in PG show that all of the volatiles contained within the strawberry flavors were significant with methyl(E)-3-phenylprop-2-enoate and 2,3-butandione reporting the highest and lowest R-indices respectively. These results show that the omission of 2,3-butandione was hardest to detect and the omission of methyl(E)-3-phenylprop-2-enoate was the easiest to discriminate.

Table 2. R-Index Values Associated with the Orthonasal Assessment of Each n-1 and n-0.5 Sample in Comparison to the Original Strawberry Flavors

Volatile	*n-1 PG (%)*	*n-0.5 PG (%)*	*n-1 Water (%)*
2,3-Butandione	62[a]	51	49
Butanoic acid	68[a]	52	48
Gamma-decalactone	69[a]	64[a]	53
Ethyl butanoate	76[a]	62[a]	54
Ethyl hexanoate	69[a]	69[a]	60[a]
Methyl dihydrojasmonate	69[a]	47	39
4-Hydroxy-2,5-dimethyl-3-furanone	72[a]	52	60[a]
Methyl(E)-3-phenylprop-2-enoate	79[a]	55	49
cis-3-Hexen-1-ol	76[a]	53	60[a]

[a] R-index of 59% or higher indicated a significant difference, *p*-value <0.05.

The results for n-0.5 testing in PG show that the omission of three volatiles – gamma-decalactone, ethyl butanoate and ethyl hexanoate – was significant when the volatile was removed. These results suggest that these three compounds are the most significant character-impact flavor compounds since even a 50 % reduction leads to sample discrimination. Interestingly, volatiles with the highest R-indices in n-1 testing were not always significantly different in n-0.5 testing. This suggests the presence of intra-modal interactions between the volatiles to account for the differences.

This approach could be used to identify which components within a flavor could be reduced in concentration without impacting on consumer perception. The reduction would provide efficiency savings and cost optimization to the manufacturer.

The results for n-1 testing in mineral water show that the omission of three volatiles – ethyl hexanoate, 4-hydroxy-2,5-dimethyl-3-furanone and cis-3-hexen-1-ol – was significant when half of the volatile concentration was removed. This result showed that it is harder to detect volatile omission once the flavor is diluted in water and due to the different partitioning of the individual compounds from these two media.

Retronasal Testing

The R-index values of the n-1 retronasal omission experiment are presented in Table 3. Each line of Table 3 corresponds to an omission test comparing the original flavor model with an omission sample omitting one volatile in three conditions; no tastants, with sucrose and with citric acid. The R-index is a measure of the relative importance of each individual volatile within the strawberry flavors delivered retronasally. In contrast to the orthonasal results, there were no significant results observed when no tastants were present.

In the presence of sugar, the omission of 4-hydroxy-2,5-dimethyl-3-furanone was significantly detected suggesting an enhancement effect between the congruent sweet taste (*14*) and caramel-like aroma. Caramel not only smells of candy but is also a thermal degradation product of sucrose strengthening the congruency effect between the aroma and taste (*15*). This enhancement made the omission of 4-hydroxy-2,5-dimethyl-3-furanone more noticeable when the original flavor model and omission sample were compared.

In the presence of acid, the omission of butanoic acid, ethyl butanoate, methyl dihydrojasmonate and methyl(E)-3-phenylprop-2-enoate was significantly detected suggesting an enhancement effect between the citric acid taste and the detected volatiles. Citric acid is prevalent in most fruits, which have sour and sweet as their most prominent tastes (*16*). Ethyl butanoate has a pineapple-like odor and methyl dihydrojasmonate has a floral-like citrus aroma (*17*) which emphasizes the congruency between citric acid and fruit. Butanoic acid also appears to demonstrate congruency with citric acid, presumably due to both being organic acids. Methyl(E)-3-phenylprop-2-enoate, has been described as having a strawberry, sweet, cinnamon odor (*17*). Although cinnamon is not necessarily congruent with citric acid, sweet and strawberry could be considered complementary.

Table 3. R-Index Values Obtained When Comparing the Full Strawberry Aroma against the n-1 Strawberry Sample in the Presence of Water, Sucrose (2% v/v) or Citric Acid (0.05% v/v)

Volatile	*Water (Blank) (%)*	*Water + Sucrose (%)*	*Water + Citric Acid (%)*
2,3-Butandione	54	42	55
Butanoic acid	46	56	64[a]
Gamma-decalactone	49	52	53
Ethyl butanoate	48	57	64[a]
Ethyl hexanoate	47	52	55
Methyl dihydrojasmonate	50	41	59[a]
4-Hydroxy-2,5-dimethyl-3-furanone	55	61[a]	49
Methyl(E)-3-phenylprop-2-enoate	44	50	64[a]
cis-3-Hexen-1-ol	54	54	47

[a] R-index of 59% or higher indicated a significant difference, p-value <0.05.

Orthonasal versus Retronasal Sensitivity

Although the retronasal olfactory system is responsible for our ability to identify the flavor of food (*18*), it has been previously shown that orthonasal olfaction can be more sensitive at both threshold and suprathreshold levels (*19*). In controlled experiments, orthonasal olfaction has been shown to be more sensitive for detecting (*20*) and identifying aromas (*21*). In this study, orthonasal olfaction supported the existing literature and was generally more sensitive at noticing the removal of single volatiles. This higher orthonasal sensitivity can be explained by the increase in volatile concentration delivered to the olfactory receptors via the orthonasal route in comparison to the retronasal route. The intensity perceived by a panellist is directly proportional to the number of molecules that reach the olfactory receptors. The concentration of volatiles in the breath during food consumption is much lower than the concentration in the headspace above a sample solution (*22*). Aqueous systems such as the strawberry flavor, are consumed within a few seconds, therefore the major factor affecting the concentration of volatiles delivered to the receptors is the transfer of volatiles from the aqueous phase to the gas phase (*23*). One limitation of this study was the measurement of the volatile concentration in the oral and nasal cavities; this would have been difficult to achieve in these sensory experiments with consumers.

A second hypothesis is that this difference in sensitivity is due to differences in how the ortho- and retronasal information is cognitively processed. In previous studies, psychophysical, electrophysiological and imaging data show clear differences in the perception and processing of ortho- and retronasal olfactory information (*24*). For this particular study, the retronasal assessment of the strawberry flavor could have been more complicated for assessors as they had the expectation they were likely to receive a strawberry taste-aroma profile. The absence of congruent tastants, which would enhance flavor perception (*25*), may have confused assessors and resulted in poor discrimination.

One further limitation of this study is that due to scheduling, different groups of assessors were used for each sensory session and the difference observed between R-index values could be due to variation in sensitivities between assessors. To limit this, the assessors were recruited from the same environment and age class and it is unlikely that differences in sensitivity can explain all of the difference observed.

Same-Different Approach in Omission Studies

This study is the first to use the same-different approach with a surety rating and calculate R-indices in the context of omission testing. It successfully identified volatile components in the strawberry flavor whose omission was more easily detected than others and enabled two different routes of volatile perception to be compared. The same-different approach utilizing R-indices constitutes an improved, effective and robust sensory protocol for use in omission experiments. Most notably, the calculation of R-indices with this approach allowed the relative importance of the different volatiles within a flavor mixture to be determined.

Conclusions

This study addressed a number of areas identified in omission research in which improvements could be made, both in terms of the sensory methodology adopted and manipulation of omission samples. The same-different approach constitutes an effective robust approach for sensory omission testing. This is of interest as it not only determines the key volatiles, but also their relative importance within a flavor mixture.

In this study, the same-different approach was used to compare perception between ortho- and retronasal routes and confirms that orthonasal perception does not necessarily represent retronasal perception. This finding has implications for the analysis of flavor mixtures used in food and beverage products. We recommend that retronasal as well as orthonasal analysis should be carried out, as the perception of the flavors can vary significantly between the two delivery routes.

References

1. Greger, V.; Schieberle, P. *J. Agric. Food Chem.* **2007**, *55* (13), 5221–5228.
2. House, K. A.; Acree, T. E. *Food Qual. Pref.* **2002**, *13* (7-8), 481–488.
3. Grosch, W. *Chem. Senses* **2001**, *26* (5), 533–545.
4. Schieberle, P.; Hofmann, T. *J. Agric. Food Chem.* **1997**, *45* (1), 227–232.
5. Reiners, J.; Grosch, W. *J. Agric. Food Chem.* **1998**, *46* (7), 2754–2763.
6. Wagner, R. K..; Grosch, W. *J. Am. Oil Chem. Soc.* **1998**, *75* (10), 1385–1392.

7. Rousseau, B.; Stroh, S.; O'Mahony, M. *Food Qual. Pref.* **2002**, *13* (1), 39–45.
8. Hautus, M. J.; Collins, S. *Percept. Psychophys.* **2003**, *65* (6), 844–860.
9. O'Mahony, M. *J. Sens. Stud.* **1992**, *7*, 1–47.
10. Seidenfeld, M. A.; Hanzlik, P. J. *J. Pharmacol. Exp. Ther.* **1932**, *44* (1), 109–121.
11. Irwin, R. J.; Stillman, J A.; Hautus, M. J.; Huddleston, L. M. *J. Sens Stud.* **1993**, *8* (3), 229–239.
12. Aust, L. B.; Gacula, M. C.; Beard, S. A.; Washam, R. W. *J. Food Sci.* **1985**, *50* (2), 511–513.
13. Bi, J.; O'Mahony, M. *J. Sens. Stud.* **2007**, *22*, 713–720.
14. Green, B. G. Oral chemesthesis: An integral component of flavors. In *Flavors Perception*; Roberts, D. D., Ed.; Wiley: U.K., 2004; pp 151−171.
15. Schifferstein, H. N.; Verlegh, P. W. *Acta Psychol.* **1996**, *94*, 87–105.
16. Pfeiffer, J. C.; Hort, J.; Hollowood, T. A.; Taylor, A. J. *Percept. Psychophys.* **2006**, *68*, 216–227.
17. Burdock, G. A. *Fenaroli's Handbook of Flavors Ingredients*, 5th ed.; Taylor & Francis Group: Boca Raton, 2010; pp 575−603.
18. Shepherd, G. M. *Nature* **2006**, *444* (7117), 316–321.
19. Bojanowski, V.; Hummel, T. *Physiol. Behav.* **2012**, *107* (4), 484–487.
20. Hummel, T.; Heilmann, S.; Landis, B. N.; Redem, J.; Frasnelli, J.; Smal, D. M. *Flavour Fragrance J.* **2006**, *21* (1), 42–47.
21. Heilmann, S.; Hummel, T. *Behav. Neurosci.* **2004**, *118* (2), 412–419.
22. Deibler, K. D.; Acree, T. E.; Lavin, E. H.; Taylor, A. J.; Linforth, R. Flavor release measurements with retronasal aroma simulator. In *Flavors*; Rothe, M., Ed.; Eigenverlag: Bergholz Rehbrucke, Germany,2000
23. Linforth, R.; Martin, F.; Carey, M.; Davidson, J.; Taylor, A. J. *J. Agric. Food Chem.* **2002**, *50* (5), 1111–1117.
24. Negoias, S.; Visschers, R.; Boelrijk, A.; Hummel, T. *Food Chem.* **2008**, *108* (4), 1247–1254.
25. Green, B. G.; Nachtigal, D.; Hammond, S.; Lim, J. *Chem. Sens.* **2012**, *37* (1), 77–86.

Chapter 8

Evidence for Perceptual Interaction Phenomena To Interpret Typical Nuances of "Overripe" Fruity Aroma in Bordeaux Dessert Wines

Panagiotis Stamatopoulos,*[,1,2] Eric Frérot,[3] and Philippe Darriet[1,2]

[1]Univ. de Bordeaux, ISVV, EA4577 Œnologie, F-33140 Villenave d'Ornon, France
[2]INRA, ISVV, USC 1366 Œnologie, F-33140 Villenave d'Ornon, France
[3]Firmenich SA, R&D Division, 1 Route Des Jeunes, CH – 1211 Genéve 8, Switzerland
*E-mail: panagiotis.stamatopoulos@u-bordeaux.fr.

Bordeaux dessert wines can present a complex flavor, recalling overripe fruits, featuring citrus, especially orange notes. It was not possible to identify the volatile compounds responsible for these typical notes using traditional strategies, i.e. analysis of wine extracts by GC-olfactometry (GC-O). Thus a new strategy, combining both analytical and sensory approaches, was applied to fractions obtained by semi-preparative HPLC of wine extracts. Then omission tests, using preparative GC with cryotrapping, were used to identify fractions presenting overripe fruit nuances, particularly orange notes. Finally, multidimensional GC/MS-O was applied to these fractions to identify a new lactone, specific to dessert wines. While this compound presents minty and fruity nuances, it was found to contribute, together with other wine volatile compounds, to some typical nuances in Bordeaux wines, particularly Sauternes, via perceptual interaction phenomena.

Introduction

Sweet wines correspond to an incomplete fermentation, leaving a certain proportion of grape sugar that has not been transformed into alcohol. Wines are arbitrarily judged to be semi-sweet, sweet, and very sweet (liquoreux) according to their sugar concentration: up to 20 g/L, 20-36 g/L, and above 36 g/L, respectively (*1*). Semi-dry and sweet winemaking is fairly similar to dry winemaking, except that the grapes must have sufficient sugar content and fermentation must be stopped before completion, either naturally or by a physical or chemical process. Very-sweet winemaking is different, as the required sugar concentrations cannot be attained during normal grape maturation. Specific processes, such as drying, freezing and noble rot are required to concentrate the juice and certain winemaking steps are unique to these wines.

Bordeaux dessert wines are produced by the effect of the Botrytis cinerea fungus on perfectly ripe grapes. These exceptional wines can only be made under specific conditions, so production is limited. The Sauternes-Barsac region is certainly one of the most highly-esteemed areas for noble rot sweet wines. The development of noble rot requires a particular climate, ideally with morning fogs to promote fungal growth, followed by warm afternoon sunshine to concentrate the grape juice, over a relatively long period of 2 to 4 weeks. In the Bordeaux vineyards, these meteorological conditions correspond to the establishment of a high-pressure ridge extending north-east from the Azores anticyclone (*2*). Noble rot also develops rapidly in the Gironde region when a short period of rain, caused by oceanic depressions, is followed by a sunny, dry spell (low humidity, 60%), with north to north-east winds, generally associated with the presence of an anticyclone in north-eastern Europe.

Recent studies (*3*) demonstrated the key role of certain volatile thiols, furanones, and aldehydes in the aromatic perception of Bordeaux sweet wines. The development of noble rot on Semillon and Sauvignon Blanc grapes contributes significantly to increased levels of these compounds. The results presented verify the empirical practices and confirm the hypothesis that Botrytis cinerea fungus amplifies the aromas of Sauternes sweet wines. It was demonstrated that Botrytis cinerea increased the concentration of key volatile compounds, such as lactones and phenylacetaldehyde. An increase was also measured in concentrations of aroma precursors, like S-cysteine and S-glutathione conjugates due to Botrytis cinerea development on grape (*4–6*).

The goal of this study was to add to knowledge about the compounds involved in the typical "overripe fruity" aroma present in noble rot sweet wines. Some specific volatile compounds with "overripe fruity" notes were identified and correlated with their descriptors by associating analytical (identification of volatile compounds by gas chromatography) and sensory (preparative gas chromatography, reconstitution and omission tests) techniques. Given the complexity of the construction of sensory images, emphasis was placed on studying perceptual interaction phenomena (antagonism, synergy, and perceptual blending). In particular, aromatic reconstitution provided an important research tool for understanding the phenomena involved in building complex aromas.

Materials and Methods

Experimental details of wine samples, GC-O, multidimensional GC/MS-O, and sensory experiments for identifying and assessing the sensory impact of the compounds responsible for this "overripe" orange aroma were described in (7) by Stamatopoulos and co-workers. The preparative GC system consisted of a Hewlett-Packard (HP) 5890 Series II gas chromatograph (Agilent Technologies, Palo Alto, United States), equipped with a flame ionization detector (FID), a flame photometric detector (FPD), a sniffing-port (ODO-1 from SGE, Ringbow, Australia), and a Gerstel preparative trapping device (Preparative Fraction Collector, Gerstel, Mülheim an der Ruhr, Germany), connected via a heated (230°C) transfer line. The preparative device consisted of an eight-port zero-dead volume valve in a heated interface and seven 1-µL glass U-tube traps (six sample traps and one waste trap), immersed in a coolant (liquid N2). The preparative device was also connected to a Gerstel 505 controller to establish the trapping zones. The compounds were separated using a type HP5 "megabore" (30 m × 0.53 mm × 1.5 µm) fused-silica capillary column from Agilent (Agilent Technologies, Palo Alto, United States).

Results and Discussion

An Alternative Method for Studying Sensory Interactions in Wine

Wine is a very complex aromatic matrix. Over a thousand compounds have been identified in wine, at concentrations ranging from a few hundred mg to a few micrograms or even nanograms. At least a hundred volatile compounds can be detected in the headspace above wine in a glass. Given the possible perceptual interactions among them, omission tests were necessary to study the real impact of a specific compound or mixture of compounds on wine aroma. The principle of omission tests is based on eliminating one or more odoriferous zones to study the impact of those compounds on perception of a wine's aroma.

A crude extract of a typical dessert wine was analyzed by preparative GC-O. One large odoriferous zone identified was reminiscent of overripe-orange aroma (Table 1). No odoriferous zones with the same flavor were identified when a dry white wine extract was analyzed, demonstrating that this zone was specific to dessert wines. Once the odoriferous zones in dessert wines had been identified, omission tests were performed. The traps were assessed by the panel, who described the various aromas in our reconstitution and omission tests. Finally, triangular tests on blotter strips were performed to validate the differences between the samples.

Several traps were used during the sensory analysis session (Figure 1). The first trapped the whole dessert wine extract (0-82.66 minutes) to evaluate its odor characteristics. The descriptors cited by the panel for this trap were overripe fruits, citrus, floral, honey, and baked sugar, similar to those mentioned for the crude extract of the "typical" dessert wine. The second trap, corresponding to the 0-36 and 40-82.66 minute retention times, the descriptors cited were honey, creamy, yeasty, and spicy. Finally, for the last trapping zone, at 36-40 minutes, the descriptors were overripe orange and dried apricot.

Table 1. Distribution of Odoriferous Zones Found in an Extract of a Typical Dessert Wine Analyzed by GC-O (HP-5 Column)

L.R.I	R.T	Odoriferous zones
1141	25,1	Honey/nutty
1166	26,3	Burnt sugar
1176	26,8	Mushroom
1345	34,6	Buttery
1380	35,8	Coconut
1394-1434	36,8-38,5	Overripe fruit/Orange
1471	40,1	Candyfloss
1478	40,4	Fruity
1500	41,3	Candyfloss
1546	43,2	Apricot
1555	43,5	Baked fruit
1578	44,4	Meaty
1655	47,4	Coconut

L.R.I : Linear Retention Index; R.T : Retention Time

The panel found a significant difference between traps one and two, highlighting the importance of fraction three in the expression of dessert wine aroma. Our omission experiment demonstrated that dessert wine aroma was mainly reminiscent of honey, yeast, and spices if the overripe orange-dried apricot zone was removed. Thus, it seems that the compounds responsible for the specific overripe-orange aroma zone would be "key" contributors for the nuances of overripe fruit in dessert wine aromas. Recombining traps two and three reproduced the initial odor of the dessert wine extract.

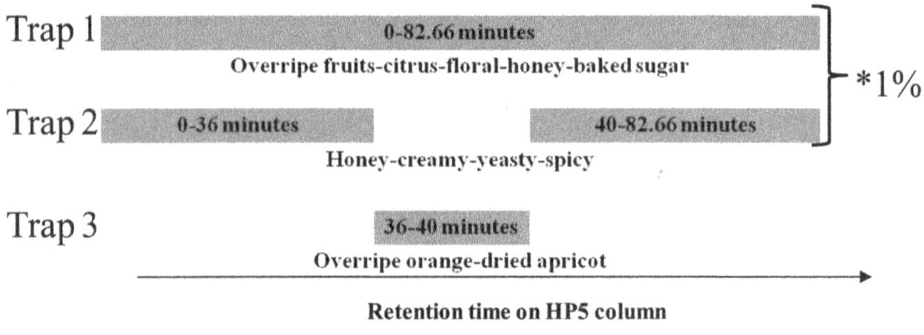

Figure 1. Omission and recombination tests using wine extract.

Characterizing the Compounds Associated with Key Aromatic Fractions Reminiscent of Overripe-Orange Aromas. Identification and Quantification of 2-Nonen-4-olide.

Extracts of Bordeaux dry white and dessert wines were initially fractionated by semi-preparative HPLC, using ethanol and water as solvent, making it possible to describe the characteristics of each aromatic fraction by direct olfaction without the presence of any malodorous or toxic solvents. Different proportions of water/ethanol were used to obtain 50 fractions with different sensory attributes. Selected dry white and dessert wine extracts were analyzed by HPLC to constitute fractions with different aromatic descriptors. Comparative sensory evaluation of the fractions revealed clear sensory differences between dessert-wine and dry-white-wine extracts, particularly in the ripe fruit register (Table 2). Sensory analysis of samples were fractions 37 and 38 had been omitted confirmed the importance of these in the perception of overall dessert wine aroma. In contrast, the fractions 37 and 38 had only a minor impact on dry white wines (Table 3). This approach facilitated the identification of new compounds and highlighted the olfactory differences between wines by focusing on less complex matrices. To characterize the compounds associated with this aroma, fraction 37, the most significant from a sensory point of view, was re-extracted with dichloromethane and the concentrated extract was injected on both a polar (BP20) and non-polar capillary (HP5), to identify the odoriferous zones corresponding to overripe orange. It was not possible to identify the key odoriferous zones with overripe fruit and orange aromas using a BP20 type capillary. However, on HP5, a typical orange aroma lasting one minute was perceived in a chromatographic zone situated between 1414 and 1443 LRI (Table 4). Then, MDGC-O-MS was used to analyze this complex matrix and identify the compound(s) associated with this aroma.

Using MDGC chromatography, the odoriferous zone was targeted by its olfactory characteristics on the GC1 olfactometry port. Then, an adequate 3 min cut was performed. Separating the extract on the analytical column (GC2, BP20 capillary) coupled with both the second olfactometry port and a mass spectrometry detector revealed the presence of several odoriferous zones, potentially related to the overripe-orange aroma. Odoriferous zones with coconut, spicy clove, and ripe fruity nuances were identified, corresponding to 3-methyl-4-octanolide (coconut), eugenol (spicy), and γ-nonalactone (ripe fruity), respectively. A fourth odoriferous zone with minty and fruity aromas corresponded to an unknown compound. MDGC analysis of the same extract, repeated on a non-polar capillary BP1, confirmed the presence of the same odoriferous zone. By coupling the capillary to the MS detector, it was possible to obtain a clear mass spectrum at the retention time of this odoriferous zone. On the basis of the MS data obtained by E.I. and chemical ionization with methane, the spectra indicated a molecular mass of M=154.099 ($[M+H]^{+}$=155) and the associated chromatographic peak was identified as a lactone: 2-nonen-4-olide. The mass spectra of the 2-nonen-4-olide standard and the one obtained from the wine extract presented the same characteristics (Figure 2). To our knowledge, this was the first time that this compound had been identified in wine. Previous studies reported the presence of 2-nonen-4-olide as a fruity flavor compound in mushrooms (*8*). Also, 2-nonen-4-olide has also been cited as a modifying and intensifying compound for flavoring (*9*).

Table 2. Sensory Evaluation of HPLC Fractions from Various Dry White and Dessert Wines

HPLC fractions	*Dessert wines*			*Dry white wines*	
	Liq-CS 2005	Liq-CDS 2004	Liq-Cds 2006	Sec-CM 2008	Sec-SB 2007
35	Solvent	Fruity	Floral	Citrus	Citrus
36	Moldy	Odorless	Citrus	Thiols	Odorless
37	Ripe orange	Ripe orange	Ripe orange	Citrus	Odorless
38	Ripe orange/ moldy	Ripe orange	Ripe orange	Floral	Fruity
39	Mushroom	Apricot	Moldy	Green	Cherry
40	Cherry	Odorless	Spicy	Solvent	Cherry
41	Banana	Banana	Banana	Banana	Solvent

Table 3. Results of Triangular Tests Using Aromatic Reconstitutions

Wine	*Matrix*	*Samples compared*		*Difference*
Dessert wine	Hydroalcoholic solution	AR (1-50)	AR (1-36+39-50)	**,
Dry white wine	Hydroalcoholic solution	AR (1-50)	AR (1-36+39-50)	-,

**, Significance level 0.1% ; -, no significance difference.

2-nonen-4-olide was quantified in 15 dessert wines from the Bordeaux region and 8 dry white wines from different appellations in France including Chablis and Alsace. Clearly, 2-nonen-4-olide was not detected in dry white wines, but only in dessert wines obtained from botrytized grapes, thus confirming the hypothesis that this compound is specific to dessert wines. The concentrations of 2-nonen-4-olide in dessert wines depended on the vintage, with a mean value around 6 µg/L. Typical and non-typical examples of dessert wines were also noticed to have different 2-nonen-4-olide concentrations (Figure 3). A triangular test with increasing ranges of concentrations of synthetic (±)2-nonen-4-olide in dilute alcohol solution found a perception threshold of 4.3 μg/L for 50% of tasters. The odor descriptors for (±)-2-nonen-4-olide range from fruity/minty flavors at low concentrations (5μg/L) to more buttery/coconut aromas at higher concentrations (100μg/L). 2-nonen-4-olide has one asymmetrical carbon in position 4, indicating the possibility of two enantiomers, R and S.

Table 4. Distribution of Odoriferous Zones Found in Fraction 37 of a Typical Dessert Wine Extract Analyzed by GC-O (HP-5 column)

L.R.I	R.T	Odoriferous zones
1263	31.4	Fruity
1346	34.7	Floral
1360	35.3	Citrus
1414-1443	38.3-38.9	Overripe orange
1602	41.4	Plastic
1791	48.8	Citrus

L.R.I : Linear Retention Index; R.T : Retention Time

Evidence for Perceptual Interaction Phenomena Involving 2-Nonen-4-olide

Three experienced assessors selected two different types of dessert wines. One was a typical Bordeaux dessert wine, with a desirable odor, reminiscent of overripe fruit, particularly orange peel, while the other was a non-typical Bordeaux dessert wine, with a fresh, fruity aroma. After fractionation of each extract by reversed-phase HPLC, several reconstitution tests were performed. Eight samples of aromatic reconstitutions in dilute alcohol solution (12% vol., pH 3.9) were presented (Table 5). In the first sensory analysis session, the fractions corresponded to "typical" Bordeaux dessert wine and the volatile compounds (3-methyl-4-octanolide, eugenol, γ-nonalactone, and 2-nonen-4-olide) were supplemented at concentrations assayed in that type of wine. In the second session, the fractions corresponded to those of a "non-typical" Bordeaux dessert wine and the concentrations of the compounds added were those assayed in that type of wine. In the third session, the "non-typical" Bordeaux dessert wine matrix (sample PAR) was supplemented with fractions "37+38" and concentrations of the volatile compounds corresponding to "typical" Bordeaux dessert wine. Each sensory session was performed in duplicate on different days. Overall aroma, overripe-orange aroma, and orthonasal appreciation of typicality were evaluated on a 0-10 point, non-structured scale, where 0 = no odor perceived and 10 = high intensity. Concerning typicality experiments, the scale was 0 = bad example of Bordeaux dessert wine and 10 = good example of Bordeaux dessert wine. For aromatic reconstitutions, fractions were retained and added individually or blended together to approach mixtures close to the original wine extract, adding double-distilled ethanol and micro-filtered water to obtain 12 % (v/v) ethanol.

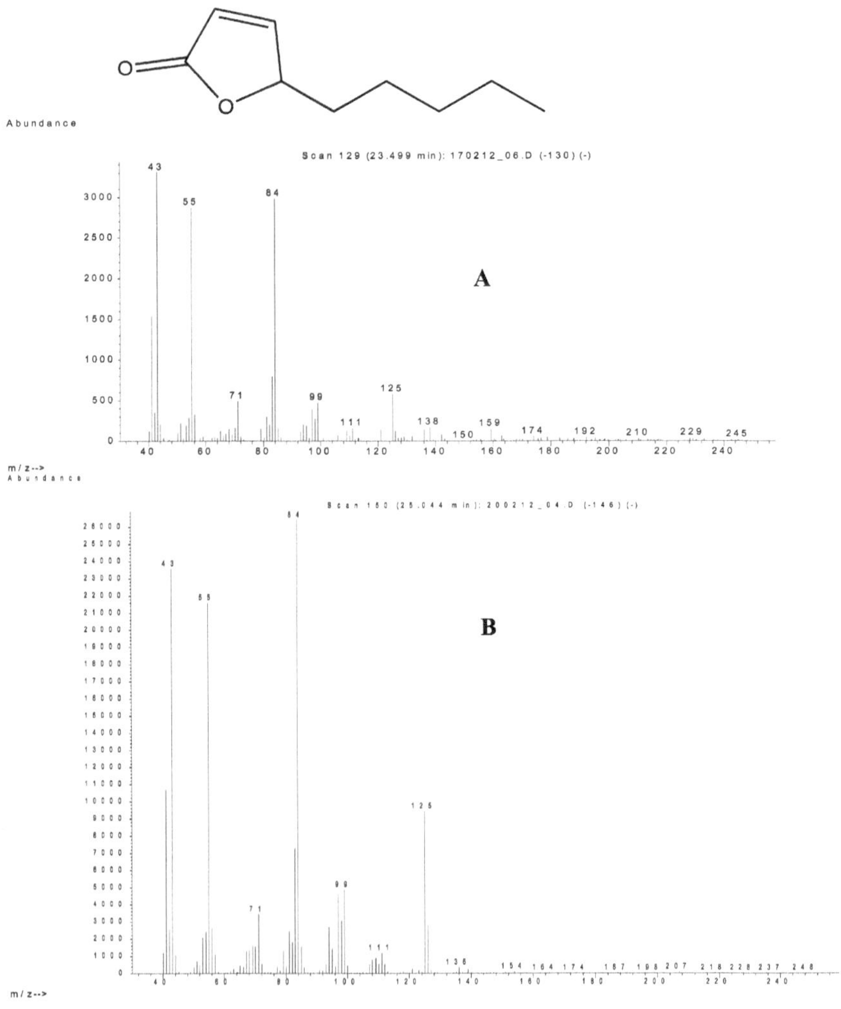

Figure 2. Chemical formulas and mass spectra of 2-nonen-4-olide in wine (A) and as a pure compound (B).

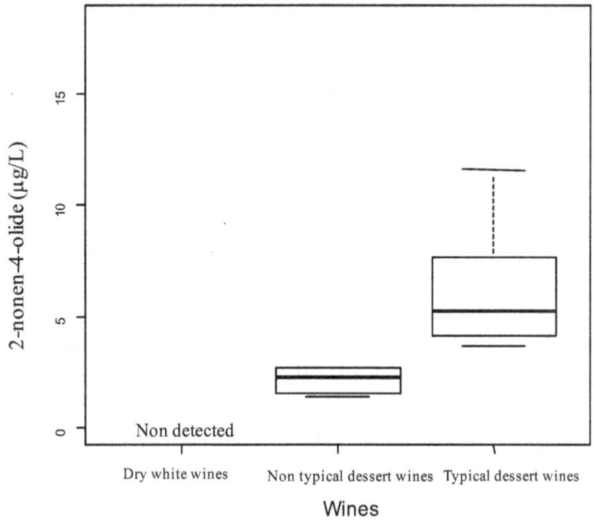

Figure 3. Concentration in µg/L of 2-nonen-4-olide in different types of white wine.

Table 5. Composition of Various Samples Used in Reconstitution and Omission Analyses

Samples	*HPLC fractions*			*Compounds*			
	1-36	37+38	39-50	3-methyl-4-octano-lide	eugenol	2-nonen-4-olide	γ-nonalac-tone
TAR	+	+	+	-	-	-	-
37+38	-	+	-	-	-	-	-
PAR	+	-	+	-	-	-	-
PAR4C	+	-	+	+	+	+	+
PAR4C-WL	+	-	+	-	+	+	+
PAR4C-E	+	-	+	+	-	+	+
PAR4C-N	+	-	+	+	+	-	+
PAR4C-γN	+	-	+	+	+	+	-

TAR: Total aromatic reconstitution; 37+38: Aromatic reconstitution with fractions 37 and 38; PAR: Partial aromatic reconstitution; PAR4C: PAR plus the four compounds; PAR4C-WL: PAR plus the three compounds without Whiskey Lactone; PARC-E without Eugenol; PARC-N without 2-nonen-4-olide; PARC-γN without γ-nonalactone; +: Addition; - : Omission.

The goal of the first experiment was to illustrate the impact of specific compounds, identified and assayed in dessert wines, on the overripe fruity character. In particular, the object of the study was to focus on the recognition of overripe-orange expression, as well as Bordeaux dessert wine typicality. Preliminary experiments comparing the total (TAR) and partial (PAR) aromatic reconstitutions supplemented with only one selected compound, revealed that the addition of one compound modifies significantly the perception of the overripe-orange aroma, resulting a lower appreciation of the aroma.

Then the first sensory experiment consisted in a direct comparison of the sample PAR (Partial Aromatic Reconstitution) supplemented with the four compounds identified and presumably associated with the overripe fruity odoriferous zone, at concentrations found in the typical dessert wine fractions. Under these conditions, the results indicated a clear recognition of overripe-orange fruity expression in PAR supplemented with the four compounds (PAR4C; mean value=6.23), much higher than in PAR fractions alone (mean value=4.04) and similar to that of the Total Aromatic Reconstitution sample (TAR) (mean value=6.73) (Figure 4). The aromatic reconstitution of fractions 37 and 38 presented a mean value of 7.43, confirming our preliminary finding that these fractions presented typical overripe-orange notes.

The evaluation of dessert wine typicality did not reveal any significant difference between the total reconstitution (TAR) alone or supplemented with the four compounds (PAR4C), with mean scores of 6.48 and 6.26, respectively (Figure 5).

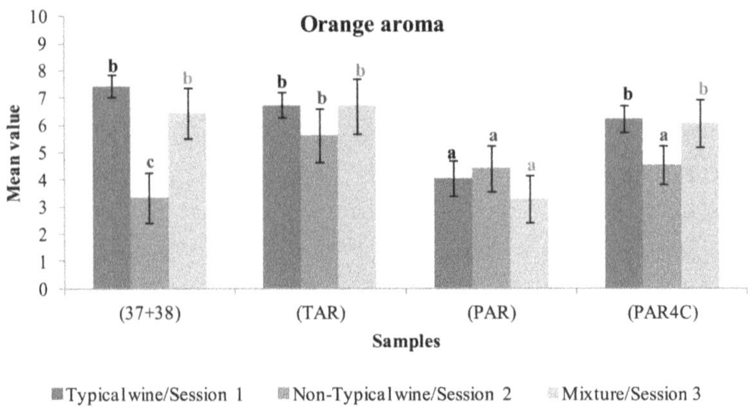

Figure 4. Overripe-orange aroma assessment in the reconstitution supplemented with four compounds.

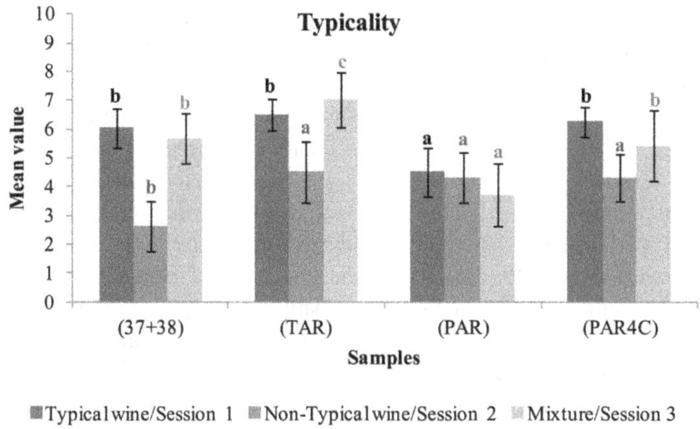

Figure 5. Typicality assessment for the reconstitution supplemented with four compounds.

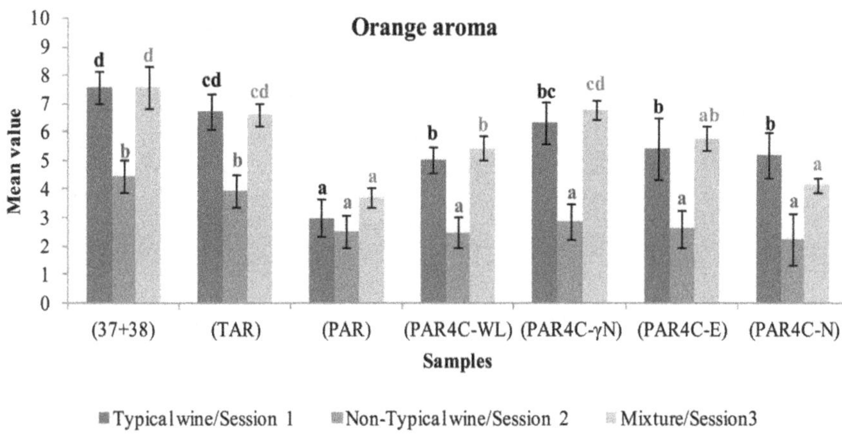

Figure 6. Overripe-orange aroma assessment in omission trials.

In order to study the impact of each compound on the perception of "overripe orange" aroma, omission tests eliminated each of the four compounds, in turn. Under these conditions, PAR had a value of 2.98. When PAR fractions were supplemented with 3 of the 4 compounds, the orange flavor was always perceived at a much higher value and, particularly when only γ-nonalactone was omitted, the results were very close to those of TAR and fractions 37+38 assessed separately. Thus, sample 37+38 presented a mean value of 7.56 and the total reconstitution (TAR) a mean value of 6.74 (Figure 6). In contrast, the total reconstitution without fractions 37 and 38 had the lowest mean value of all: 2.98. In the omission experiments, only sample PAR4C-γN was not significantly different from the total reconstitution (Figure 6). This revealed the direct impact of 2-nonen-4-olide, eugenol, and 3-methyl-4-octanolide on the overripe orange flavor in the Bordeaux dessert wines and the minor contribution of γ-nonalactone to its expression, at the concentrations considered. Also the existence of perceptual interactions between oak-derived compounds and the newly identified lactone, 2-nonen-4-olide, in Bordeaux dessert wines was confirmed, supporting previous findings on synthetic perception at brain level. Also the major impact of oak-derived compounds, such as eugenol and 3-methyl-4-octanolide, on this orange aroma was revealed.

Moreover, the results concerning the typicality of Bordeaux dessert wines revealed a significant difference ($p<0.01$) between samples 37+38 and TAR and samples PAR, PAR4C-E, and PAR4C-N (Figure 7), highlighting the impact of these compounds on wine typicality. First of all, the omission of fractions 37 and 38 had a clear negative impact, confirming the importance of these fractions for wine typicality. The omission of eugenol and 2-nonen-4-olide clearly had a negative impact on Bordeaux dessert wine typicality (Figure 7).

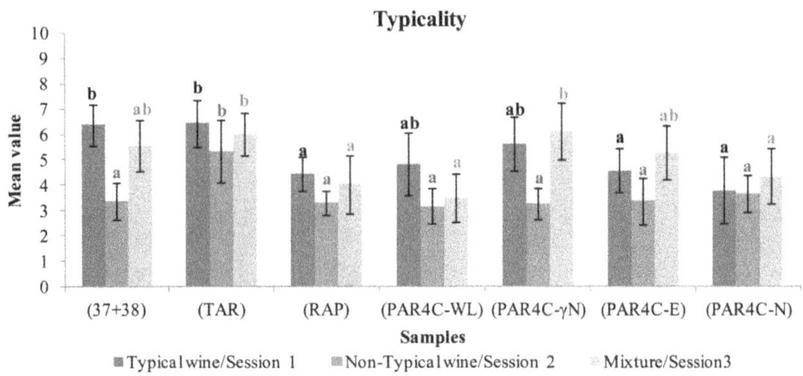

Figure 7. Typicality assessment in omission trials.

Session 2

The PAR sample was supplemented with the four compounds mentioned above, at the concentrations found in the non-typical wine fractions. The sensory test consisted of comparing the perception of orange flavor nuances detected in total and partial reconstitution samples, as well as in the reconstitution of fractions 37 and 38. A mean value of 4.54 was measured and a significant difference with the Total Aromatic Reconstitution (TAR) was noted (Figure 4). In contrast, lower values were measured and no significance difference was observed for the Bordeaux dessert wine typicality of the samples, with the exception of the sample containing the aromatic reconstitution of fractions 37 and 38, which scored the lowest mean value of 2.62, significantly different from the other samples (Figure 5).

The results of the sensory experiment with the omission of one compound out of four revealed a significance difference between samples 37+38 and TAR and samples PAR, PAR4C-WL, PAR4C-γN, PAR4C-E, and PAR4C-N (Figure 6). Samples 37+38 and TAR scored the highest mean values for overripe-orange aroma: 4.46 and 3.96, respectively, while samples PAR, PAR4C-WL, PAR4C-γN, PAR4C-E, and PAR4C-N scored the lowest mean values (Figure 6). The addition of the compounds to the PAR4C sample had a minor impact, due to the very low concentrations added.

The results concerning the typicality of Bordeaux dessert wines indicated a significant difference ($p<0.001$) between samples TAR, 37+38, and PAR, PAR4C-WL, PAR4C-γN, PAR4C-E, and PAR4C-N (Figure 7). TAR had the highest mean score for Bordeaux dessert wine typicality (Figure 6). This experiment confirmed not only that that the absence of specific compounds had a clear negative impact on wine typicality but also that, when sample 37+38 was added to the PAR of a non-typical wine, it was rated as a typical example of Bordeaux dessert wine. Only the association of compounds found in these specific samples produces a good example of Bordeaux dessert wine.

Session 3

In this sensory experiment, perception of the overripe-orange aroma was studied in the non-typical wine matrix, supplemented with the target compounds at concentrations found in the typical Bordeaux dessert wine, as previously reported. The partial aromatic reconstitution supplemented with the four compounds (PAR4C) scored a mean value of 6.06, with no significant difference compared to the total aromatic reconstitution. When the four compounds were omitted (PAR), the overripe orange perception had the lowest mean value 3.28 (Figure 6). Significant differences in terms of Bordeaux dessert wine typicality were observed between the total aromatic reconstitution alone and supplemented with the four compounds (Figure 7). Thus, a different matrix may influence the perception of wine typicality.

Paradoxically, the omission of eugenol had no significant impact on the total aromatic reconstitution under these conditions. The TAR and PAR4C-E samples had similar scores for overripe-orange aroma after statistical analysis (Figure 6). Under these conditions, the 37+38, TAR and PAR4C-γN samples had the highest mean values: 7.58, 6.62, and 6.79, respectively (Figure 6).

In contrast, a significant difference ($p<0.001$) in terms of Bordeaux dessert wines typicality was noted between sample PAR4C-N and the others, confirming the impact of 2-nonen-4-olide on wine typicality (Figure 7), as its omission resulted in a lower score for wine typicality. However, the omission of eugenol had no clear effect on wine typicality, in contrast to our first sensory session, where eugenol seemed to play a key role. The different matrix had a significant impact on eugenol expression, both in the overripe-orange aroma and wine typicality.

To sum up, 3-methyl-4-octanolide and 2-nonen-4-olide were found to play an important role in the expression of overripe-orange flavor, while eugenol and γ-nonalactone only had a minor effect. Finally, the specific matrix of a non-typical Bordeaux dessert wine had a minor impact on expression of the overripe-orange aroma, probably due to the specific molecules present and the concentrations added.

A previous study (*10*) reported that the strength of Cabernet Sauvignon "berry" character was very strongly correlated with concentrations of oak lactones and eugenol. Lactones similar to those in oak are known as active aroma compounds in many fruits (*11*), so the relationship between cis-3-methyl-4-octanolide concentrations and berry-aroma intensity is not surprising. Previous findings confirmed the impact of oak-derived compounds on the fruity aroma of wines, but the research reported here revealed the contribution of a novel compound, 2-nonen-4-olide, to the overripe-orange flavor in Bordeaux dessert wines.

Conclusion

This work built on research by Sarrazin (*3*), who demonstrated the existence of a typical sensory concept and identified various "key compounds" in Bordeaux dessert wines. However, the characteristic overripe fruit aroma, specifically overripe-orange, had not previously been elucidated. It was already clear that the perception of these aromatic notes was associated with a combination of molecules in the olfactory system of the tasters, via sensory interaction phenomena. This experiment used a new approach, involving preparative gas chromatography, as well as omission and recombination tests, to assess the importance of certain volatile compounds associated with the overripe-orange aroma. Our findings led to the identification of a new lactone specific to noble rot dessert wines. While this molecule presents a fruity/minty aroma, it contributes to the overripe-orange aroma, via perceptual interactions with other volatile compounds through a perceptual blending.

References

1. Ribéreau-Gayon, P.; Dubourdieu, D.; Donéche, B.; Lonvaud, A. *Traité d'Œnologie – Microbiologie du vin, Vinifications*, 5th ed.; Dunod: Paris, 2004; pp 598–599.
2. Dubourdieu, D. La vinification des vins liquoreux de pourriture noble. *Rev. Fr. d'Œnologie* **1999**, *176*, 32–35.
3. Sarrazin, E. *Recherches sur l'arôme des vins liquoreux de pourriture noble issus des cépages Sémillon et Sauvignon blanc.* Ph.D Thesis, Université de Bordeaux 2, Bordeaux-France. 2007.
4. Sarrazin, E.; Dubourdieu, D.; Darriet, P. Characterization of key-aroma compounds of botrytized wines, influence of grape botrytization. *Food Chem.* **2007**, *103*, 536–45.

5. Thibon, C.; Cluzet, S.; Mérillon, J. M.; Darriet, P.; Dubourdieu, D. 3-Sulfanylhexanol Precursor Biogenesis in Grapevine Cells: The Stimulating Effect of Botrytis Cinerea. *J. Agric. Food Chem.* **2011**, *59*, 1344–51.
6. Thibon, C.; Dubourdieu, D.; Darriet, P.; Tominaga, T. Impact of noble rot on the aroma precursor of 3-sulfanylhexanol content in Vitis vinifera L. cv sauvignon blanc and semillon grape juice. *Food Chem.* **2009**, *114*, 1359–64.
7. Stamatopoulos, P.; Frérot, E.; Tempére, S.; Pons, A.; Darriet, P. Identification of a New Lactone Contributing to Overripe Orange Aroma in Bordeaux Dessert Wines via Perceptual Interaction Phenomena. *J. Agric. Food Chem.* **2014**, *62*, 2469–78.
8. Jong, S. C.; Birmingham, J. M. Mushrooms as a source of natural flavor and aroma compounds. In *Mushroom biology and mushroom products*; World Society for Mushrooms Biology and Mushrooms Products. First International Meeting Proceedings, August 23−26, 1993, Hong Kong; The Chinese University Press: Hong Kong, 1993; pp 345−366.
9. Kindel, G.; Vossing, T.; Looft, J.; Wohrle, I.; Lages, R.; Loges, H. Novel use of Nonenolide. U. S. Patent Application US 2007/0297993A1.
10. Spillman, P. J.; Sefton, M. A.; Gawel, R. The contribution of volatile compounds derived during oak barrel maturation to the aroma of a chardonnay and cabernet sauvignon wine. *Aust. J. Grape Wine Res.* **2004**, *10*, 227–235.
11. Gatfield, I. L.; Sommer, H. The enzymatic and fermentative production of lactones and their use in natural flavors. In *Recent Developments in Flavor and Fragrance Chemistry: Proceeding of the 3rd International Haarman and Reimer Symposium*; Hopp, R., Mori, K., Eds.; VCH Publishers: New York, 1993; pp 291−304.

Chapter 9

Serial Dilution Sensory Analysis (SDSA) Applied To Exploring Sensory Attributes Essential for Food Aroma

T. Aishima,* K. Iizuka, and K. Morita

Chemometrics and Sensometrics Laboratory, 1-197 Sengen- cho, Omiya-ku, Saitama 330-0842, Japan
***E-mail: aishima@chemsensmetrix.com.**

A new methodology to identify sensory attributes essential for characterizing aroma notes of liquid food has been developed by combining serial dilution and with subsequent descriptive sensory analysis. To demonstrate the effectiveness of this new methodology, the serial dilution sensory analysis (SDSA) was applied to canned coffee beverage and soy sauce. In SDSA, sensory scores in all attributes decreased in accordance with dilution. However, score percentages calculated for individual sensory attributes based on their total in every dilution suggested meaningful changes. In coffee, proportions of the coffee note itself and the burnt note decreased drastically according to dilution but those of sweet, green and dusty notes increased inversely. By diluting soy sauce, proportions of alcoholic and burnt notes decreased but those of fruity and sweet increased. Partial least squares regression (PLSR) analysis quantitatively differentiated positive and negative effects of individual sensory attributes on characterizing original aroma notes.

Introduction

Regardless of plant, animal or metabolites of microorganisms, their odors are composed of many volatile compounds each smelling similarly or differently. From both of the scientific and industrial points of view, many efforts have been made to identify a compound or group of compounds responsible for characterizing aroma notes in various food from the beginning of aroma research

(*1*). Although various skepticisms have been expressed afterwards, the concept of odor activity value (OAV) or odor unit was proposed in the mid-1960s to quantify the aroma intensity of individual components contained in an aroma extract (*2*). In the 1980s gas chromatography- olfactometry (GC-O) dilution analysis was introduced, first as combined hedonic response measurement (CHARM) analysis (*3*) and then aroma extract dilution analysis (AEDA) (*4*). Since then, GC-O dilution analysis has been used to determine the most potent odorants in scores of foods. Especially AEDA has widely been applied because of its facile concept and versatility. However, our nose does not sense individual aroma components separately but rather simultaneously senses a mixture of numerous aroma notes, where complicated mutual interactions may take place.

As widely known, diluting beverage or liquor with water not only makes the aroma strength weaker but the aroma note itself changes or deteriorates. This change in aroma note caused by dilution supports the correctness of the odor unit concept because a profile of components contained in the dilution should be the same as that of the original food. In other words, change in the aroma note caused by dilution indicates that significant profile changes may have occurred due to the absence of aroma characters derived from some volatiles with low odor unit or low concentration.

The descriptive sensory analysis (*5*), that has been used frequently to describe sensory properties of food, can quantify individual sensory attributes in food by fully taking account of mutual interactive effects existing between attributes derived from chemicals. Of course, information obtained from sensory analysis cannot be directly related to instrumental data or chemicals but it will supply supportive knowledge to assess the significance of components identified by instrumental analysis. In this study, we applied SDSA to coffee and soy sauce in order to demonstrate the effectiveness of this new methodology.

Premise of SDSA

The introduction of the odor unit and GC-O dilution values to aroma research for quantitatively evaluating the significance of individual components in mixtures has proven useful for solving problems related to aroma quality, especially in the measurement of inappropriate, off- and taint odors. However, these methodologies have intrinsic limitations that are derived from the methodology itself. The skepticism or criticism against the principle of the odor unit and AEDA may be summarized as follows. First, results obtained from these two methodologies do not consider any interactive effects. Second, there is an intrinsic dependency on pretreatments. In other words prior to GC or GC-MS analysis in AEDA, volatiles should be extracted from the food matrix and concentrated using some physical or chemical treatment that more or less gives significant effect on the composition of the resulting concentrate (*6*). A solution to these criticism may be to have no sample pretreatment prior to analysis and to applying sensory analysis, because this analyzes food aroma by fully taking account of interactive effects.

The odor unit is calculated by dividing the concentration of an aroma compound in food by its threshold concentration (*2*). By utilizing the concept of odor unit, the basic concept of SDSA is illustrated in Figure 1. Hypothetical aroma profiles composed of six components (A, B, C, D, E and F) in original food and its contracted profile in three-fold dilution are comparatively described. In the instrumental analysis by using GC or GC-MS, peak areas of individual compounds simply decreased according to dilution and so the resulting profile would be simply contracted but it keeps the original pattern. Although no interactive effects have been considered to simplify the olfaction process, if the concept of odor unit is correct, the initial food aroma is recognized as an integrated aroma character derived from six different compounds but the number of recognizable attributes decreased to only four (A, C, D and F) after dilution. This clearly suggests that sample dilution not only makes the aroma profile contract in size but also changes the profile itself due to the absence of two aroma attributes (B and E).

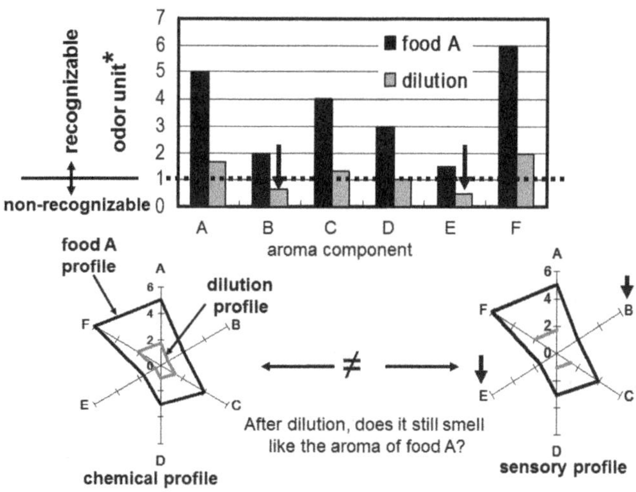

Figure 1. Basic concept of SDSA.

The quantitative descriptive sensory analysis was developed in the mid-1970s (*5*). Although the analytical resolution of this sensory method is much lower, the basic idea of descriptive sensory analysis is similar to that of GC analysis as schematically shown in Figure 2 (*7*).

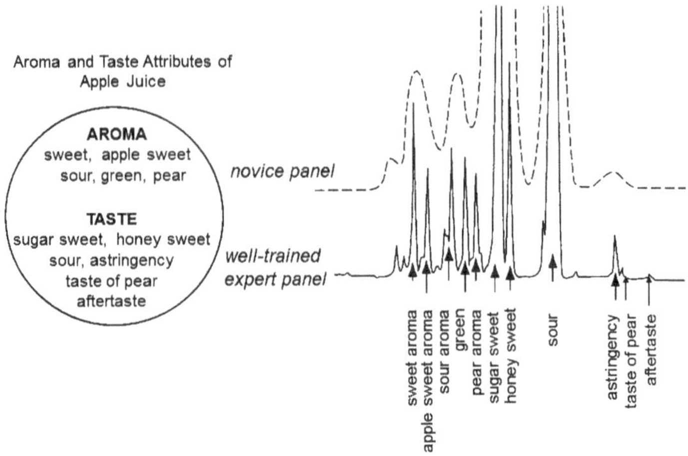

Figure 2. Basic concept of descriptive sensory analysis exemplified by chromatographic separation. (reproduced with permission from reference (7))

The whole process of descriptive sensory analysis is shown in Figure 3. In the first step of this method, all sensory attributes found in samples are listed by sniffing and tasting by individual members of a trained panel composed of not many but not less than 6 members (*8*). In the second step, sensory attributes that are commonly recognized by all members are selected through intensive discussion that is modulated by a panel leader who takes the role of an active listener and acts as a facilitator without swaying the opinion of individually panelists. The panel leader's primary aim is to establish consensus among the panelist by presenting a reference such as fruit, vegetable, spice or ingredients, if needed (*8*, *9*). In the descriptive sensory analysis, the first two steps for selecting attributes may correspond to identification of peak components in GC-MS. The third step corresponds to calibration where firstly a trial evaluation is carried out for a set of samples. In the evaluation, the strengths of attributes selected in the preceding steps are quantified using a category or line scale. After that, scores given by individual panelists along with scores averaged based on the whole panel are immediately fed back to them for tuning scales. Finally in the fourth step, each panelist individually quantifies strengths of every attribute. Then, scores given for each attribute by a panel are averaged to obtain sensory scores.

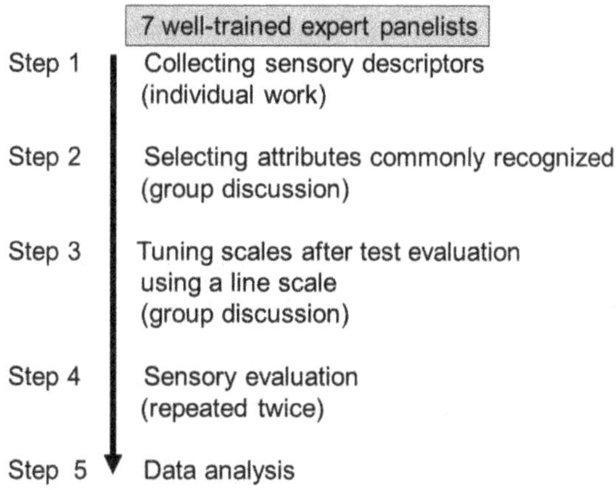

Figure 3. Procedure for descriptive sensory analysis.

Thus, we have combined the sample dilution in AEDA and sensory descriptive analysis instead of using GC or GC-MS as shown in Figure 4. According to the principle of SDSA, we can expect to obtain the quantified aroma strengths in individual attributes, not necessarily corresponding to a specific compound though, by fully taking account of whole interactive effects. Further, attributes essential for characterizing the original aroma note could be identified if the original aroma note would be more or less still preserved even in the highly diluted solution.

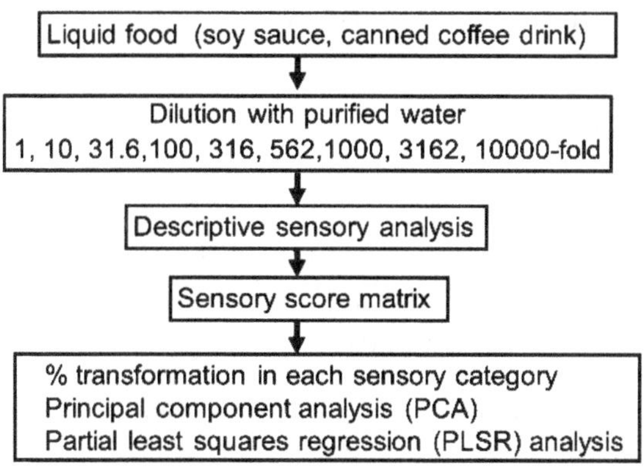

Figure 4. Procedure for SDSA.

Materials and Methods

Samples

The canned coffee beverage "Black" (UCC Ueshima Coffee, Kobe, Japan) and deep-colored type soy sauce (Kikkoman, Tokyo, Japan) were purchased from the local market. As shown in Figure 4, canned coffee and soy sauce were diluted 1, 10, 31.6, 100, 316, 562, 1,000, 3,162 and 10,000-fold, *i.e.*, logarithms: 0, 1, 1.5, 2, 2.5, 2.75, 3, 3.5 and 4, with purified water (Japanese pharmacopoeia standard: Fujimi Seiyakusyo, Osaka, Japan).

Sensory Evaluation

Sensory data have been collected through the steps shown in Figure 3. A trained sensory panel composed of 7 women aged 30's to early 60's, experienced in conducting the descriptive sensory analysis for more than 10 years, was applied. In the whole steps of SDSA, 40 ml of each sample at room temperature, *ca.* 23°C, placed in a 100 ml white styrene cup and covered with a transparent plastic lid was served to the sensory panel in the descending order of dilution values. In the preliminary sessions, when a sample was served, panelists opened the lid and sniffed the headspace aroma to generate aroma descriptors and then one scoop of sample solution was put into the mouth using a 7 ml volume plastic spoon and a 0.5 ml coffee spoon for diluted and undiluted samples, respectively, to generate taste and flavor descriptors. Strengths of attributes identified as common in all dilutions in preliminary sessions were quantified using a line scale arranging three anchors, the center for moderate and 10% distances from each end for "very weak" and "very strong", respectively. After evaluating a sample solution, two minutes interval was placed before starting to evaluate the next sample. During the interval, panelists rinsed their mouth with mineral water. This sensory evaluation was repeated twice and replicates were averaged (*10*).

Data Analysis

Obtained sensory scores for each attribute in every dilution were transformed into percentage proportions based on their total in sensory categories, *i.e.*, aroma, taste and flavor, to qualitatively compare their profiles. Multidimensional sensory data were analyzed by principal component analysis (PCA) for graphically displaying mutual relationships of samples and attributes on a two dimensional plane. The partial least squares regression (PLSR) analysis was applied to analyze contributions of attributes to generate the original aroma note using Unscrambler ver. 9.7 (CAMO Software AS, Oslo, Norway).

Results

Coffee

Although no information on the sample had been supplied to the sensory panel in advance, they unambiguously identified the aroma note of coffee even in the highest or 10,000-fold dilution. Fifteen attributes, nine for aroma including coffee aroma itself, three for taste and three for flavor, were commonly identified in all dilutions in the preliminary sessions (Table 1).

Table 1. Sensory Attributes for Canned Coffee

category	attribute	description
aroma (11)	alcoholic	alcohol, fermented
	fruity	canned peach, plum
	sweet aroma	cotton cake, caramel
	roasted	roasted cereals
	burnt	burnt rice or rice cake
	steamed	steamed cereals
	woody	cedar or cypress timber
	yoghurt	lactic fermentation
	dusty	dust, old warehouse
	miso aroma	fermented soy bean
	soy sauce aroma	*soy sauce* aroma
flavor (5)	roasted flavor	roasted cereals
	sweet flavor	cotton cake, caramel
	steamed soy bean	steamed soy bean
	miso flavor	fermented soy bean
	soy sauce flavor	*soy sauc* e flavor
taste (5)	sweet	
	salty	
	sour	
	umami	
	bitter	

Sensory scores in all attributes decreased in accordance with dilution values as shown in Figure 5. Next, sensory profiles of dilutions were illustrated based on score percentages calculated for individual sensory categories, such as aroma, taste and flavor, based on their total of scores. Thus, sensory profiles of percentage proportions for each dilution clearly depicted significant differences in their aroma characters as shown in Figure 6. In accordance with the increase in dilution values, percentages of burnt and coffee aroma notes decreased drastically but those of sweet, green and dusty notes increased inversely. Sensory attributes showing larger percentages in highly diluted samples may be derived from compounds with large odor unit values but may not necessarily contribute to the aroma note of coffee (*11*).

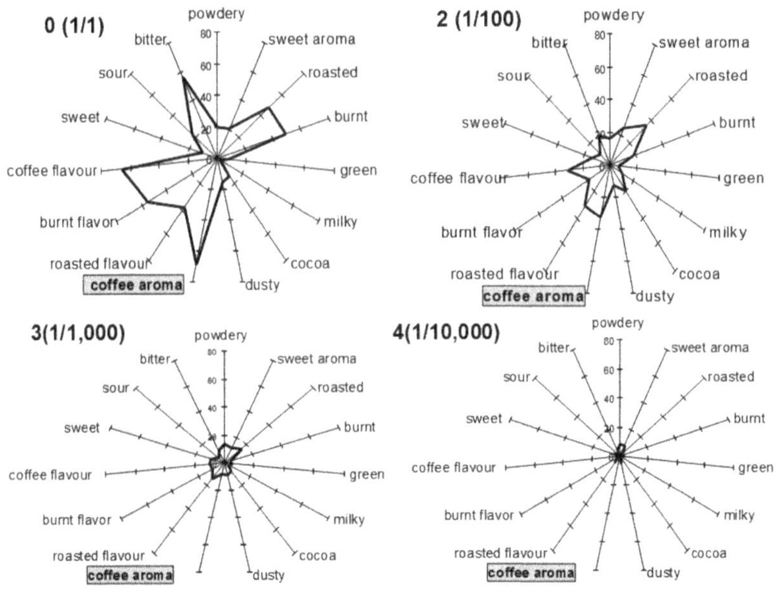

Figure 5. Sensory profiles of undiluted coffee (0 or 1/1) and its serial dilutions.

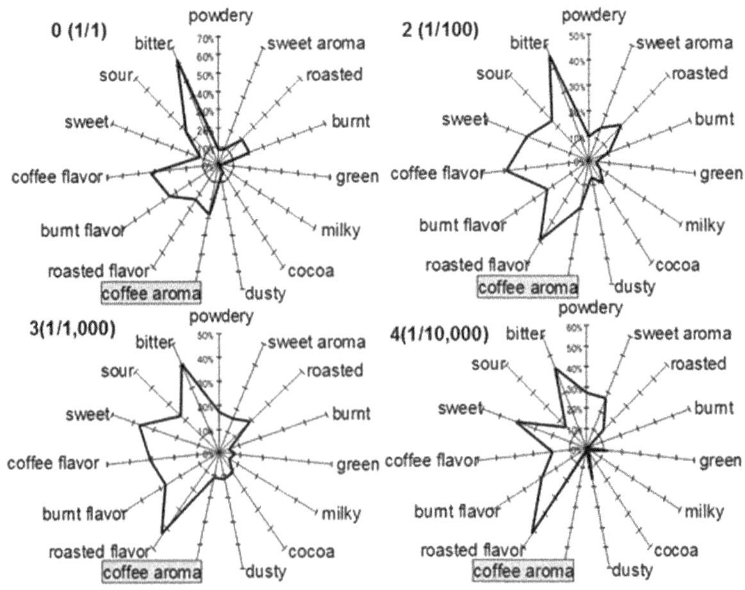

Figure 6. Profiles of sensory attribute proportions (%) in undiluted coffee and its serial dilutions.

The PCA biplot shown in Figure 7 suggests that sensory characters in dilutions have changed continuously in accordance with dilution values. Against our expectation, however, the closest attribute to undiluted coffee was burnt flavor but the coffee aroma note was located closely to the 10-fold dilution instead. On the other hand, sweet and green notes were located nearby highly diluted coffee.

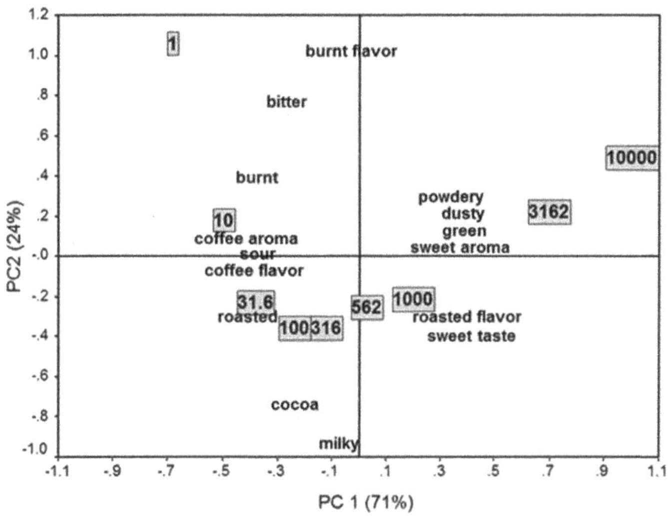

Figure 7. PCA biplot for coffee. Numbers in squares indicate dilution values.

The PLSR model calculated by using the coffee aroma as criterion and eight other aroma attributes as predictors was highly predictive (R^2=0.991) as shown in Figure 8. Both PLS loadings and regression coefficients for each predictor attribute clearly suggests high contribution of burnt and roasted notes to the coffee aroma. However, attributes indicating their significant contribution to the coffee aroma do not necessarily occupy larger proportions in sensory profiles, as indicated in Figure 6. A strong contribution of burnt and roasted notes to the aroma note of coffee suggested by PLSR strongly supports the importance of the roasting process in coffee production.

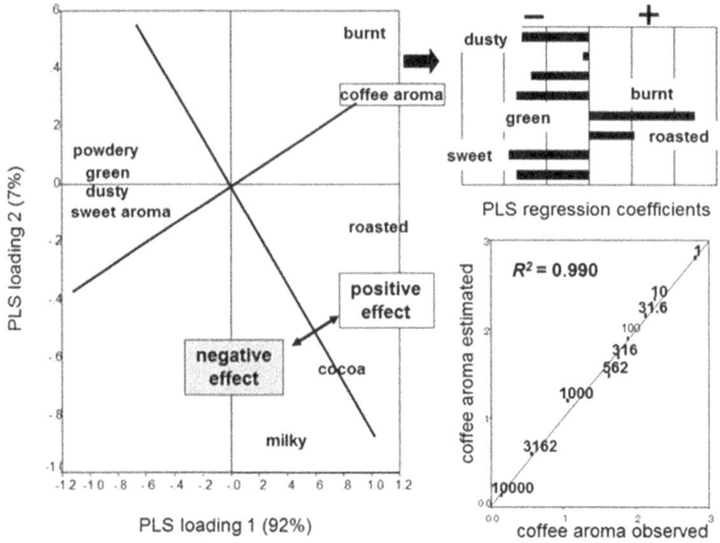

Figure 8. PLSR loading plots and model fit for coffee.

Soy Sauce

In the preliminary sessions, the sensory panel detected the unique aroma note of soy sauce even in the highest dilution. Twenty-one sensory attributes, 11 for aroma including soy sauce aroma, 5 for taste and 5 for flavor shown in Table 2 were identified in these sessions.

Sensory scores in all attributes decreased in accordance with dilution values as shown in Figure 9. According to dilution, proportions of alcoholic and burnt notes decreased rapidly but those of fruity and sweet notes increased inversely (Figure 10). On the other hand, some attributes, *i.e.*, yoghurt and *miso* or soybean paste, initially increased but started decreasing afterwards. In fermented soy products, *i.e.*, soy sauce and *miso*, a very strongly sweet smelling furanone derivative produced by yeast fermentation was identified and the authors claimed that it should be the character impact compound of soy sauce aroma (*12*) but increase in the percentage of sweet aroma in diluted soy sauce makes the soy sauce note weaker. Surely, we never associate a sweet aroma with soy sauce.

Table 2. Sensory Attributes for Soy Sauce

category	attribute	description
aroma (9)	powdery	physical sensation
	sweet aroma	cotton cake, caramel
	roasted	roasted cereals
	burnt	burnt rice or rice cake
	green	fresh leaf, green vegetables
	milky	chocolate
	cocoa	cocoa powder (van Houten)
	dusty	dust, old warehouse
	coffee aroma	coffee aroma
flavor (3)	roasted flavor	roasted cereals
	burnt flavor	burnt rice or rice cake
	coffee flavor	coffee flavor
taste (3)	sweet	
	sour	
	bitter	

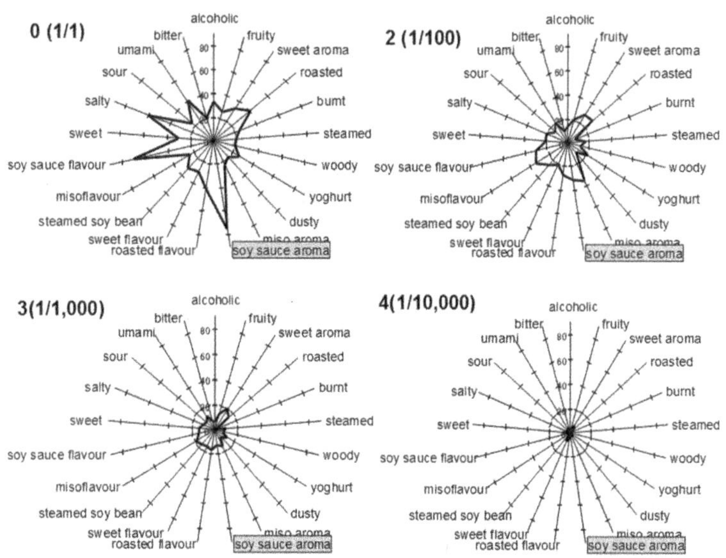

Figure 9. Sensory profiles of undiluted soy sauce and its serial dilutions.

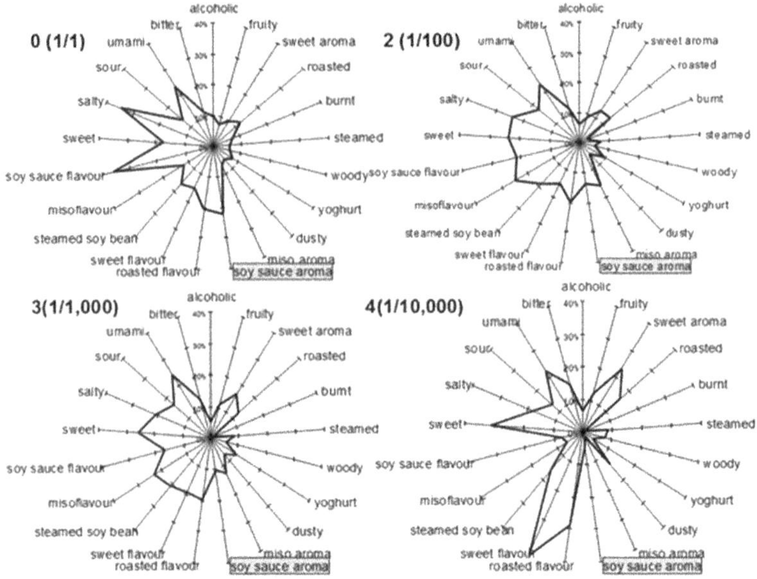

Figure 10. Profiles of sensory attribute proportions (%) in undiluted soy sauce and its serial dilutions.

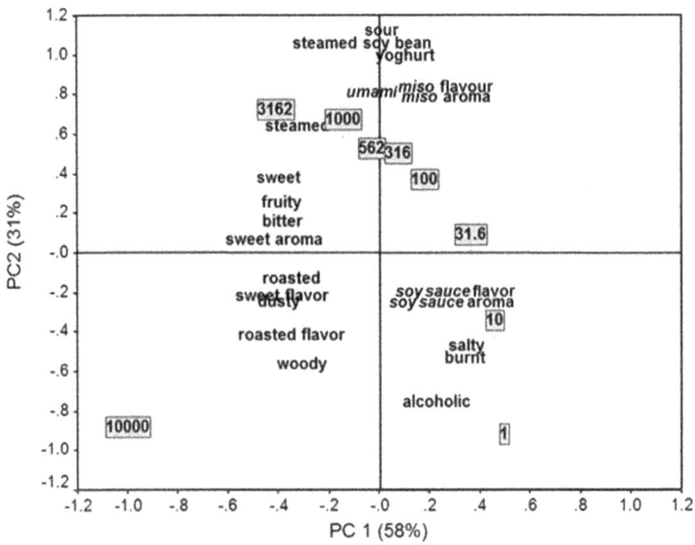

Figure 11. PCA biplot of sensory data of soy sauce dilutions.

The PCA biplot shown in Figure 11 suggests that whole sensory characters in dilution have changed continuously in accordance with dilution values. Similar to the biplot for coffee, the closest attribute to the original soy sauce was located close to the10-fold dilution. On the other hand, the alcoholic aroma note was located in the vicinity of undiluted soy sauce but the soy sauce aroma note itself was found close to the 10-fold dilution.

The PLSR model calculated using soy sauce aroma as criterion and 10 aroma attributes as predictors was highly predictive (R^2=0.988). The PLS loading plot shown in Figure 12 clearly suggests a high contribution of burnt and alcoholic notes to the aroma note of soy sauce. However, attributes with significant contributions to the soy sauce aroma do not necessarily occupy larger proportions in the sensory profiles shown in Figures 10. High contributions of alcoholic note derived from yeast fermentation in the *moromi* or mash period and a burnt note comes from the pasteurization process of raw soy sauce. High contributions of these two attributes strongly suggest the significance of these two processes for the brewing of soy sauce, as it has been widely recognized by soy sauce manufacturers in Japan.

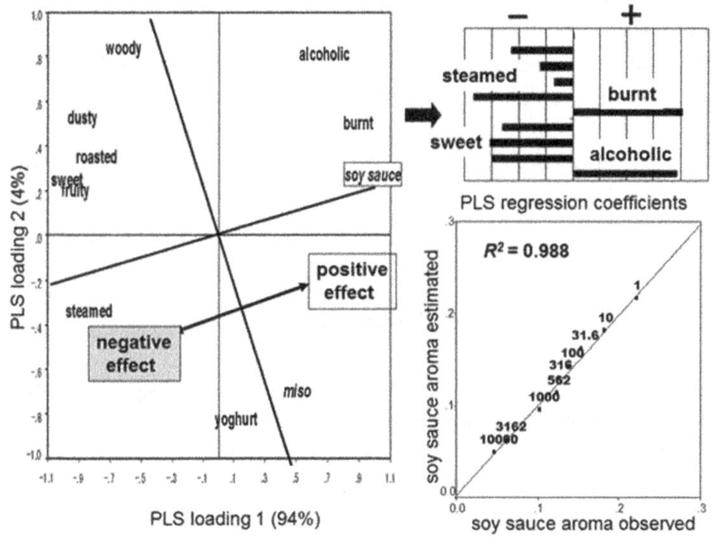

Figure 12. PLS loadings, regression coefficients and model fit for soy sauce.

Discussion

Steven's law (*13*) describes the relationship between a single chemical or physical stimulus and its perceived intensity and is given as $R = kC^n$, where R, C, n and k are the perceived sensation intensity, the stimulus concentration, the growth rate of the perceived intensity and a constant of proportionality, respectively. As indicated, the basic idea of the odor unit does not account for differences in n for individual components. However, results obtained from this study are nevertheless consistent with Steven's law, whereby the perceived intensity in the aroma of the volatile compounds decreased with increasing dilution. This decrease varied according to *n*, which is intrinsic to the individual compounds, because profiles composed of multiple sensory attributes were noticeably deformed. On the other hand, however, if the intensities in all components had decreased with the same curve or *n*, the profile would have been simply contracted.

Current analytical methods are mostly reductionistic in nature, where efforts have mainly been focused on measuring the significance of individual constituents by ignoring interactive effects. Flavor research is no exception because of the extreme complexity in nature. However, skillful techniques to determine complex sensory interactions such as additive, synergistic and suppressive effects by utilizing the information obtained from CHARM or AEDA have been reported (*6*).

Every element in the real world exists more or less under mutual interaction. According to the long history of aroma research, we realized that no single method makes it possible to illustrate whole picture of the aroma constituents in a food. Every method reveals some facet of the aroma profile but the profile is strongly determined by the biases introduced by the methodology itself (*6*). There is no universal approach to the instrumental analysis of aroma. No current instrument except for the human nose can analyze food aroma undisturbed by fully taking account of whole interactions. Although, sensory data cannot directly be related to aroma compounds in food, the results described here may be utilized as an aid to conduct aroma research effectively because SDSA can supply researchers with quantitative and qualitative information that makes it possible to conduct more focused research.

References

1. Patton, S.; Josephson, D. A method for determining significant of volatile flavor compounds in foods. *Food Res.* **1957**, *22*, 316–318.
2. Guadagni, D. G.; Buttery, R. G.; Harris, J. Odour intensities of hop oil components. *J. Sci. Food Agric. Technol.* **1966**, *17*, 142–144.
3. Acree, T. E.; Barbard, J.; Cummingham, D. G. In *Analysis of Volatiles, Methods and Applications*; Schreier P., Ed.; Walter de Gruyter: New York, 1984; pp 251−168.
4. Ullrich, F.; Grosch, W. Z. Identification of the most intense odor compounds formed during the autooxidation of linoleic acid. *Z. Lebensm.-Unters. Forsch.* **1987**, *184*, 277–282.
5. Stone, H.; Sidel, J.; Oliver, S.; Woolsey, A.; Singleton, R. C. Sensory evaluation by quantitative descriptive analysis. *Food Technol.* **1974**, *28* (11), 24–34.
6. Reineccius, G. In *Food Flavour Technology*; Taylor, A. J., Ed.; CRC Press: Boca Raton, FL, 2002; pp 210−251.

7. Aishima, T. Correlating sensory attributes to gas chromatography-mass spectrometry profiles and e-nose responses using PLS regression analysis. *J. Chromatogr. A* **2004**, *1054*, 39–46.
8. Sensory Evaluation Division of the Institute of Food Technologists Sensory evaluation guide for testing food and beverage products. *Food Technol.* **1981**, *35* (11), 50–59.
9. Meilgaaed, M; Civille, G. V.; Carr, T. *Sensory Evaluation Techniques*, 3rd ed.; CRC Press: Boca Raton, FL, 1999; pp 166−168.
10. Poste, L. M.; Mackie, D. A.; Butler, G.; Larmond, E. *Laboratory Methods for Sensory Analysis of Food*; Research Branch Agricultural Canada Publication 1864/E: Ottawa, ON, 1991.
11. Semmelroch, P.; Grosch, W. Studies on Character Impact Odorants of Coffee Brews. *J. Agric. Food Chem.* **1996**, *44*, 537–543.
12. Nunomura, N.; Sasaki, M.; Yokotsuka, T. Shoyu (soy sauce) flavor components. Acidic fractions and the characteristic flavor components. *Agric. Biol. Chem.* **1980**, *44*, 339–351.
13. Stevens, S. S. On the psychophysical law. *Psychol. Rev.* **1957**, *64*, 153–181.

Chapter 10

Non-Homeostatic Intake of Snack Foods: Molecular Triggers and Effects on Brain Activity Pattern

Tobias Hoch,*,[1] Andreas Hess,[2] and Monika Pischetsrieder[1]

[1]Department of Chemistry and Pharmacy, Food Chemistry Division, Friedrich-Alexander-Universität Erlangen-Nürnberg (FAU), Schuhstr. 19, 91052 Erlangen, Germany
[2]Institute of Experimental and Clinical Pharmacology and Toxicology, Friedrich-Alexander-Universität Erlangen-Nürnberg (FAU), Fahrstr. 17, 91054 Erlangen, Germany
***E-mail: tobias.hoch@fau.de.**

Craving for special types of food like snack food can tremendously influence our energy balance. The result: obesity due to a non-homeostatic, hedonic food intake, i.e. an intake of energy independent of hunger and satiety. The intake of potato chips – an often craved highly palatable snack food – has a great influence on whole brain activity pattern. Especially the reward system as well as circuits regulating food intake, sleep and locomotor activity are affected. Furthermore, we could show that the fat and carbohydrate content is a main contributor to the palatability of potato chips. These first steps of the identification of the molecular triggers and the corresponding effects on brain activity pattern of the non-homeostatic intake of highly palatable snack food are reviewed in this chapter.

Introduction

Hedonic food intake, eating for pleasure without a physiological need, can overrule the homeostatic energy balance and therefore, in the long run, lead to hyperphagia, i.e. an energy intake beyond satiety (*1*). Besides chocolate and sweets, savory snacks like potato chips are very attractive to many people and largely contribute to energy intake (*2*, *3*). Hedonic hyperphagia can be influenced by the emotional state, mental health conditions or sleep deprivation (*4*). Furthermore, molecular food composition and energy density are decisive factors for the induction of hedonic hyperphagia (*5*, *6*). Many physiological systems are involved in the regulation of food intake, like distinct brain systems, hormones, dopamine, melanocortins, opioids or endocannabinoids (*7–12*). In our work, we investigated the influence of the intake of potato chips on whole brain activity pattern of *ad libitum* fed rats. Furthermore, we analyzed the main contributors to the palatability of potato chips by using a two choice preference test with rats.

Influence of Snack Food on Feeding Related Behavior and Whole Brain Activity Pattern of Rats

We could show that the intake of the snack food potato chips has a decisive influence on feeding related behavior and whole brain activity pattern of *ad libitum* fed rats (*13*). To investigate the links between food intake and behavior or brain activity we developed a protocol using manganese-enhanced magnetic resonance imaging (MEMRI) (*14*, *15*). As displayed in Figure 1, we conducted our study on a 4-week based schedule with male Wistar rats (Charles River, Sulzfeld, Germany, n=16 per group) which had access to standard chow pellets and tap water *ad libitum* over the whole course of the study. The schedule started with a one week habituation phase (HP) in which the animals were familiarized with the study conditions. In the following one week training phase (TP), the rats got in contact with the respective test food potato chips or powdered standard chow presented additionally to the standard feed. The test food was provided throughout the whole TP 24 hours a day for 7 days. In the following intermediate phase (IP, one week), the test food was removed to simulate withdrawal. For the upcoming manganese phase (MnP), rats were implanted with osmotic pumps, filled with the contrast agent manganese chloride. Activation of cells/neurons is accompanied by an influx of Ca^{2+}, particularly at the synapse. Manganese is transported using Ca^{2+} transport systems like voltage gated Ca^{2+} channels. Therefore, also Mn^{2+} is transported into the neuropil of activated neurons. In contrast to Ca^{2+}, Mn^{2+} accumulates in the cells with only a slow release over weeks and acts as MRI contrast agent due to its paramagnetic character. Therefore, the integral activity of distinct brain areas can be measured by MEMRI (*15*). Application of manganese chloride has a toxicological drawback. Single injections of the required dose led to a suppressed food intake and activity in general. Eschenko et al. (2010) (*16*) suggested the use of implanted osmotic pumps, which release the contrast agent continuously at a rate of 1 µL/h over 7 days. This application method had no toxic or behavioral side effects and enabled an integral measurement of structure specific whole brain activity pattern during the intake of potato chips over 7 days.

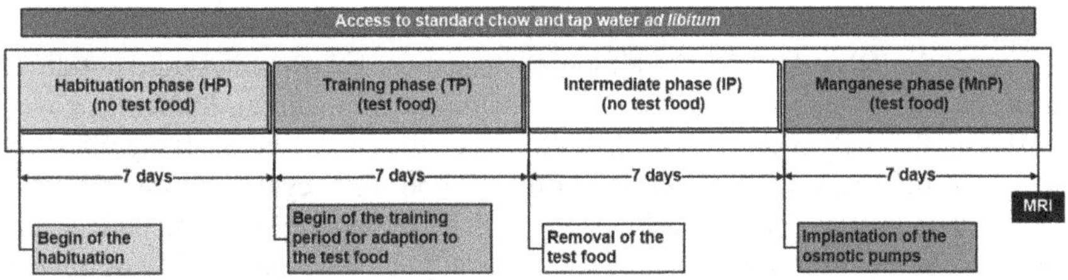

Figure 1. Study design for the investigation of the influence of snack food intake on whole brain activity pattern by manganese-enhanced magnetic resonance imaging (MEMRI). Crushed potato chips and powdered standard chow were used as test food. Reproduced with permission from reference (13).

During the study, we evaluated behavioral data of 16 rats in 4 cages per group. Test food intake was measured on a daily basis and locomotor activity was recorded continuously over the 7 days of TP and MnP in one-hour-bins.

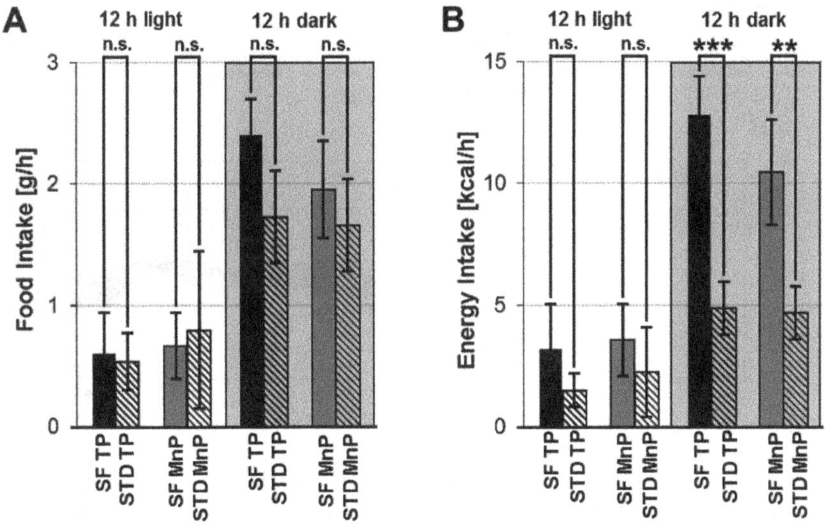

*Figure 2. Test food (A) and energy (B) intake of the animals of the snack food group (SF, snack food crushed potato chips) and the standard chow group (STD, powdered standard chow). Both parameters are shown separately during the 12 hours light and 12 hours dark cycle as well as during training phase (TP) and manganese phase (MnP). The mean ± standard deviation of 16 animals in each group as an average over 7 days of the respective period is shown. *** $p<0.001$, ** $p<0.01$, *$p<0.05$, n.s.: not significant. Reproduced with permission from reference (13).*

The average amount of ingested test food, which was available *ad libitum* over 7 days of the TP and the MnP additionally to standard chow pellets, did not differ between the two groups snack food and standard chow during the light as well as during the dark cycle (Figure 2A). However, the energy intake is significantly higher in the snack food group during the 12-hour dark cycle (Figure 2B). This elevated energy intake of snack food is independent of hunger and satiety and therefore solely of hedonic nature. Consequently, the high palatability of the test food leads to hedonic hyperphagia, an energy intake beyond physiological needs. Hedonic hyperphagia due to the snack food potato chips but not due to powdered standard chow could be shown during the training phase, where the animals got in contact with the test foods for the very first time. Additionally, this phenomenon could be observed during the MnP with implanted osmotic pumps in which the respective test food was already known from the TP (Figure 2B).

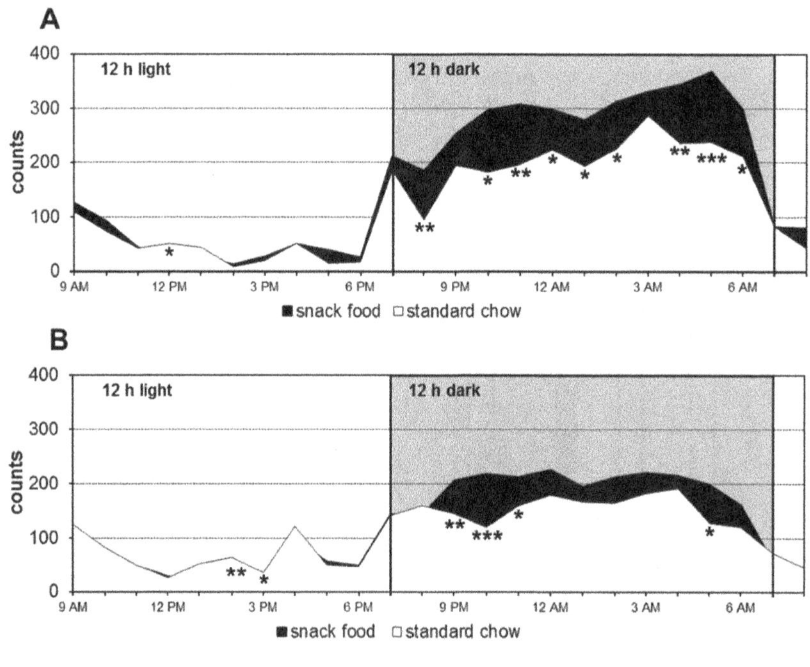

*Figure 3. Feeding related locomotor activity during training phase (A) and manganese phase (B) with access to snack food (potato chips) or powdered standard chow. Data are presented as the mean of 16 animals over 7 d per group. *** p<0.001, ** p<0.01, * p<0.05. Reproduced with permission from reference (13).*

As a second behavioral readout, we measured the feeding related locomotor activity of the animals simultaneously with the food and energy intake during the training phase (Figure 3A) and the manganese phase (Figure 3B). During their resting times, the animals normally stay in the back of the cage. To take up the respective test foods, the animals had to move to the front of the cage, which was recorded by cameras installed above each cage. Pictures were taken every 10 seconds, and the movement of the animals in(to) the "snack food or powdered standard chow area" of the cage was registered and counted. Counts were defined as "one animal shows locomotor activity near the food dispensers on one picture". It could be observed that locomotor activity was relatively low during the light cycle of the day in both groups and during both phases of the study. During the dark cycle, locomotor activity was much higher in general with significant differences between the food groups. In the TP, animals of the snack food group showed a significantly elevated feeding related locomotor activity at nine time points (one-hour bins) (Figure 3A). During the MnP, the animals of the snack food group also showed elevated feeding related locomotor activity, significant at four time points. Elevated feeding related locomotor activity can be interpreted as a high attraction of the provided test food, because the animals had to move from the back of the cage to the front in order to take up the respective test food. During both phases, the snack food potato chips caused a significantly higher feeding related locomotor activity than the powdered standard chow control. Therefore, it can be concluded that *ad libitum* access to the high caloric snack food potato chips leads to an elevated energy intake and a higher feeding related locomotor activity to access the test food. This underlines the particular attraction of the snack food potato chips.

Imaging of whole brain activity pattern of the animals of both food groups snack food (potato chips) and standard chow resulted in significantly differently activated brain regions. The spatial distribution of the voxel based morphometric (VBM) analysis is displayed in Figure 4 showing brain areas, which are significantly activated (colored in red) or deactivated (colored in blue) by snack food intake. For this purpose, statistics have been calculated on the basis of every voxel with a size of 109x109x440 µm. For detailed methods see Hoch et al. (2013) (*13*).

Further characterization of the significantly differently activated brain regions was performed by a structure based densitometric analysis by using an in-house 3D digital version of the rat brain atlas by Paxinos and Watson (*17*). Therefore, the grey values of each labeled brain structure (166 in total) were analyzed by overlaying the greyscale pictures with a digital rat brain atlas to measure the "brightness" of the distinct brain areas. The higher the activity in the respective brain area was, the more manganese accumulated and consequently the brighter is the picture in this area. Figure 4A shows an example of an overlay of the digital rat brain atlas with one of the smallest analyzed brain structures (VTA, ventral tegmental area, volume 0.7914 mm3, 152 voxel) highlighted in yellow. The average MEMRI image data of 16 animals of one food group is shown in the background of Figure 4A and underlines the very good resolution of 109x109x440 µm. The analysis led to significantly differently activated brain structures comparing the standard chow with the snack food group. These brain areas could be mainly assigned to the functional groups reward, food intake, (REM) sleep and locomotor activity (Table 1).

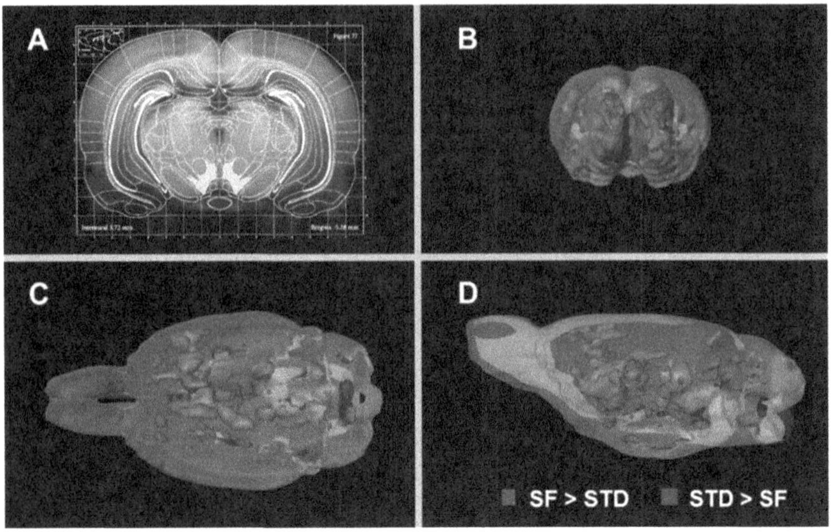

Figure 4. In (A) the overlay of a slice of the reconstructed dataset with the corresponding atlas slice (Bregma -5.28 mm) from the rat brain atlas by Paxinos and Watson (17) is shown with one of the smallest analyzed regions (VTA) marked in yellow. In (B), (C) and (D), significantly differently activated brain areas due to the intake of snack food (SF, potato chips) and standard chow (STD) are displayed in a rat brain surface. Areas colored in red are significantly (p<0.05) activated by the snack food whereas areas colored in blue are significantly (p<0.05) deactivated by the snack food. The results are displayed in axial (B), horizontal (C) and sagittal (D) view. Reproduced with permission from reference (13). (see color insert)

Table 1. Number of Significantly Differently Activated Brain Areas in the Respective Functional Groups

Functional Group	*# brain areas potato chips vs. standard*
Reward	27
Food intake	11
Sleep	11
Locomotor activity	6

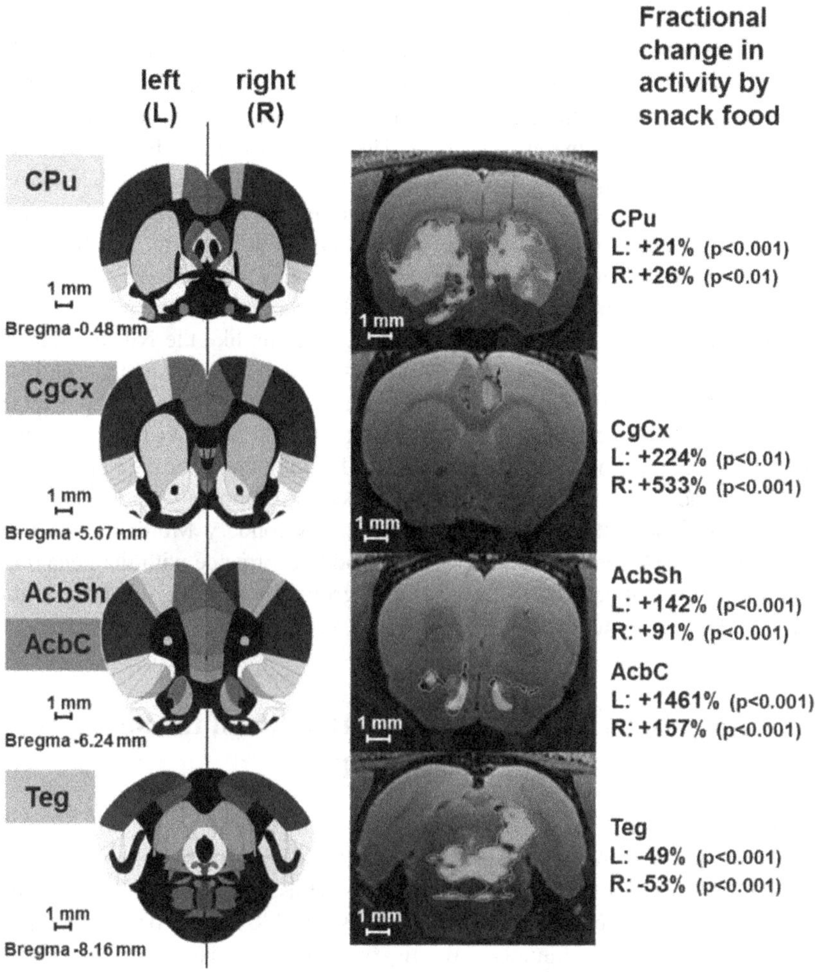

Figure 5. In the left column, the localization of five brain areas is shown on slices of the rat brain atlas by Paxinos and Watson (17). The middle column shows the significantly differently activated areas (VBM) in an overlay with the corresponding standard T2 weighted MRI anatomy and atlas labels. In the right column, the respective fractional change in activity by snack food intake is shown. CPu = Caudate Putamen, CgCx = Cingulate Cortex, AcbSh = Shell region of the Nucleus Accumbens, AcbC = Core region of the Nucleus Accumbens, Teg = Tegmental Nuclei. Reproduced with permission from reference (13). (see color insert)

Evaluation of the fractional change in activity due to the intake of snack food showed that some brain areas are much stronger activated. Examples are shown in Figure 5. The Caudate Putamen (CPu) is significantly activated by snack food to an extent of 21 % (left hemisphere, L) to 26 % (right hemisphere, R). This brain structure plays a decisive role in the reward circuit as well as in the regulation of locomotor activity. The Cingulate Cortex (CgCx) and especially the Nucleus Accumbens (Acb) with the shell (AcbSh) and the core (AcbC) substructures are key structures of the brain reward system. These areas were significantly activated to an extensive degree. CgCx was activated between 200 and over 500 %. Acb was even activated up to nearly 1500 %. This observation underlines the great impact of snack food intake on the brain reward system. Furthermore, the Tegmental Nuclei (Teg) were significantly deactivated by snack food. This structure serves as an example for brain areas responsible for the regulation of sleep.

Furthermore, brain areas mediating satiety signals like the Nucleus Tractus Solitarius, the Arcuate Nucleus, the Dorsomedial Hypothalamus and the Paraventrical Thalamic Nucleus Anterior were down regulated by the intake of potato chips. In contrast, further structures of the circuits regulating food intake like the Infralimbic Cortex, the Lateral Hypothalamus or the Septum were significantly upregulated by snack food intake. Additionally, brain structures of the locomotor system like the Primary and Secondary Motor Cortex were activated by snack food intake, which is in good accordance with the behavioral data where the animals of the snack food group show a significantly higher feeding related locomotor activity. For details, see Hoch et al. (2013) (*13*).

Molecular Determinants of the Palatability of the Snack Food Potato Chips

The snack food potato chips has a decisive influence on several physiological circuits in the brain. As described above, besides others, the brain reward system is highly affected by snack food intake (*13*). Thus, the question arises, which properties or components of snack food are responsible for this phenomenon. To answer this question, we aimed at investigating the molecular determinants of the palatability of the snack food potato chips in a second study (*18*). Therefore, we used the test foods shown in Figure 6 to address the single main components of potato chips. All mixtures are based on a 1:1 mixture of potato chips and powdered standard chow to minimize the influence of the texture on the palatability testing.

To simulate a snacking situation, the respective test foods were presented in short time windows (10 minutes) 3 times per day (Figure 7A). Furthermore, the test foods were presented additionally to *ad libitum* available standard chow pellets and tap water to simulate snacking independent of homeostatic needs. Two test foods were presented in each session at the same time to evaluate the preference for one of the test foods (Figure 7B). As readout, we measured the respective test food intake during the 10-minute tests and the feeding related locomotor activity with one count defined as “one rat takes food from one food dispenser”. Cameras were placed above the cages taking pictures every 10 seconds as shown in Figure 7C. Both readout parameters showed a very high correlation between all tested conditions ($r = 0.9204$, $R^2 = 0.8471$, $p<0.001$). Therefore, only the results regarding “food intake” are shown below.

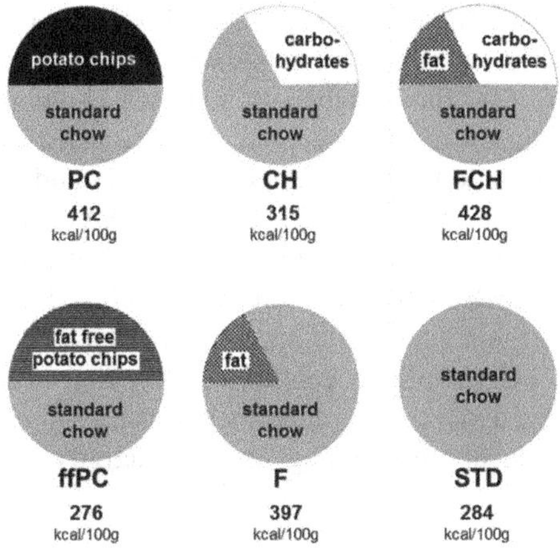

Figure 6. Composition of the used test foods and their respective energy contents. Reproduced with permission from reference (18).

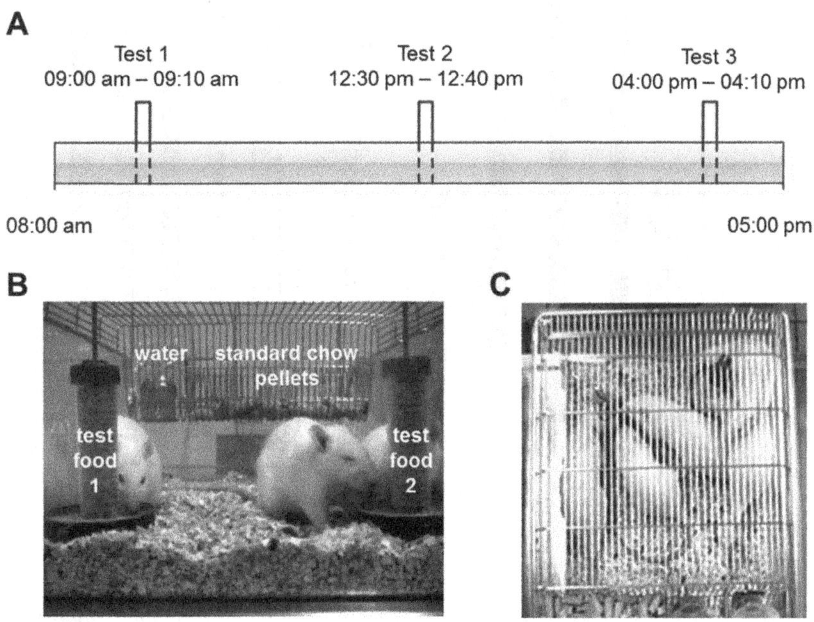

Figure 7. The schedule of the preference tests is shown in (A). In (B), the setup of the two choice preference test is displayed with both test foods in front and standard chow pellets as well as tap water in the back of the cage. In (C), a picture for the evaluation of the counts is shown, taken from a camera placed above each cage. Reproduced with permission from reference (18).

At first, we tested standard chow (STD) in both food containers to evaluate possible place preferences of the animals during two-choice preference testing. As shown in Figure 8A, food from both standard chow food dispensers was taken in equal amounts and no place preference occurred. As next results, the potato chips test food (PC) was preferred over STD and over its single main components carbohydrates (CH) and fat (F) (Figure 8A). Only the mixture of fat and carbohydrates similar to potato chips (FCH) led to a comparable test food intake as the whole snack food. Therefore, we concluded that the mixture of fat and carbohydrates is the main contributor to the palatability of potato chips. The outstanding role of FCH could be confirmed by testing the components CH, F and FCH vs. STD and against each other. FCH was highly significantly preferred over all other components underlining its great activity to induce food intake (Figure 8B).

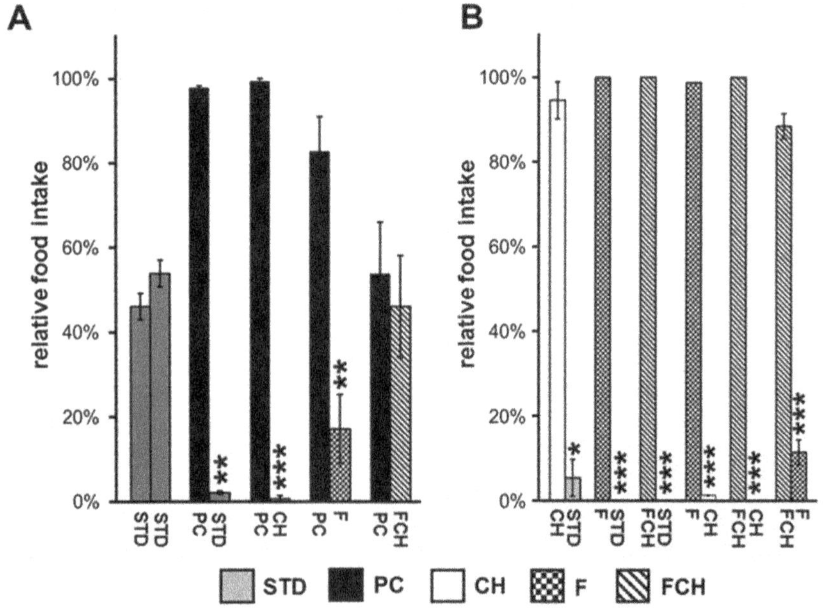

*Figure 8. Relative food intake during two-choice preference tests. (A) Standard chow (STD) vs. STD as well as potato chips test food (PC) vs. the components carbohydrates (CH), fat (F) and the mixture of fat and carbohydrates as present in potato chips (FCH). (B) Tests of the single main components of PC vs. STD and against each other. Mean ± standard deviation of the relative food intake of the independent animal groups (cages) is shown. ***p<0.001, ***p<0.01, *p<0.05. Reproduced with permission from reference (18).*

Furthermore, we tested the influence of energy density and other features on the palatability of the snack food potato chips by using a test food containing fat free potato chips (ffPC, Figure 6) as an additional test food. As shown in Figure 9, ffPC was preferred over STD and CH although ffPC has the lowest energy density of all test foods. Consequently, the texture, mouth feeling or other components of ffPC might have a measurable influence on the palatability (Figure 9).

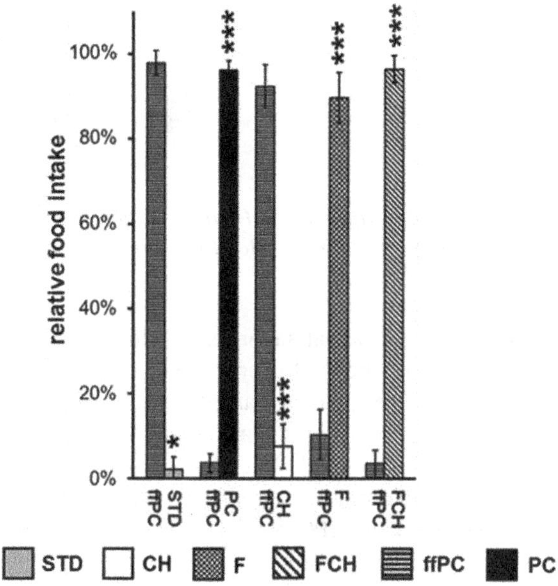

*Figure 9. Relative food intake during two-choice preference tests, investigating the influence of energy density and texture on the palatability of potato chips. The test foods standard chow (STD), carbohydrate (CH), fat (F), mixture of fat and carbohydrate (FCH), fat free potato chips (ffPC) and potato chips (PC) were used. Mean ± standard deviation of the relative food intake of the independent animal groups (cages) is shown. ***p<0.001, *p<0.05. Reproduced with permission from reference (18).*

Thus, we could show with our two-choice preference tests that potato chips are a highly palatable test food under snacking conditions. Potato chip test food (PC) was preferred over the single main components of potato chips fat (F) and carbohydrates (CH). Only the mixture of fat and carbohydrates (FCH) reached the same palatability as PC (Figure 10A). This leads to the conclusion that the specific mixture of both main components fat and carbohydrates is the most important determinant of the intake of potato chips (Figure 10B). The palatability of the mixture of fat and carbohydrates may be caused by its energy density. However, the tests with fat free potato chips showed that in addition to the energy contents, other features such as texture, flavor or further minor components may trigger non-homeostatic intake of snack food such as potato chips.

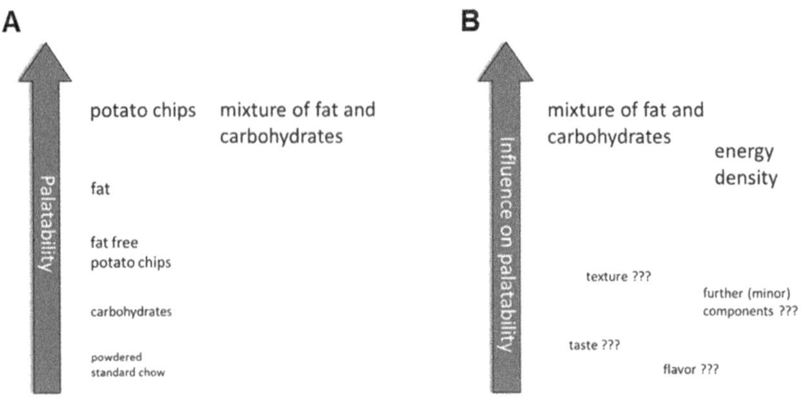

Figure 10. Ranking of the investigated test foods (A) and summary of the factors influencing palatability during the reported two-choice preference tests (B).

In summary, we could show that savory snack food, such as potato chips, induce non-homeostatic intake beyond energy needs as well as feeding related locomotor activity. The intake is accompanied by a strong activation of brain reward circuits. Additionally, brain areas related to food intake, locomotor activity and sleep were affected by snack food. In elaborated preference tests, we could show that potato chips intake is mainly caused by the fat and carbohydrate content of the snack food. However, from ffPC, there is evidence that also other features, such as texture, flavor or minor components have an influence on the observed behavior and modulation of brain activity, which is to be investigated in further studies.

References

1. Berthoud, H.-R. Homeostatic and non-homeostatic pathways involved in the control of food intake and energy balance. *Obesity* **2006**, *14* (S8), 197S–200S.
2. Drewnowski, A.; Kurth, C.; Holden-Wiltse, J.; Saari, J. Food preferences in human obesity: carbohydrates versus fats. *Appetite* **1992**, *18*, 207–221.
3. Slining, M. M.; Mathias, K. C.; Popkin, B. M. Trends in food and beverage sources among US children and adolescents: 1989-2010. *J. Acad. Nutr Diet* **2013**, *113*, 1683–1694.
4. Sharma, A. M.; Padwal, R. Obesity is a sign - over-eating is a symptom: an aetiological framework for the assessment and management of obesity. *Obes Rev.* **2010**, *11*, 362–370.
5. McCrory, M. A.; Fuss, P. J.; Saltzman, E.; Roberts, S. B. Dietary determinants of energy intake and weight regulation in healthy adults. *J Nutr.* **2000**, *130*, 276S–279S.
6. Kenny, P. J. Common cellular and molecular mechanisms in obesity and drug addiction. *Nat. Rev. Neurosci.* **2011**, *12*, 638–651.

7. Lenard, N. R.; Berthoud, H.-R. Central and peripheral regulation of food intake and physical activity: pathways and genes. *Obesity* **2008**, *16* (Suppl 3), S11–22.
8. Berthoud, H.-R. Metabolic and hedonic drives in the neural control of appetite: who is the boss? *Curr. Opin. Neurobiol* **.2011**, *21*, 888–896.
9. Pandit, R.; Jong, J. W. de; Vanderschuren, L. J. M. J.; Adan, R. A. H. Neurobiology of overeating and obesity: the role of melanocortins and beyond. *Eur. J. Pharmacol.* **2011**, *660*, 28–42.
10. Alsiö, J.; Olszewski, P. K.; Levine, A. S.; Schiöth, H. B. Feed-forward mechanisms: addiction-like behavioral and molecular adaptations in overeating. *Front. Neuroendocrinol.* **2012**, *33*, 127–139.
11. Jarosz, P. A.; Sekhon, P.; Coscina, D. V. Effect of opioid antagonism on conditioned place preferences to snack foods. *Pharmacol. Biochem. Behav.* **2006**, *83*, 257–264.
12. DiPatrizio, N. V.; Astarita, G.; Schwartz, G.; Li, X.; Piomelli, D. Endocannabinoid signal in the gut controls dietary fat intake. *Proc. Natl. Acad. Sci. U.S.A.* **2011**, *108*, 12904–12908.
13. Hoch, T.; Kreitz, S.; Gaffling, S.; Pischetsrieder, M.; Hess, A. Manganese-enhanced magnetic resonance imaging for mapping of whole brain activity patterns associated with the intake of snack food in ad libitum fed rats. *PLoS one* **2013**, *8*, e55354.
14. Koretsky, A. P.; Silva, A. C. Manganese-enhanced magnetic resonance imaging (MEMRI). *NMR Biomed.* **2004**, *17*, 527–531.
15. Lin, Y. J.; Koretsky, A. P. Manganese ion enhances T1-weighted MRI during brain activation: An approach to direct imaging of brain function. *Magn. Reson. Med.* **1997**, *38*, 378–388.
16. Eschenko, O.; Canals, S.; Simanova, I.; Beyerlein, M.; Murayama, Y.; Logothetis, N. K. Mapping of functional brain activity in freely behaving rats during voluntary running using manganese-enhanced MRI: Implication for longitudinal studies. *Neuroimage* **2010**, *49*, 2544–2555.
17. Paxinos, G.; Watson, C. *The Rat Brain in Stereotaxic Coordinates*; Elsevier: Amsterdam 2007.
18. Hoch, T.; Pischetsrieder, M.; Hess, A. Snack food intake in ad libitum fed rats is triggered by the combination of fat and carbohydrates. *Front. Psychol.* **2014**, *5*, 250.

Chapter 11

The Role of Temporal Dominance of Sensations (TDS) in the Generation and Integration of Food Sensations and Cognition

S. Fiszman*

Physical and Sensory Properties of Food and Consumer Science Group IATA-CSIC, Agustin Escardino 7, 46980 Paterna (Valencia), Spain
***E-mail: sfiszman@iata.csic.es.**

In recent years, the oral processing of food has regained the attention of food technologists and researchers. The complex process of mastication, in which food is minced into smaller particles and mixed with saliva until a bolus that can be swallowed safely and comfortably is formed, can be situated in a space that involves structure, lubrication and time. As temporality is a key factor in all the changes described, a sensory method called Temporal Dominance of Sensations (TDS) appeared a few years ago. TDS has made it possible to evaluate a sequence of sensations generated by the oral trajectory during mastication and has been applied to a number of food items. Several examples of the possibilities that TDS offers for relating the perceived sequence of sensations with food composition and structures will be examined. As a result, it will be seen how TDS can contribute to understanding the drivers of liking, which decide the quality of the food items and consumer acceptance.

Temporal Dominance of Sensations

The Temporal Dominance of Sensations (TDS) method was created at the Centre Européen des Sciences du Goût in 1999 and was first presented to the scientific community (*1*) at the Pangborn Symposium. Since then, an increasing number of papers have used this method to analyze a large number of food items and eating and drinking situations, and relate the results to numerous factors in order to gain a deeper understanding of the cognitive processes surrounding the act of eating.

The method consists of identifying which of a list of attributes is the sensation perceived as dominant at each moment of consumption of a food product. The raw data file records the scores of each attribute over the time allowing to get the sequence of dominant sensations elicited which is recorded and then analyzed (*2*).

Dominant has been defined as "the most striking perception at a given time" (*3*), or "the most intense sensation" (*4*). One selected attribute is considered dominant until the subject changes his/her selection or terminates the evaluation. On some occasions, as panelists indicated that they perceived no dominant sensation, a "new" descriptor called "Nothing dominates" has been added to the list, thus blocking the previous dominant sensation recorded (*5*) and avoiding computing dominance of the last attribute selected when really the sensation had ceased and no new dominance had appeared. When the method was first proposed, the assessors were asked to score the intensity of the selected attribute. However, intensity ranking was found not to be really necessary (*6*), as the results from numerous studies show that dominance rates alone can provide important temporal information.

Regarding special training of the assessors, a short introduction to the concept of temporality in the perception of attributes is considered useful (*7*), and a first simulated session to check whether the method has been understood is also recommended.

TDS curves (dominance rate versus time of consumption) are not related with intensity but with the number of times (number of assessors) that an attribute has been cited as dominant at a given time. It is naturally conceived that the greater the intensity of an attribute, the more it will be perceived and cited by the panelists, so the higher its dominance rate will be (*8*). If a certain product is characterized by several high-intensity attributes, however, the panelists cannot cite all of them at the same time and are obliged to make a choice, which will have an impact on the final dominance rate. These authors also commented that the converse holds true, as TDS does not allow some discriminating (but really not dominant) attributes of a sensory profile to appear significant, since it does not permit the evolution of weak intensity sensations to be known.

Since the duration of mastication up to swallowing differs from one subject to another, the sensory perception time scales differ as well (*9*). In order to take this fact into account in computing the TDS curves, the data from each subject are usually normalized according to individual mastication durations, so the X-axis shows values from X=0 (clock starting, first score) to X=100 (swallowing or clock stop). Swallowing does not always end the tests, since it could be interesting to know something about the aftertaste after swallowing, so the end-point of the test could be to click on the Stop button when no more sensations are perceived.

In order to obtain more meaning, clarity and understanding from the TDS curve, two lines are usually drawn. The first, called the chance level, is the dominance rate that an attribute can obtain by chance. Its value, P0, is equal to 1/p, p being the number of attributes. The second, called the significance level, is the minimum value this proportion should reach if it is to be considered significantly higher than P0. It is calculated using the confidence interval of a binomial proportion based on a normal approximation (*3*). Some studies have considered TDS curves consistent at panel level when the individual attribute curves exceed the chance level, while others have considered them consistent when they are above the significance level (*10*).

TDS performance has been compared or used in combination with other sensory descriptive methods using trained panelists. Some examples are: with key attribute sensory profiling for food with contrasting textural layers such as fried and microwaved fish sticks (*7*), and with quantitative descriptive analysis (QDA) for yoghurts formulated with different fat contents or gelatin and starch concentrations (*11*), or to compare the potential of yeast strains as starter cultures for dry fermentation of coffee beans (*12*). A sequential approach, i.e. QDA then TDS using the same set of trained assessors to measure the sensory properties of commercial blackcurrant squashes has also been employed (*13*).

TDS has also been used in combination with consumer-based techniques. A free sorting task, sensation scoring and a focus group have been used in combination to investigate the sensation of freshness in yoghurt and yoghurt-like products (*14*): the aim of this study was to determine the key sensory attributes underlying freshness, and their temporal order was determined by TDS. Other combinations, such as free sorting and sensory profiling, have been used to describe the taste of bottled drinking waters containing different amounts of minerals (*15*).

Due to the time-dependent character of the evaluation, several studies have compared TDS with other temporal techniques such as Time Intensity (TI). Both of these have been used, together with QDA, to assess perceptions of bitterness in different white wines (*16*). A further study analyzed a number of flavor attributes of the same wines, which differed in the skin contact time of the crushed grapes during fermentation (*17*). In this latter study, TDS analysis showed that the perception of astringency dominated over the bitter sensation and also revealed further subtle differences in sourness perceptions.

In an attempt to obtain a real-time flavor profile of beers (*18*), three different techniques were used to study the evolution of four characteristic flavor attributes of beer. These techniques were TI, TDS and drinking profile (DP). In the DP technique, in addition to the flavor attributes, scores were given for "linger", providing an idea of the "length, finish, and aftertaste of the main characteristic flavors of beer". The attributes were scored on a 10-point scale (0=none to 9=extremely strong) at five successive sips, waiting 45 s between the sips and scoring moments. The TI technique provided useful information for studying the kinetics of particular attributes, and showed the main differences in the beer profiles, but proved very time-consuming. TDS and DP provided useful information when used as complementary techniques, but DP was also very time-consuming.

As TDS appeared relatively recently, a number of studies have searched for the best experimental scenario this method requires.

Regarding the number and order of the attribute list, a study (*6*) found that panelists tended to use a relatively constant number of attributes per test, whatever the number of attributes in the list. The same study showed that panelists were also able to use different types of attributes in the same list with no impact on the number of selections of each attribute. A list with a maximum of 10 attributes was recommended. Finally, a tendency to select the attributes at the top of the list earlier in the sequence than the attributes at the bottom of the list was detected, leading to a recommendation to balance the attribute order across the panelists (i.e. each panelist has a different attribute order, but the same order for all the product evaluations).

A complete protocol has been presented (*19*) to assess panel performance in TDS experiments in terms of discrimination ability and agreement, both at panel and subject levels, together with the corresponding data analyses and keys for interpretation. It is based on the analysis of variance (ANOVA) framework commonly used for assessing panel performance in the usual context of descriptive analysis.

A multiple-sip TDS approach was recently proposed to study the influence of sucrose replacement by low-calorie sweeteners in orange juice (*20*). Using TDS over three consecutive sips, each lasting 20 s, made it possible to identify differences in the dynamics of the sensory characteristics of the juices that had not been identified through single-sip measurements.

Linking to Liking

It has been suggested that studies correlating perceived temporal sensations and consumer acceptance would be of great interest. The dynamic nature of TDS may contribute to the understanding of liking drivers (*21*). Because of its temporal character, in certain scenarios TDS is considered more appropriate than static descriptive analysis for explaining consumer responses. This is probably the case in products with an oral trajectory that leads to huge changes in structure and, consequently, in flavor release. In fact, given that liking can also be evaluated temporally during consumption, some researchers have attempted to measure it through a temporal procedure (*22, 23*). The following section discusses some examples of how TDS has contributed to ascertaining drivers of liking in several foods.

Biscuit Quality

Looking for a better understanding of biscuit eating quality, TDS has been used in combination with other techniques (*24*). This work assessed the oral texture perception of biscuits with two different amounts of fat and two different amounts of wheat fiber. The attributes selected for the TDS task were *Hardness*, *Crispness*, *Crunchiness*, *Pastiness*, *Fat mouthfeel*, *Grittiness*, and *Dry mouthfeel*. In addition, the consumers (n=100) scored their overall liking, liking for the texture, liking for the flavor, and perceived adequacy of the levels of four of the attributes (*Hardness*, *Dry mouthfeel*, *Fat mouthfeel* and *Pastiness*) for each sample on bipolar just-about-right (JAR) scales ranging from 1=much too little to 5=much too much, with 3=just about right.

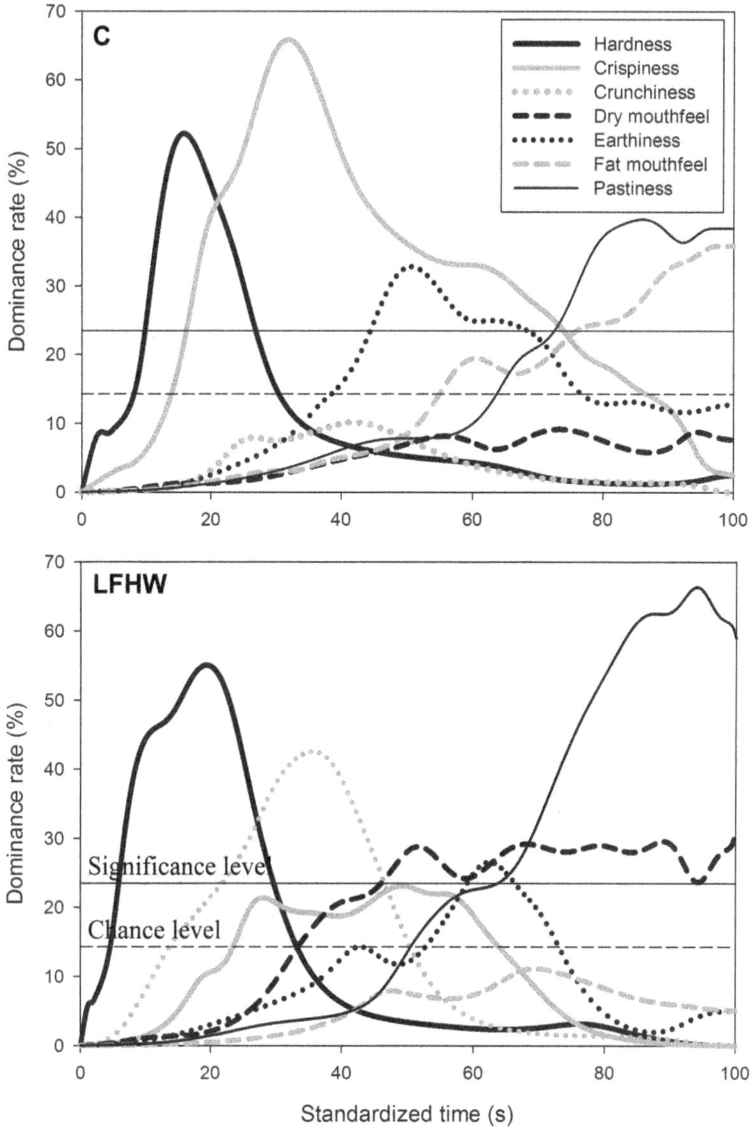

Figure 1. Normalized Temporal Dominance of Sensations (TDS) curves for different biscuit formulations. C, control sample; LFHW, low-fat, high-wheat fiber sample.

The TDS results showed that *Hardness* was the first dominant attribute of all the formulations during the mastication process, probably because it depended on the first bite. The dominance of the other parameters seemed to depend more on the fat and fiber levels, as *Crispness* appeared with high-fat biscuits and *Crunchiness* with high-fiber, low-fat ones, whilst both attributes were perceived in intermediate formulations. In the high-fiber formulations, *Grittiness* and *Dry mouthfeel* were selected as dominant later in the chewing phase. At the end of the mastication all the biscuits were perceived as Pasty: at this point all the textural contrasts had already disappeared. *Fat mouthfeel* was also perceived with both high-fat and low-fat biscuits, with or without the addition of a low level of fiber (Figure 1). A penalty analysis based on the JAR scales showed that excessive hardness and an excessively dry mouthfeel were the most penalizing sensory characteristics and caused a significant drop in biscuit acceptability.

These results showed that as fat and fiber levels modulated the dominance of the sensations experienced during oral assessment of the biscuit, knowing the occurrence and intensity of these characteristics would be a valuable tool for assessing a biscuit's eating quality, providing insights which could give clear pointers for biscuit reformulation.

Ice Cream

In-mouth texture largely determines the acceptability of ice cream, making it a key quality factor. Ice cream consumption involves special in-mouth handling which seeks to melt the solid, frozen cream delicately by tongue movements involving several oral structures. As a result, the product slowly melts and becomes a smooth, creamy viscous liquid as its temperature increases.

Time is therefore an important issue in the sensory perception of ice cream but its evaluation has barely been considered (*21*). On a macro scale, the ice cream texture perceived is determined by the microscopic features of the structure. In turn, microstructure is determined by complex molecular interactions. Being an emulsion (fat droplets) and a foam (air bubbles) in a continuous phase (sugar solution) which is partially frozen (ice crystals), emulsification and stabilization of the ingredients becomes essential. Although egg, dairy cream, and milk proteins act as naturally occurring stabilizers, the addition of hydrocolloid mixtures is a normal practice in industrial ice cream manufacture.

In the following example, TDS was performed on six ice cream samples (*21*). They were formulated very differently in order to analyze the texture-stabilizing effect of each ingredient. They all contained the same amount of sugar and one of the six combinations of ingredients: only milk, milk plus cream and egg yolk, milk plus hydrocolloid mix, milk plus cream and hydrocolloid mix, milk plus egg yolk and hydrocolloid mix and milk plus cream, egg yolk and hydrocolloid mix. *Iciness, Coldness, Creaminess, Roughness, Gumminess,* and *Mouth Coating* were assessed in the TDS task. In addition, two consumer (n=100) tests were performed. In the first, *Overall Liking* and liking for *Appearance*, *Texture* and *Flavor* were scored on 9-point box-scales labelled from 1=dislike extremely to 9= like extremely. In the second, a Check-all-that-apply (CATA) questionnaire was answered by the consumers, who described the samples by selecting appropriate attributes from a given list. The texture attributes presented were *Rough surface, Creamy, Appearance, Crystallized, Soft, Hard, Smooth, Creamy texture, Coarse texture, Gummy, Elastic, Cold, Melt easily, Melt slowly, Fatty, Aerated, Dense, Easy to spoon* and *Fatty residual sensation.*

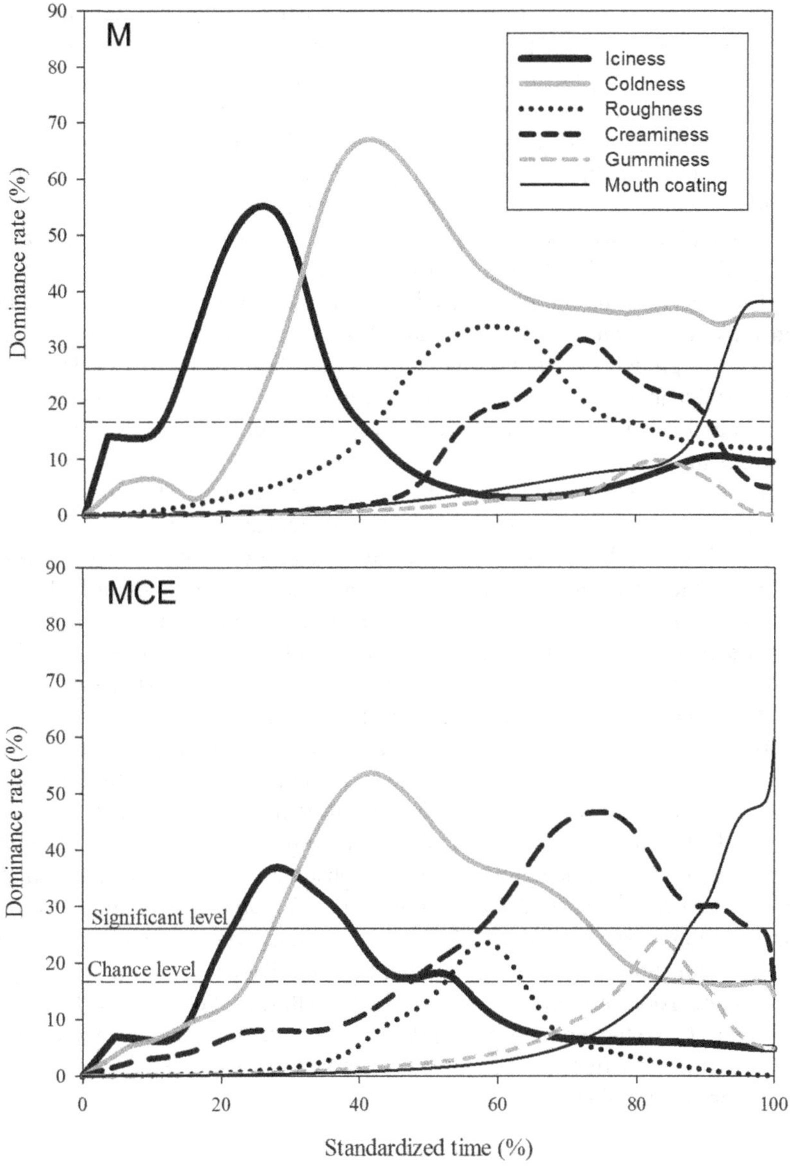

Figure 2. Normalized Temporal Dominance of Sensations (TDS) curves for the different ice cream formulations. M, only milk and sucrose; MCE, milk, dairy cream, egg yolk and sucrose.

It is interesting that *Iciness* was included as a different attribute from *Coldness* since the panel was able to distinguish between three different sensations (*Iciness*, *Coldness* and *Roughness*) that are all related to the presence of ice crystals. The strong direct relationship between big ice crystals and the development of a coarse and/or icy texture (*Iciness*) is well known. In contrast, when the structure of the ice cream is well stabilized the ice crystals remain small and the sensation was just called *Coldness*. In general, finer structures produce sensory properties such as *Creaminess* and *Smoothness*, with coldness having less initial impact (Figure 2).

The sensation of *Gumminess*, associated with the perception of a fairly cohesive substance that could almost be chewed, appeared with a significant dominance rate at 60% of the consumption time for the sample containing cream, egg yolk and hydrocolloid mix. The excessive cohesiveness of this sample was attributed to overstabilization of the formulation.

The consumer liking test showed that the more "traditional" sample, made with milk, cream and egg yolk, obtained the highest overall liking score. In general, the products containing cream were the best liked, which also agrees with their dynamic profiles. They were perceived as creamy earlier than the other samples and it is well know that creaminess is a determining factor for hedonic responses in this product category.

The least liked samples were those containing no cream. The measurements of liking for some specific attributes help in understanding the results for these last samples, as all the scores were quite low, particularly for texture, where they performed very poorly. Their TDS profiles also had very distinct dynamic properties: they were perceived as less creamy in the mouth and some of them were perceived as having iciness, coldness and roughness as the dominant sensations, with coldness lasting until the end of the consumption process, in line with their low texture appreciation and overall liking scores.

Multi-factorial analysis of CATA questions makes it possible to work with different variables to obtain a concise representation that considers all the information together, linking it to sample positioning. The purpose of this test is to gain a better understanding of which sensory descriptors define the samples and are responsible for the hedonic response of consumers. Overall liking was highly correlated to *Smoothness*, *Melting slowly* and *Creamy texture,* and negatively correlated to *Coarse texture*, *Cold*, *Crystallized*, and *Rough surface*. Also, *Easy to spoon* came up as highly correlated to liking and Hard as opposed to liking.

In conclusion, it could be said that hydrocolloids (and cream or to a lesser extent egg) tuned the temporal perception of the ice cream attributes, reducing the first impact of sensations such as iciness and coldness. They also favored an early perception of creaminess.

Other Products

Coffee

The impact of "crema" (the smooth, dense, hazelnut-brown foam on top of a freshly brewed espresso) characteristics on in-mouth sensory perception and their link with the release of a pleasant roasted coffee aroma has been investigated using TDS (*25*). Espresso coffees with different foam characteristics and similar above-cup and in-mouth flavor sensory profiles were evaluated by instrumental (Proton Transfer Reaction Mass Spectrometry (PTR-MS) of headspace and nose-space) and sensory (TDS) techniques. A number of volatile compounds released in the crema headspace contribute to the pleasant freshness note of roasted coffee. The standard quantity of espresso crema was shown to be associated with the optimum release of pleasant high volatiles, both in the above cup headspace and in-mouth. However, the TDS study demonstrated that an increased amount of crema was associated with increasing dominance of the roasted attribute during coffee consumption.

Olive Oil

The use of two types of olive oils was found to affect the hedonic responses to tomato purée (*26*). Descriptive analysis and TDS methods were used to assess the impact of two Italian extra-virgin olive oils with different sensory properties on the perceived sensory profiles of pureed tomatoes. Both descriptive data and TDS curves showed that the olive oils strongly influenced the sensory properties of the tomato samples by modifying the intensity and the dominance rate of their characteristic attributes (suppression of sourness and watery character) and/or by contributing new sensations, peculiar to each oil (bitterness and grassy flavor in one case, and unripe fruit flavor in the other). These modifications of the sensory properties of tomatoes induced by the oils affected consumer liking and perceived freshness.

Cheese

TDS and TI analyses were used in the development of a Mozzarella cheese with reduced sodium content and good acceptability (*27*). Through these analyses, it became possible to describe the sensorial profile of the reformulated products and to optimize the recipes. Since the use of other salts for sodium chloride replacement contributes unpleasant tastes such as metallic or bitter, it seemed very appropriate to evaluate the temporal occurrence of the dominant sensations in the new low-sodium Mozzarella cheeses. By means of TDS, the temporal dominance of sweet, sour, salty, bitter, umami, spicy, and off-flavor was determined and was used to conclude that it was possible to produce Mozzarella cheese with up to a 54% reduction in sodium content, using a mixture of NaCl, KCl and monosodium glutamate, without affecting liking.

Chewing Gum

Liking scores resulting from asking consumers only once whether they liked a chewing gum (static liking, SL) and those obtained when asking repeatedly during consumption (dynamic liking, DL) were compared (*28*). Three different mint chewing gums were evaluated by two groups of consumers at home using an internet application. In the SL task, consumers were prompted to rate their liking only after 5 min of chewing, during which time they were presented with a series of curious facts to read on the screen as a background task. In the DL task, the consumers were asked to rate the samples every 45 s over a period of 10 min, with a maximum of 10 s in which to answer, while performing the same background task. The study found that for all the samples, the SL ratings were significantly higher than the DL results after 5 min of consumption. These higher values were probably closer to the consumers' first impression than to their preference after 5 min, showing that discrimination among the samples was smaller in the SL than in the DL task. The analysis of the results also showed the moment at which preference became discriminant among the samples. Additionally, an inversion pattern of preference in some samples was found in the DL, and could not have been found by the SL technique. The use of a dynamic approach to study preference seems more pertinent when preference during the whole of a food product's performance needs to be validated.

Temporal Liking and TDS

Liking is commonly measured as a single integrated response to the overall eating experience, but is likely to vary during food intake. The temporal aspects of hedonic assessment have been investigated (*22*) by comparing the results of two methods: one which elicited any change in liking during consumption, and another which scored liking at four predefined points during the consumption time. The results of this exploratory study showed that at a panel level, the two methods showed similar dynamics of liking for two of the three wheat flakes studied but detected poor consensus among the subjects. In addition, an analysis of the individual responses showed many different patterns of dynamic liking. Finally, correlation coefficients between overall liking scores and dynamic liking data, calculated for each subject, suggested the importance of the beginning of the consumption event in the overall liking score (i.e., the time-averaged response). The authors state that better understanding of the dynamics of liking will provide a deeper understanding of the determinants of the overall liking score and help to identify the key moments of the consumption experience.

In the same line of research, recently a new approach has been proposed (*23*). This study assessed six flavored fresh cheese samples. During the first session, the consumers completed a classic liking test (9-point hedonic scale) for the six products, presented monadically. The consumers received no instructions regarding the moment to give their response, but had to click on the "In mouth" button on the screen when putting the sample into their mouths. This triggered a timer that recorded the time at which the subject scored his/her liking and stopped when the "I confirm my choice" button was clicked. During the second session, the same products were evaluated by a temporal liking test: the consumers had to taste each sample and indicate their liking throughout the tasting on an ordinal hedonic scale of nine boxes until they no longer felt any sensation after swallowing the product. During the third session, the consumers performed a TDS task with the six products, without any training. The fourth session was identical to the first except that the hedonic scale appeared on the screen 1 min after the subjects had begun to taste the product, forcing them to wait a minute before giving their answer. The results of the study showed that temporal liking was more discriminative than classic liking, and that classic liking scores may have been given before swallowing, whereas waiting for 1 min before asking for a liking score decreases discrimination of the products. Importantly, in the opinion of the authors, combining temporal liking and TDS data obtained from the same consumers made it possible to suggest the attributes, and perception times, that were responsible for liking or disliking the product.

References

1. Pineau, N.; Cordelle; S.; Schlich, P. Temporal dominance of sensations: a new technique to record several sensory attributes simultaneously over time. *Abstract Book of 5th Pangborn Sensory Science Symposium*; July 20–24, 2003, Boston, MA; p 121.
2. Le Révérend, F. M.; Hidrio, C.; Fernandes, A.; Aubry, V. Y. Comparison between temporal dominance of sensations and time intensity results. *Food Qual. Preference* **2008**, *19*, 174–178.
3. Pineau, N.; Schlich, P.; Cordelle, S.; Mathonnière, C.; Issanchou, S.; Imbert, A.; Rogeaux, M.; Etiévant, P.; Köster, E. Temporal Dominance of Sensations: Construction of the TDS curves and comparison with time–intensity. *Food Qual. Preference* **2009**, *20*, 450–455.
4. Labbe, D.; Schlich, P.; Pineau, N.; Gilbert, F.; Martin, N. Temporal dominance of sensations and sensory profiling: a comparative study. *Food Qual. Preference* **2009**, *20*, 216–221.
5. Meillon, S.; Viala, D.; Medel, M.; Urbano, C.; Guillot, G.; Schlich, P. Impact of partial alcohol reduction in Syrah wine on perceived complexity and temporality of sensations and link with preference. *Food Qual. Preference* **2010**, *21*, 732–740.
6. Pineau, N.; Goupil de Bouillé, A.; Lepage, M.; Lenfant, F.; Schlich, P.; Martin, N.; Rytz, A. Temporal Dominance of Sensations: What is a good attribute list? *Food Qual. Preference* **2012**, *26*, 159–165.
7. Albert, A.; Salvador, A.; Schlich, P.; Fiszman, S. Comparison between temporal dominance of sensations (TDS) and key-attribute sensory profiling for evaluating solid food with contrasting textural layers: Fish sticks. *Food Qual. Preference* **2012**, *24*, 111–118.
8. Meillon, S.; Urbano, C.; Schlich, P. Contribution of the Temporal Dominance of Sensations (TDS) method to the sensory description of subtle differences in partially dealcoholized red wines. *Food Qual. Preference* **2009**, *20*, 490–499.
9. Lenfant, F.; Loret, C.; Pineau, N.; Hartmann, C.; Martin, N. Perception of food oral breakdown: The concept of sensory trajectory. *Appetite* **2009**, *53*, 659–667.
10. Di Monaco, R.; Su, C.; Masi, P.; Cavella, S. Temporal Dominance of Sensations: A review. *Trends Food Sci. Technol.* **2014**, *38*, 104–112.
11. Bruzzone, F.; Ares, G.; Giménez, A. Temporal aspects of yoghurt texture perception. *Int. Dairy J.* **2013**, *29*, 124–134.

12. Evangelista, S. R.; Ferreira Silva, C.; Pedrozo da Cruz Miguel, M. G.; Souza Cordeiro, C.; Marques Pinheiro, A. C.; Ferreira Duarte, W.; Freitas Schwan, R. Improvement of coffee beverage quality by using selected yeasts strains during the fermentation in dry process. *Food Res. Int.* **2014**, *61*, 183–195.
13. Ng, M.; Lawlor, J. B.; Chandra, S.; Chaya, C.; Hewson, L.; Hort, J. Using quantitative descriptive analysis and temporal dominance of sensations analysis as complementary methods for profiling commercial blackcurrant squashes. *Food Qual. Preference* **2012**, *25*, 121–134.
14. Bouteille, R.; Cordelle, S.; Laval, C.; Tournier, C.; Lecanu, B.; This, H.; Schlich, P. Sensory exploration of the freshness sensation in plain yoghurts and yoghurt-like products. *Food Qual. Preference* **2013**, *30*, 282–292.
15. Teillet, E.; Schlich, P.; Urbano, C.; Cordelle, S.; Guichard, E. Sensory methodologies and the taste of water. *Food Qual. Preference* **2010**, *21*, 967–976.
16. Sokolowsky, M.; Fischer, U. Evaluation of bitterness in white wine applying descriptive analysis, time-intensity analysis, and temporal dominance of sensations analysis. *Anal. Chim. Acta* **2012**, *732*, 46–52.
17. Sokolowsky, M.; Rosenberger, A.; Fischer, U. Sensory impact of skin contact on white wines characterized by descriptive analysis, time–intensity analysis and temporal dominance of sensations analysis. *Food Qual. Preference* **2015**, *39*, 285–297.
18. Vázquez-Araújo, L.; Parker, D.; Woods, E. Comparison of temporal–sensory methods for beer flavor evaluation. *J. Sens. Stud.* **2013**, *28*, 387–395.
19. Lepage, M.; Neville, T.; Rytz, A.; Schlich, P.; Martin, N.; Pineau, N. Panel performance for Temporal Dominance of Sensations. *Food Qual. Preference* **2014**, *38*, 24–29.
20. Zorn, S.; Alcaire, F.; Vidal, L.; Giménez, A.; Ares, A. Application of multiple-sip temporal dominance of sensations to the evaluation of sweeteners. *Food Qual. Preference* **2014**, *36*, 145–153.
21. Varela, P.; Pintor, A.; Fiszman, S. How hydrocolloids affect the temporal oral perception of ice cream. *Food Hydrocolloid* **2014**, *36*, 220–228.
22. Sudre, J.; Pineau, N.; Loret, C.; Martin, N. Comparison of methods to monitor liking of food during consumption. *Food Qual. Preference* **2012**, *24*, 179–189.
23. Thomas, A.; Visalli, M.; Cordelle, S.; Schlich, P. Temporal Drivers of Liking. *Food Qual. Preference* **2015**, *40*, 365–375.
24. Laguna, L.; Varela, P.; Salvador, A.; Fiszman, S. A new sensory tool to analyse the oral trajectory of biscuits with different fat and fibre contents. *Food Res. Int.* **2013**, *51*, 544–553.
25. Barron, D.; Pineau, N.; Matthey-Doret, W.; Ali, S.; Sudre, J.; Germain, J. C.; Kolodziejczyk, E.; Pollien, P.; Labbe, D.; Jarisch, C.; Dugas, V.; Hartmann, C.; Folmer, B. Impact of crema on the aroma release and the in-mouth sensory perception of espresso coffee. *Food Funct.* **2012**, *3*, 923–930.
26. Dinnella, C.; Masi, C.; Zoboli, G.; Monteleone, E. Sensory functionality of extra-virgin olive oil in vegetable foods assessed by Temporal Dominance

of Sensations and Descriptive Analysis. *Food Qual. Preference* **2012**, *26*, 141–150.

27. Rodrigues, J. F.; Gonçalves, C. S.; Pereira, R. C.; Carneiro, J. D. S.; Pinheiro, A. C. M. Utilization of temporal dominance of sensations and time intensity methodology for development of low-sodium Mozzarella cheese using a mixture of salts. *J. Dairy Sci.* **2014**, *97*, 4733–4744.
28. Galmarini, M. V.; Symoneaux, R.; Visalli, M.; Zamora, M. C.; Schlich, P. Static vs. dynamic liking in chewing gum: A new approach using a background task and a natural setting. *Food Qual. Preference* **2015**, *40*, 381–386.

Chapter 12

Experimental Approaches To Better Understand the Retention of Aroma Compounds in Oro-Naso-Pharyngeal Cavities

I. Déléris,[*,1,2] A. Saint-Eve,[1,2] M. Kauffmann,[1,2] G. Feron,[3] and I. Souchon[1,2]

[1]INRA, UMR 782 Génie et Microbiologie des Procédés Alimentaires, 1 ave L. Brétigniéres, F-78850 Thiverval-Grignon, France
[2]AgroParisTech, UMR 782 Génie et Microbiologie des Procédés Alimentaires, 1 ave L. Brétigniéres, F-78850 Thiverval-Grignon, France
[3]INRA, UMR 1324 Centre des Sciences du Goût et de l'Alimentation, 17 rue Sully, F-21065 Dijon, France
[*]E-mail: isabelle.deleris@grignon.inra.fr. Tel: +33 (0)1 30 81 54 39. Fax: +33 (0)1 30 81 55 97.

Better understanding the persistence of aroma compounds during food consumption has constituted a challenging issue in food science for a long time. Due to the complexity and the diversity of the phenomena involved, it has often been studied through *in vitro* approaches. The main objective of the present study was to propose a global approach to address this topic in *in vivo* conditions. Four simple experimental protocols were developed to differently expose the compartments of naso-oro-pharyngeal cavities to flavored gaseous samples. Assumptions on possible mechanisms (mass transfer, dilution, interactions with mucosa and/or saliva, *etc.*) were proposed to explain the shapes of release kinetics that were observed. Release differences appeared to be dependent on the physicochemical properties of volatile molecules, on the physiological characteristics of individuals (notably saliva properties) and on the compartment of the naso-oro-pharyngeal cavities that was considered. These achievements constitute a first step to progress in the understanding of relationships that exist between aroma release and perception.

Introduction

Consumer choice and preferences for foods are largely driven by their sensory properties. Among the different perceptions, aroma perception is particularly important since it occurs before consuming the product (orthonasal perception), continues during the oral phase of the consumption (retronasal perception) and can sometimes persist until several minutes after the swallowing occurs (*1*). It has been identified that some aroma compounds, for instance menthol and menthone responsible for the mint aroma or estragole and dehydro-ar-ionene responsible for the liquorice aroma, have delayed release patterns that could explain longest perception. Depending on products, sensory persistence phenomena can have different hedonic connotations and could impact food appreciation, either positively or negatively (for example, persistence can be a positive characteristic during coffee or wine consumption but is negatively considered in the case of garlic consumption). In addition to product acceptance by consumers and to the pleasure induced by food consumption, recent studies highlighted the contribution of food flavouring on consumer behavior (food processing, food intake, satiation and sometimes satiety) (*2–4*). Relationships between perception and release are largely described in literature but the underlying mechanisms at the origin of release dynamics are still not clearly understood. The complexity of aroma perception is mainly due to the various types of mechanisms (physicochemical, physiological, neurobiological, cognitive, *etc.*) that are involved at different time and space scales. The dynamics of phenomena probably also contribute to complexity (*5–7*). This is notably mentioned in studies focusing on differences between orthonasal and retronasal perceptions (differences in perception thresholds, in air flow rate levels and directions, in compound solubility within the mucus layer and in neuronal connections between both pathways) (*5*, *6*, *8–10*).

When dealing with the persistence of aroma compounds, several patterns of aroma release can be identified, with either large peaks spreading over time or several secondary peaks occurring after each new swallowing event with a rapid return to zero between each (*11*). The determining role of some physicochemical properties of aroma compounds, such as volatility, hydrophobicity or solubility, on release persistence has been highlighted using instrumental and/or Quantitative Structure-Activity Relationships (QSAR) modeling approaches on model or real food products (*12–14*). However, the mechanisms involved are probably much more complex. Even if aroma compound properties (*12*, *13*, *15*, *16*) and product characteristics (*17–19*) can account for the diversity in the shape of release kinetics, the high inter-individual variability that exists on physiological parameters (salivation, chewing, velum position, breath flow rate, *etc.*) remains an important factor to explain the differences that were observed (*20–23*). The role of the contact area between nasal mucus and air has also been mentioned (*24*). Concerning saliva, even if its role is assumed, it remains unclear, notably because artificial saliva is often used, in spite of the fact that it does not mimic the real one. The variety and complexity of the phenomena involved probably largely explain the wide range of results available in literature, their dependence on the nature of aroma compounds and/or on food products and the difficulty to draw clear conclusions.

Improving the understanding of mechanisms needs a better knowledge of the characteristics of saliva and of oral, nasal and pharyngeal mucosa as well as the development of specific systems to study their properties. However, the design of such systems requires to simplify biological models so that the mechanisms and/or interactions of interest can readily be studied. Extensive literature detailed the characterization of drug transport into (*i.e.*, penetration) and through (*i.e.*, permeation) different types of mucosa (*e.g.*, oral, nasal, oesophageal, intestinal) in *ex vivo* (*25–27*) or *in vitro* experimental set-ups (*28–31*). Briefly, *ex vivo* experiments were carried out by using fresh or frozen mucosal specimens sandwiched between donor and receptor compartments of vertical or horizontal diffusion cells (*26*, *27*). Similar *in vitro* set-ups using cell cultures are also largely described in the literature (*29*, *32*, *33*). The cumulative amount of drug in the receptor compartment allowed the estimation of permeation profile through mucosa while drug amount/concentration in mucosal tissue resulting from penetration reflected compound bioavailability within mucosa. In the field of toxicology, notably in relation to alcohol ingestion or smoking, some studies focused on the impact of specific molecules such as ethanol (*34*), menthol (*35*, *36*) or of complex mixtures such as cigarette smoke (*37*), on the integrity of oral or nasal mucosa. In the field of pharmaceutics, studies focusing on the oral cavity largely mentioned that several factors, such as the contact area, the degree of permeability or the degree of lubrication by saliva, can vary, depending on the structure and type of mucosa (gingival, buccal, sublingual, *etc.*) and thus can impact drug availability (*38–41*). In the field of food science, works on interactions between oro-pharyngeal lubricated mucosa and aroma compounds remained limited. But, the existence of interactions between aroma compounds and nasal, oral and/or pharyngeal mucosa and/or mucus constituents and/or salivary proteins has been assumed in lots of studies to explain the shape of release kinetics (*12*, *20*, *42–45*). Results obtained with aroma model aqueous solutions, for which no retention effect due to solution constituents occurred (*46*, *47*) and/or with original methodologies (*20*) tend to confirm this assumption.

The main goal of the present study is to better understand interactions between aroma compounds and naso-oro-pharyngeal mucosa from an overall point of view. The originality of the proposed approach relies on the development of a simple *in vivo* experimental strategy to control the exposure of the different compartments to aroma compounds. This will help in identifying the location of interactions and in determining the respective contributions of aroma compound properties and individual physiology.

Material and Methods

Aroma Compounds

Molecules (diacetyl, 2,5-dimethylpyrazine, ethyl propanoate, (*Z*)-3-hexen-1-ol, hexanal, menthone, 2-nonanone and menthol) were purchased from Sigma Aldrich (France) (food grade quality) (Table 1).

Table 1. Aroma Compound Characteristics. Chemical Formulae and Structures and Physicochemical Properties from Literature and Experimental Mean Air/Water K_{aw} and Air/Saliva K_{as} Partition Coefficients (37°C) with Associated Standard Deviations.

Aroma compound	Solution n°	Chemical formula	Chemical structure	MW (g/mol)	LogP[1]	PTR-MS[2] fragmentation[3]: main *m/z* peaks (relative abundance)	Partition coefficients ($\times 10^{-3}$) K_{aw} (air/water) est. (25°C)[1]	K_{aw} (air/water), exp. (37°C)	K_{as} (air/saliva), exp. (37°C)
Diacetyl	2	$C_4H_6O_2$	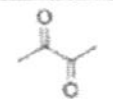	86.09	-1.34	87 *(100)*	0.54	1.3± 0.2	0.32± 0.07*
2,5-Dimethylpyrazine	2	$C_6H_8N_2$	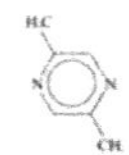	108.14	0.63	109 *(100)*	0.14	0.03± 0.02	0.12± 0.02
Ethyl propanoate	1	$C_5H_{10}O_2$	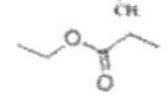	102.13	1.21	75 *(100)*; 103 *(20)*; 57 *(14)*	15.9	14.9 ± 1.4	12.9 ± 5.5
(Z)-3-Hexen-1-ol	1	$C_6H_{12}O$		100.16	1.61	55 *(100)*; 83 *(39)*	0.63	0.78 ± 0.07	0.76 ± 0.12
Hexanal	2	$C_6H_{12}O$		100.15	1.78	55 *(100)*; 83 *(89)*; 43 *(87)*	8.64	16.7 ± 1.5	11.5 ± 4.6
Menthone	3	$C_{10}H_{18}O_2$	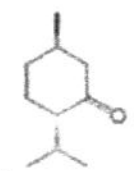	154.25	3.05	81 *(100)*; 155 *(41)*; 137 *(34)*	5.74	10.4 ± 1.0	11.5± 4.9
2-Nonanone	1	$C_9H_{18}O_2$	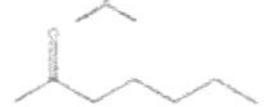	142.24	3.14	143 *(100)*; 41 *(20)*; 69 *(18)*	11.1	19.6 ± 8.4	9.7 ± 1.4
Menthol	3	$C_{10}H_{20}O$	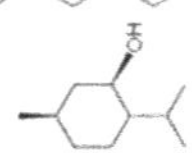	156.27	3.40	83 *(100)*; 55 *(41)*; 57 *(37)*	1.04	4.4± 0.4	3.1± 0.4*

[1]: estimation with EPI Suite™ program
[2]: PTR-MS: Proton Transfer Reaction-Mass Spectrometry
[3]: experimental data
*: stars mean significant difference between K_{aw} and K_{as} for a given compound (Mann and Whitney test, $p<0.05$).

They were selected as they belong to several chemical classes, have different physicochemical properties and differ in their release behaviors (persistence) (preliminary tests, not shown). Individual concentrated stock solutions were prepared in polypropylene glycol (Sigma Aldrich, France) and used throughout the study.

Experimental Determination of the Air/Water (K_{aw}) and Air/Saliva (K_{as}) Partition Coefficients of Aroma Compounds

Two mL of aroma compound solutions (300 mg/kg final concentration) were prepared, either in water (Evian, France) or in pooled human saliva.

Aqueous solutions of aroma compounds were prepared extemporaneously from concentrated stock solutions (20-fold dilution).

Pooled human saliva was collected from 33 persons under stimulated conditions (*48*). Volunteers exhibited no known illness at the time of collection and did not declare any olfactory and gustatory dysfunctions. They were asked to chew a piece of paraffin sealing film (Parafilm®M, Brand GMBH+CO KG, Wertheim, Germany) for 5 min and then to spit out their saliva at regular intervals (every 30 s). Saliva samples were stored in iced-vessel (0°C) during the collection period and finally mixed together to obtain whole pooled saliva, which was aliquoted in 1.5 mL Eppendorf tube. Aliquots were stored at -80°C. For the preparation of flavored samples, concentrated aroma solutions were directly diluted in saliva aliquots, extemporaneously defrosted.

The K_{aw} and K_{as} partition coefficients of aroma compounds were determined at 37°C by the Phase Ratio Variation method (PRV)(*49*). Different volumes of aroma compound solutions (0.05, 0.2, 0.5 and 2.0 mL) were placed in closed vials (22.4 mL, Chromacol, France) and incubated under stirring at 37°C for 1 hour. Preliminary tests confirmed that this duration was sufficient enough to allow thermodynamic equilibrium to be established without any aroma loss. Then, 2 mL aliquots of the headspace above the solutions were sampled with an automatic headspace CombiPal sampler (CTC Analytics, Switzerland) and injected into a gas chromatograph (GC-FID HP6890, Agilent Technologies, Germany) equipped with an HP-INNOWax polyethylene glycol semi-capillary column (30 m length, 0.53 mm internal diameter, with a 1 µm-thick film) and a flame ionization detector (FID). The temperatures of the gas chromatograph injector and detector were set at 250°C. The oven program duration was 27.5 min, starting at 50°C, with 4°C/min up to 150°C, then 20°C/min up to 200°C. The carrier gas was helium (average velocity of 57 cm/s at 50°C).

FID peak areas were measured using the Hewlett–Packard Chemstation integration software. A non-linear regression was applied to determine K_{aw} and K_{as} coefficients (*50*). All experiments were performed in triplicate to validate measurement repeatability. Comparison between K_{aw} and K_{as} was performed with Mann and Whitney test and Conovan-Iman procedure (multiple paired comparison) ($p<0.05$) (XlStat, Addinsoft).

Table 2. Median Values of the Physiological Characteristics of Panelist Classes Determined by HAC (Hierarchical Ascendant Classification Performed on the Physiological Data Set) and Associated Quartiles

	Panelist classes	*A*		*B*		*C*	
	Panelist number/class		5		2		1
	V_{Tidal} (L)	0.89	(0.51; 1.19)	0.91	(0.90; 0.93)	1.0	(0.9; 1.1)
	V_{nose}^{*} (cm^3)	9.5	(8.0; 10.2)	17.9	(17.6; 18.2)	10.4	(9.3; 11.2)
	V_{mouth}^{*} (cm^3)	41.6	(38.3; 45.5)	60.7	(59.6; 61.7)	64.8	(63.0; 66.3)
	$V_{pharynx}$ (cm^3)	31.0	(21.8; 32.5)	35.0	(31.5; 38.5)	31.2	(30.9; 31.8)
Salivary parameters at rest	Salivary flux * (g/min)	0.59	(0.58; 0.69)	0.43	(0.32; 0.54)	0.14	(0.11; 0.15)
	Antioxidant* (eq mM Trolox)	54.5	(39.9; 67.0)	71.3	(69.5; 73.2)	26.8	(26.4; 27.2)
	LipLipolysis* (mU/mL)	0.048	(0.045; 0.054)	0.069	(0.067; 0.071)	0.12	(0.11; 0.13)
	Amylase (U/mL)	247.7	(96.6; 384.6)	169.3	(143.7; 195.0)	345.4	(333.7; 366.1)
	Proteolysis* (U/mL)	5.8	(5.4; 5.8)	7.5	(5.9; 9.0)	129.9	94.6; 150.0)
	Lysozyme* (U/mL)	884.8	(391.0; 1187.4)	1327.2	(933.1; 1721.3)	1723.0	(1651.2; 1839.1)
	Proteins* (mg/mL)	0.30	(0.20; 0.40)	0.25	(0.24; 0.26)	0.36	(0.33; 037)

* : stars indicate parameters that are significantly different between classes (Kruskal and Wallis test and Conovan-Iman procedure (multiple paired comparison), $p<0.05$).

Panelists

Eight panelists (four men / four women, all members of the laboratory, 20-40 years old) were recruited for the study. They were informed about the nature of the study, gave their signed consent and received a financial compensation for their participation. They were instructed not to smoke, eat, drink, or use any persistent-flavored product for at least one hour before Proton Transfer Reaction-Mass Spectrometry (PTR-MS) or saliva collection sessions.

The volumes of the oral, nasal and pharyngeal cavities of subjects were measured with the Eccovision Acoustic Rhinopharyngometer (Eccovision, Sleep Group Solutions, North Miami Beach, FL 33162, USA). The air/product areas of cavities were calculated automatically for each individual using a specific software (*51*). The tidal volume of each individual was measured with a spirometer (Pulmo System II, MSR, Rungis, France) (*52*). Non-stimulated saliva was collected by asking volunteers to swallow the saliva present in mouth before starting and then spit every 30 s for 5 min into ice-chilled vessels. The final saliva weight was measured and flow rate was calculated as g/min. Whole saliva samples were centrifuged at 13 400 x g for 5 min at 4°C to remove cellular debris and bacteria (Eppendorf, model 5415 R, Germany). The supernatants were frozen and stored at -80°C before analysis. Protein concentration (expressed in mg/mL) was obtained by standard Bradford protein assay Quick Start (Bio-Rad, France) using bovine serum albumin (Sigma-Aldrich, France) as standard calibration. The lipolytic (lipolysis), proteolytic (proteolysis), lysozymal (lysozyme) and amylolytic (amylase) activities of individual salivas (expressed in U/mL) were determined as previously described (*48*). For all these parameters, three replicates were determined. Based on the whole set of physiological data, three classes of individuals were highlighted (Hierarchical Ascendant Classification) (Table 2). These panelist classes mainly distinguished on mouth and nose volumes and on some salivary parameters (salivary flow rate, lipase, protease and lysozyme activities and antioxidant and protein concentrations) (Kruskal and Wallis test and Conovan-Iman procedure (multiple paired comparison), $p<0.05$). The analysis of release data was thus performed on the basis of these classes.

Gaseous Sample Preparation for *in Vivo* Experiments

To really focus on the interactions of aroma compounds with mucosa and/or saliva and avoid the known influence of sample composition and/or structure on aroma release (*17–19*), gaseous samples were used. 25 mL of aqueous solutions of aroma compounds were prepared using water (Evian, France) and aroma stock solutions. They were stored in 250 mL-flasks (Schott, France), closed by caps equipped with valves, for 4 hours before measurements (at ambient temperature). Concentrations in the liquid phase were chosen to be sufficient enough so that aroma compounds can be detected by PTR-MS measurements, *i.e.* 300 mg/kg.

Three different mixtures of aroma compounds were used to reduce the number of sessions. The composition of flavored solutions were defined on the basis of two criteria: i) to optimize the sampling frequency and cycle duration during a single PTR-MS acquisition (number of ions to be detected) and ii) to avoid any ion overlapping during measurement (from fragmentation data, Table 1). Solution 1 contained 2-nonanone, ethyl propanoate and *(Z)*-3-hexen-1-ol, solution 2 diacetyl, 2,5-dimethylpyrazine and hexanal and solution 3 menthone and menthol (Table 1).

Protocols

Four inhalation protocols were defined to differently expose the compartments of the naso-oro-pharyngeal cavities (Figure 1). Samples were inhaled either through the mouth or the nose and air sampling for PTR-MS measurement was performed either in the mouth or the nose. Protocols were named according to (i) sampling location (Nose or Mouth), (ii) inhalation location (Nose or Mouth) and (iii) swallowing allowed or not (S or ns). For all protocols except N.M.ns, panelists were allowed to swallow 5 s after sample inhalation and other swallowing events were free. For N.M.ns. protocol, swallowing was not allowed for at least 2 minutes after sample inhalation, leading to saliva accumulation in mouth. In the case of M.M.S. protocol, the use of a nose clip allowed to isolate the nasal cavity from aroma compound exposure.

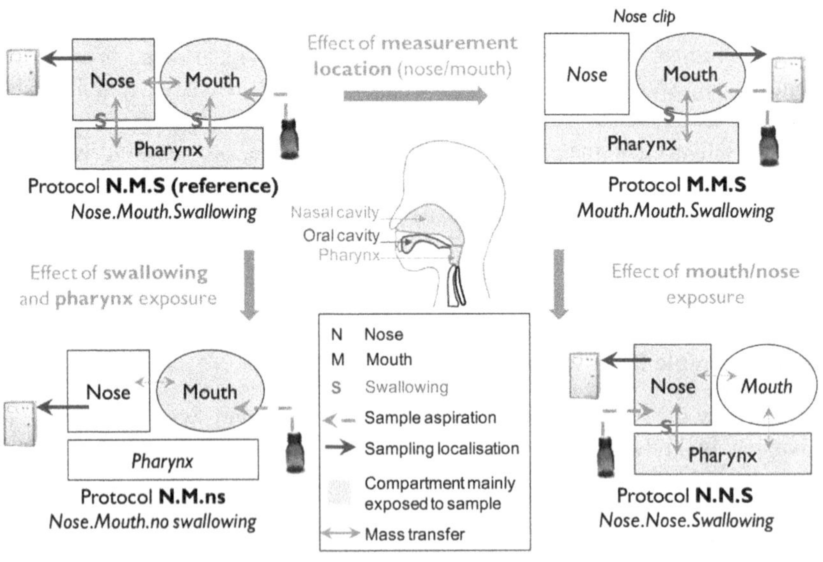

Figure 1. Schematic representation of the different protocols used in the present study to characterize the in vivo release of aroma compounds depending on inhalation and sampling locations. Protocols are named according to (1) sampling location (Nose or Mouth), (2) inhalation location (Nose or Mouth) and (3) swallowing allowed or not (S or ns).

PTR-MS Measurements of in Vivo Aroma Release Kinetics

In vivo release kinetics were measured using a High-Sensitivity PTR-MS (Ionicon Analytik, Innsbruck, Austria). The PTR-MS instrument drift tube was thermally controlled (60°C) and operated with a voltage set at 600.1 (±0.4) V and a pressure of 1.95 ± 0.01 mbar (*E/N*=156.6 ± 0.9 Td). Measurements were performed using the Multiple Ion Detection (MID) mode. Panelist breath was monitored with the signal of mass/charge ratio (*m/z*) 59, which is attributable to endogenous acetone. For *m/z* 21 (H_3O^+), 37 (H_2O-H_3O^+) and 59, the dwell time per mass was fixed to 0.05s. The mean signal for H_3O^+ ion was $11.3\times10^6 \times 2.4\times10^6$ counts per second (cps) and its variation along the measurement period was lower than 5%. Signal for H_2O-H_3O^+ ion (*m/z* 37) did not exceed 2.8% of the one of *m/z* 21 (equipment specification). From the fragmentation patterns of individual compounds (Table 1), the molecules studied were monitored at *m/z* 55 and 83 (*(Z)*-3-hexen-1-ol), *m/z* 75 and 103 (ethyl propanoate) and *m/z* 143 (2-nonanone) (solution 1), *m/z* 55 (hexanal), *m/z* 87 (diacetyl) and *m/z* 109 (2,5-dimethylpyrazine) (solution 2) and *m/z* 55 and 83 (menthol) and *m/z* 81 and 155 (menthone) (solution 3). With this selection, no fragment overlapping was observed. For these ions, a dwell time per mass of 0.1 s was selected as a compromise between sensitivity for aroma compound detection and appropriate sampling frequency with regards to the phenomena to be measured. Thus, depending on the flavoured solution that was analyzed, measurements were performed on six to eight specific *m/z* and the duration of measurement cycles ranged between 0.45 and 0.65 s.

Six sessions of 45 min were planned to obtain three replicates of release kinetics for each molecule, each protocol and each subject. During a session, subjects started with the analysis of a blank sample to get used to protocols and then had to test six samples. Nose-space or mouth-space air was sampled *via* the two inlets of a stainless nosepiece placed either in both nostrils of the assessors or in mouth, respectively. The inlet of the PTR-MS instrument was connected to the sampling device *via* a 1/16” PEEK™ tube maintained at 110°C. Room air was first analyzed for 10 s. After positioning the sampling device in nostrils or mouth, panelists were asked to breathe regularly for 30 s (breath analysis). Then, they opened the flask containing the flavored gaseous sample just in front of the nostrils or mouth and inhaled it in one breath using the previously defined protocols. For N.M.S., N.M.ns. and N.N.S. protocols, panelists were asked to keep their mouth closed and to only breathe through the nosepiece. For M.M.S. protocol, they breathed through the mouth, keeping it as closed as possible. Each assay lasted 4-5 minutes. Between each sample, panelists were asked to clean their mouth by drinking mineral water (Evian, France). Panelist breath was tested before each new measurement. All the measurements were performed within a 19-day period. These measurements led to the determination of molecule release kinetics, *i.e.* intensity I_t=f(time t), for each panelist and each protocol. As solution composition was specifically defined for the experiment and exactly known, aroma compounds can be unambiguously detected at the stated *m/z*. For this reason and to facilitate text readability, the use of compound names rather than their *m/z* was preferred hereafter.

For data handling, these experimental kinetics were divided into two main periods: the phase before (phase 0) and after product inhalation (phase 1). For each sample, the mean PTR-MS signal measured during phase 0 was subtracted from the PTR-MS signals obtained during exposure to sample. Quantitative release parameters (maximal intensity (I_{max}), which indicates the maximum concentration reached by a compound, and area under the curve (AUC), which is related to the total amount of molecule that is released) were extracted from these corrected release curves. Temporal release parameters (time at which I_{max} occurred (t_{max}), peak widths at 20% of I_{max} ($\Delta t_{20\%}$) and second half-peak width at 50% of I_{max} ($t_{50\%}$-t_{max})) were also determined. In addition, standardized release kinetics were obtained by dividing each intensity value of the curve by the corresponding I_{max} ($I_{t\_stand}=I_t/I_{max}$). Standardized areas under curve ($AUC_{stand.}$), determined from these standardised kinetics, were used as an indication of persistence behavior. Because the objective was to compare the extent of aroma release between products, the use of arbitrary units for aroma release data was sufficient for the analysis of intensity differences. As ions related to the same molecule behaved in the same way, only the results of the main product ion for each molecule is presented in the text. The comparison of release kinetics between protocols for a given ion is possible without bias. The comparison of release behaviors between ions for one protocol must be done carefully as sampling frequencies were not exactly the same for the different ions.

Statistical Analysis

Non-parametric descriptive analysis was carried out on data sets. Comparative analysis was performed using Kruskal and Wallis or Mann and Whitney tests, associated with Conovan-Iman procedure (multiple paired comparison), to highlight differences on *in vivo* release kinetics between protocols for a given molecule or between molecules for a given protocol, respectively. The level of significance was set at $p<0.05$. In addition, Spearman tests ($p<0.05$) were applied to determine potential correlations between ion release parameters extracted from *in vivo* release data and either the physiological characteristics of panelists or the air/saliva partition properties of aroma compounds (XlStat, Addinsoft).

Results and Discussion

Protocol Effect on Aroma Release Kinetics

The comparison between N.M.S. and N.M.ns protocols enabled to focus on the effect of swallowing and pharynx exposure on aroma release. With these both protocols, panelists sucked samples with the mouth, meaning that sample amount was similar between protocols and replicates for one panelist. Table 3 presents the normalized values of release parameters obtained from N.M.ns protocol (N.M.S. parameters were used as reference for normalization).

Table 3. Normalized Release Parameters for N.M.ns Protocol (Mean Values, All Panelists, All Replicates). Normalization Was Done Using Release Parameters from N.M.S. Protocol as Reference. Stars Mean Values That Significantly Differ between Protocols (Mann and Whitney Test and Conovan-Iman Procedure (Multiple Paired Comparison), $p<0.05$). Bold Characters Highlight Significant Effects while Italic Characters Refer to Non Significant Effects.

Normalized parameters for N.M.ns protocol (reference=N.M.S. protocol)	I_{max}	*AUC*	$AUC_{stand.}$	t_{max}	$\Delta t_{20\%}$	$t_{50\%}-t_{max}$
(Z)-3-Hexen-1-ol	**1.40 ***	**1.29 ***	*0.89*	*0.93*	*0.93*	*0.95*
Hexanal	*1.00*	**1.23 ***	*1.15*	**1.74 ***	*1.17*	*1.00*
Menthol	**1.14 ***	**1.31 ***	*1.14*	*0.79*	*1.27*	**1.56 ***
Menthone	**1.26 ***	**1.19 ***	*0.95*	**1.10 ***	**0.86 ***	**0.71 ***
Diacetyl	**1.30 ***	**1.27 ***	*0.90*	**1.54 ***	*1.04*	*0.93*
Ethyl propanoate	*1.33*	*1.45*	*0.95*	**1.59 ***	*1.13*	*1.00*
2,5-Dimethylpyrazine	**1.17 ***	**1.28 ***	**1.16 ***	*1.01*	*1.34*	*1.09*
2-Nonanone	**1.27 ***	**1.24 ***	**0.84 ***	**1.22 ***	**0.82 ***	**0.85 ***

Table 4. Normalized Release Parameters for M.M.S Protocol (Mean Values, All Panelists, All Replicates). Normalization Was Done Using Release Parameters from N.M.S. Protocol as Reference. Stars Mean Values That Significantly Differ between Protocols (Mann and Whitney Test and Conovan-Iman Procedure (Multiple Paired Comparison), $p<0.05$). Bold Characters Highlight Significant Effects while Italic Characters Refer to Non Significant Effects.

Normalized parameters for M.M.S protocol (reference=N.M.S. protocol)	I_{max}	AUC	$AUC_{stand.}$	t_{max}	$\Delta t_{20\%}$	$t_{50\%-tmax}$
(Z)-3-Hexen-1-ol	**1.93 ***	**2.24 ***	**0.69 ***	**1.52 ***	**1.59 ***	**1.61 ***
Hexanal	**0.31 ***	**0.62 ***	**0.30 ***	**1.47 ***	**1.81 ***	*1.13*
Menthol	**1.29 ***	**1.46 ***	**0.87 ***	*0.91*	**1.69 ***	**2.85 ***
Menthone	*1.17*	**1.22 ***	*0.85*	*1.06*	*1.02*	*1.16 **
Diacetyl	*1.09*	**1.45 ***	*0.82*	*1.16*	**1.44 ***	*1.29*
Ethyl propanoate	*0.90*	*1.19*	*0.85*	**2.14 ***	*1.04*	*0.81*
2,5-Dimethylpyrazine	**2.38 ***	**3.40 ***	**0.72 ***	*0.87*	**9.02 ***	*6.73*
2-Nonanone	**1.36 ***	**1.62 ***	**0.84 ***	*1.32*	**1.28 ***	**1.49 ***

For all molecules, except for hexanal and ethyl propanoate, the values of I_{max} and/or AUC were lower when panelists swallowed (N.M.S. protocol) than when they did not (N.M.ns protocol). In addition, a lower t_{max} was observed with swallowing for all ions, except for alcohols and 2,5-dimethylpyrazine. As almost all ions were concerned, these results could be partly explained by swallowing, which reduced the residence time of air in mouth (removal of a part of the flavored air contained in the mouth) and favor the transport of molecules into the nasal cavity (velopharynx opening) (*53*). Yet, it was difficult to explain the specific results obtained for some molecules.

Another interesting result was the particular behaviors of menthone and 2-nonanone: the parameters related to persistence ($AUC_{stand.}$, $\Delta t_{20\%}$ and $t_{50\%}$-t_{max}) were significantly higher with than without swallowing. Interactions between these molecules and lubricated pharyngeal mucosa could be assumed: the breath flow rate could be responsible for the progressive stripping of these aroma compounds retained in the lubricated pharyngeal mucosa and their transport through the nasal cavity. An opposite significant behavior could be noticed for 2,5-dimethylpyrazine, with a lower $AUC_{stand.}$ with than without swallowing. In this case, specific interactions with oral mucosa could be assumed.

By comparing M.M.S. and N.M.S. protocols, the effect of measurement location could be evaluated. Once again, samples were introduced in the mouth in both cases, meaning that quantitative comparison can be done between these protocols. Table 4presents the normalized values of release parameters obtained from M.M.S protocol (N.M.S. parameters were used as reference for normalization).

Lower values of I_{max}, AUC and peak widths ($\Delta t_{20\%}$ and $t_{50\%}$-t_{max}) and higher values of $AUC_{stand.}$ were obtained for *(Z)*-3-hexen-1-ol, menthol, 2,5-dimethylpyrazine and 2-nonanone when measurement was located in nose (N.M.S.) instead of in mouth (M.M.S.). An opposite behavior was noticed for hexanal, with highest I_{max} and AUC when measurement was performed in the nose. For diacetyl, difference between protocols only occurred on AUC and $\Delta t_{20\%}$ parameters. In the case of ethyl propanoate, protocols only differed on t_{max}. These results highlighted quite complex and different behaviors depending on molecules and general assumptions were difficult to propose. The decrease in quantitative parameters (intensities and areas under curve) as well as on peak widths could be due to air dilution by breath flow rate when sampling and measurement were not located in the same place (N.M.S. protocol). But, as it was not observed for all molecules, other mechanisms, not yet identified, are probably involved.

By comparing, N.N.S. and M.M.S. protocols, the effects of mouth or nose exposure could be explored. These two protocols were only compared on temporal parameters as sample inhalation through the mouth or the nose did not lead to similar inhaled amounts. Table 5 presents the values of normalized release parameters obtained from M.M.S protocol (N.N.S. parameters were used as reference for normalization).

Results highlighted lower values of $AUC_{stand.}$ and $\Delta t_{20\%}$ for nose exposition than for mouth exposition for all molecules, except 2-nonanone and menthone. t_{max} was shortest when nose was exposed, except for hexanal and menthone. These results could be explained by specific interactions with the lubricated mucosa in the oral cavity for all molecules, except hexanal and menthone. The effect on $t_{50\%}$-t_{max} was less obvious but when present, it confirmed a lowest persistence when only nose was exposed (*(Z)*-3-hexen-1-ol and menthol).

Table 5. Normalized Release Parameters for M.M.S Protocol (Mean Values, All Panelists, All Replicates). Normalization Was Done Using Release Parameters from N.N.S. Protocol as Reference. Stars Mean Values That Significantly Differ between Protocols (Mann and Whitney Test and Conovan-Iman Procedure (Multiple Paired Comparison), p<0.05). Bold Characters Highlight Significant Effects while Italic Characters Refer to Non Significant Effects.

Normalized parameters for M.M.S protocol (reference=N.N.S. protocol)	$AUC_{stand.}$	t_{max}	$\Delta t_{20\%}$	$t_{50\%\text{-}tmax}$
(Z)-3-Hexen-1-ol	**1.48** *	**2.08** *	**2.23** *	**2.24** *
Hexanal	**3.05** *	*1.40*	**1.91** *	*0.59*
Menthol	**1.24** *	**1.87** *	**1.61** *	**1.47** *
Menthone	*1.04*	*1.10*	*0.99*	*0.93*
Diacetyl	**1.16** *	**1.40** *	**1.26** *	*1.29*
Ethyl propanoate	**4.19** *	**1.92** *	**1.41** *	*1.30*
2,5-Dimethylpyrazine	**1.36** *	**1.26** *	**2.13** *	*1.77*
2-Nonanone	*0.92*	**1.27** *	*1.17*	*1.30*

These results clearly illustrated that aroma release and persistence involved numerous and complex phenomena that are difficult to identify. They seemed to be dependent on both the exposed cavity and molecule properties. To go further on the understanding of molecule effects, data were analyzed per protocol.

Ion Effect on Aroma Release Kinetic

The comparison of *in vivo* release kinetics between molecules highlighted different release behaviors. Some examples of typical release curves for three ions are illustrated in Figure 2 (N.M.S. protocol, one panelist).

Diacetyl was released very rapidly and presented a fast decrease after swallowing. Similar behavior was observed for ethyl propanoate and hexanal (not shown). *(Z)*-3-Hexen-1-ol, was also released rapidly once the sample was inhaled but presented a different persistence behavior: its decrease rate was slower than for diacetyl and regular variations, related to breathing frequency, can be observed (such as for almost all other ions except the ones from ethyl propanoate, diacetyl and hexanal, not shown). In the case of menthone, a 3 s-delay in the peak occurence can be observed as well as slower initial release rate and decrease rate. The release kinetic of this molecule seemed also more impacted by breath flow rate. A release behavior similar to the latter was observed for 2-nonanone (not shown).

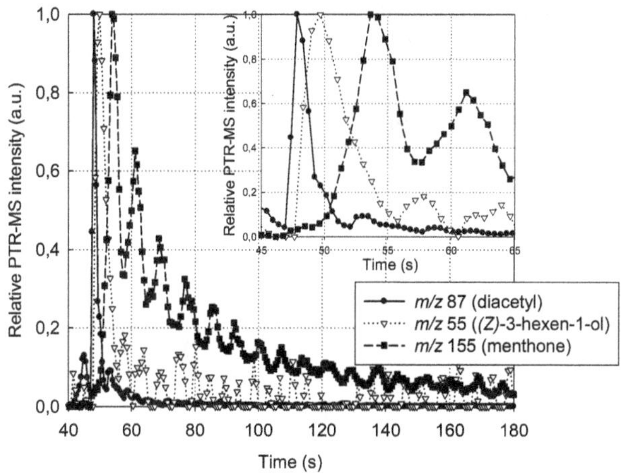

Figure 2. Example of standardized in vivo release kinetics for m/z 87 (diacetyl, solution 2), m/z 55 ((Z)-3-hexen-1-ol, solution 1) and m/z 155 (menthone, solution 3) obtained with the N.M.S. protocol and for one panelist. The release intensities were standardized using respective ion I_{max}.

When comparing the values of extracted parameters, protocol per protocol, significant differences between ions on both temporal and quantitative release parameters were highlighted, regardless of the protocols (Kruskal and Wallis test and Conovan-Iman procedure, $p<0.05$). Examples are given in Figure 3 in the case of N.M.S. protocol. It is yet important to note the slight difference in sampling frequencies between the different ions: it could partly contribute to final differences in temporal parameters between ions but cannot completely explain the discrepancies that were observed.

For example, in the case of N.M.S protocol, it was observed that ethyl propanoate, diacetyl and hexanal had the lowest $AUC_{stand.}$ and peak widths ($\Delta t_{20\%}$ and $t_{50\%}$-t_{max}) and were released sooner (t_{max}) than other molecules, meaning they were weakly persistent (Figure 3). Menthol and 2,5-dimethylpyrazine had an opposite behavior with highest values of $AUC_{stand.}$ and t_{max}. (*Z*)-3-Hexen-1-ol and menthone appeared to have an intermediate behavior, with only a delayed release (t_{max}). Differences between molecules concerning peak widths were less evident and more difficult to understand. Nevertheless, all these results tended to confirm the existence of aroma compound retention by lubricated mucosa and highlighted the main role of the physicochemical properties of molecules.

Partition Properties of Aroma Compounds and Correlation with Release Parameters

To help in understanding the origin of differences between ions on release parameters, the K_{aw} and K_{as} of aroma compounds were determined (Table 1). K_{aw} at 37°C were in agreement with literature data. K_{as} at 37°C ranged between 0.12×10^{-3} (2,5-dimethylpyrazine) and 12.9×10^{-3} (ethyl propanoate). Only a significant retention of menthol and diacetyl by saliva can be noticed (29 and 75% retention respectively). For other aroma compounds, no significant difference between K_{aw} or K_{as} was observed, meaning that no specific interaction with saliva exists for these aroma compounds in the tested conditions (thermodynamic equilibrium).

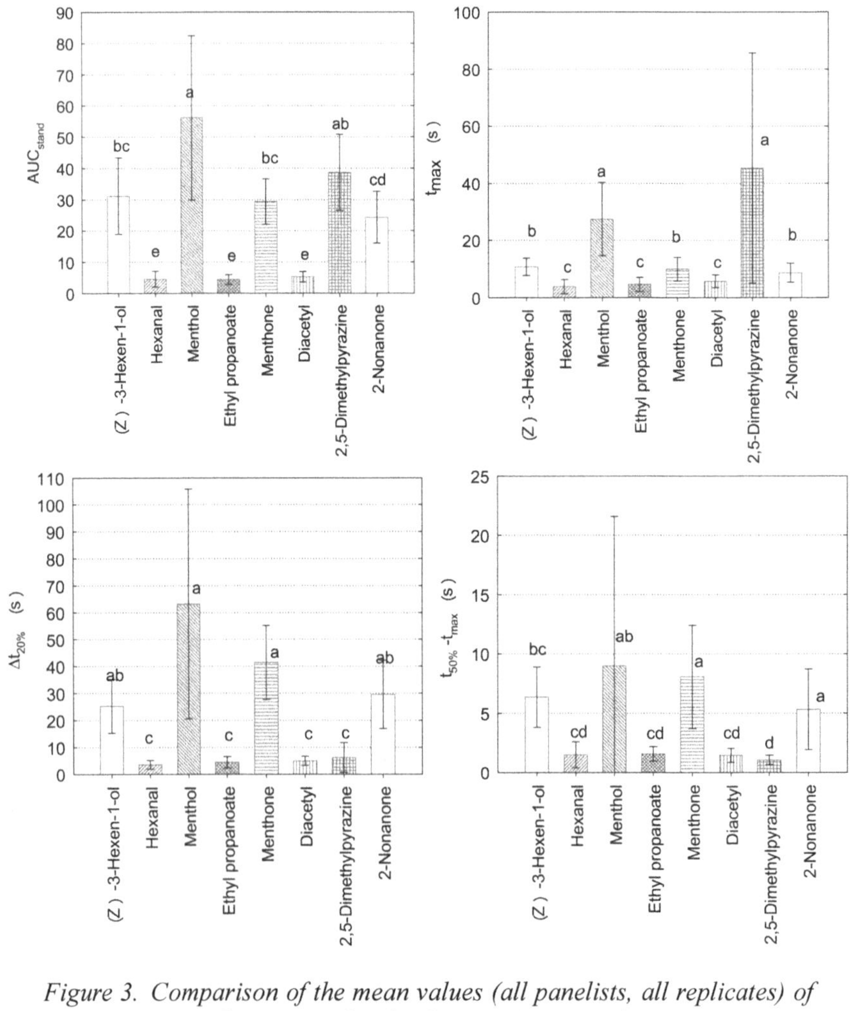

Figure 3. Comparison of the mean values (all panelists, all replicates) of release parameters between molecules for N.M.S. protocol. Letters a to e mean significant difference between ions (Kruskal and Wallis test and Conovam-Iman procedure, $p<0.05$).

To better understand the relationships between the air/saliva partition properties of molecules and release parameters, Spearman correlations between them were established ($p<0.05$) (Table 6). Even if positive correlations existed between K_{as} and I_{max} or AUC, no assumption was proposed as the amount of sample inhaled can be different between protocols.

For all protocols, the K_{as} at 37°C of aroma compounds were negatively correlated with $AUC_{stand.}$, t_{max} and $\Delta t_{20\%}$. These results highlighted that the more volatile compounds were released sooner and with lower persistence than less volatile ones.

Table 6. Correlation Coefficients between Release Parameters Extracted from *in vivo* Release Data and K_{as} Partition Coefficients (Spearman Test, p<0.05)

		I_{max}	AUC	$AUC_{stand.}$	t_{max}	$\Delta t_{20\%}$
$K_{a/s,\ 37°C}$	N.M.S.	0.63	0.58	-0.30	-0.41	-
	N.M.ns	0.65	0.61	-0.37	-0.35	-
	M.M.S.	0.57	0.43	-0.36	-0.34	-0.264
	N.N.S.	0.48	0.48	-0.27	- 0.23	-0.123

A Hierarchical Ascendant Classification applied on compound release parameters enabled to identify four groups of molecules with similar release behaviors: i) menthol, ii) menthone, *(Z)*-3-hexen-1-ol and 2,5-dimethylpyrazine, respectively, iii) 2-nonanone and iv) hexanal, diacetyl and ethyl propanoate respectively. All these data confirmed the role of the physicochemical properties of molecules on release behavior, even if it cannot explain all phenomena.

Assumptions on the Influence of Anatomy, Physiology, and/or Physicochemistry on Release Mechanisms

To propose some assumptions, the correlations between anatomical and physiological data and release parameters were studied (not shown). Negative correlations between cavity volumes (V_{mouth}, $V_{pharynx}$, V_{nose}) and $AUC_{stand.}$ (persistence) and between V_{mouth} and $t_{50\%}$-t_{max} were obtained, highlighting a probable dilution effect when cavity volume increased. But, as this effect was not observed for all molecules, other mechanisms, which compensate the dilution effect, probably occurred. Correlations between salivary and release parameters obtained with protocols involving mouth exposure (N.M.S., M.M.S. and N.M.ns) highlighted the probable role of saliva constituents: $AUC_{stand.}$ and $t_{50\%}$-t_{max} decreased with (i) an increase in lipolysis and/or proteolysis activities for ethyl propanoate, hexanal, menthone, *(Z)*-3-hexen-1-ol and 2,5-dimethylpyrazine, (ii) a decrease in amylase activity for menthone, *(Z)*-3-hexen-1-ol and 2,5-dimethylpyrazine and (iii) a decrease in salivary flow rate at rest for all molecules, except 2-nonanone.

Table 7. Summary of the Nature of Main Mechanisms That Could Be Involved and Description of Their Influence on Aroma Release Behavior in Relation with Physiological Characteristics of Panelists

	Physical mechanism		*Interaction mechanism*	
	Nature	Consequences on release	Nature	Consequences on release
V_{mouth} ↗	dilution ↗	I_{max} and AUC ↘	lubricated mucosa/air contact area ↗	I_{max} and AUC ↘
	initial release rate (to I_{max})↘	t_{max} ↘	-	-
	air renewal time in mouth ↗	$t_{50\%}$-t_{max} ↗ (persistence)	-	-
	pathway length to reach the nasal cavity↗	t_{max} ↗	-	-
$V_{pharynx}$ ↗	swallowed volume ↗	AUC ↘	lubricated mucosa/air contact area ↗	I_{max} and AUC ↘
V_{nose} ↗	air renewal time in nose ↘	$AUC_{stand.}$↗ (persistence)	-	-
salivary flow rate at rest ↗	-	-	[salivary compounds] ↘, dilution ↗	$AUC_{stand.}$ ↘ (persistence)
[salivary compounds] ↗	-	-	interaction↗, degradation ↗	$AUC_{stand.}$ ↗ (persistence)

Two kinds of mechanisms with different natures could be assumed to explain aroma release and persistence (Table 7).

First, the release behavior of aroma compounds can be due to physical and mass transfer mechanisms: the increase in cavity volumes can induce dilution effect, the stripping of a highest volume when swallowing (pharynx) and/or a slowest renewal of the air phase (nose), all implying an increase in persistence phenomena. These mechanisms are not dependent on molecule properties. For instance, these phenomena could explained the higher release intensity and renewal time and the lower t_{max} of ethyl propanoate and diacetyl in the case of panelists from class A than from class B (with higher mouth and nose volumes). Secondly, there are probably interactions between some aroma compounds and lubricated mucosa tissues cells and/or saliva. For instance, the lower persistence of *(Z)*-3-hexen-1-ol and 2,5-dimethylpyrazine for class B than for class C panelists, the latter having higher salivary concentrations, is an illustration of saliva retention. These interactions depend not only on molecule and mucosa properties, but also on cavity volumes (modification of contact area available for mass transfer): the lower persistence in mouth of *(Z)*-3-hexen-1-ol in the case of class A in comparison with class B probably results from a lower contact area between air and mucosa in the case of class A, due to smaller mouth volume.

The nature of these interactions remains still unclear even if several assumptions could be proposed on the basis of muco-adhesion theory (*54*) and on the possible enzymatic degradation of aroma compounds (*46*, *47*). Their consequences on release phenomena can be in opposition or not to the ones due to physical mechanisms, leading to final different behaviors that are observed upon molecules.

To summarize, it seems that the release behaviors of ethyl propanoate, hexanal and diacetyl were mainly influenced by stripping and dilution effects due to breath flow rate in the nasal cavity. In the mouth, phenomena are probably much more complex and include both physical (dilution, air renewal) and interaction (presence of saliva, contact area) mechanisms. These mechanisms seem to impact all molecules, except perhaps 2-nonanone and menthone. In the case of hexanal and ethyl propanoate, their specific behavior could eventually be explained by biochemical or enzymatic degradation (*46*, *47*). In the pharynx, it seems that only the release behaviors of 2-nonanone and menthone were affected, probably through both kinds of mechanisms. Nevertheless, the exact origins of release remain quite difficult to identify as it probably results from a combination of phenomena, each of them not always acting in the same way.

Conclusions

This study allowed the development of simple but original protocols to study aroma release and persistence. As already known, neither volatility nor solubility or hydrophobicity alone can explain the release behavior of aroma compounds. Phenomena are much more complex and have probably different natures. The role of cavity volumes was highlighted through both dilution and contact area effects. The roles of saliva and of swallowing were also illustrated, more specifically for two molecules, *i.e.* 2-nonanone and menthone. Further investigations on the precise role of saliva characteristics on the behavior of aroma compounds, in relation with their hydrophobicity should notably be considered. Thus, even if these results constitute a first basis, some experiments are still needed to progress in the understanding of aroma release, notably in relation with perception.

Acknowledgments

We gratefully acknowledge the panelists for their contribution to in vivo and sensory measurements. We also acknowledge Hélène Brignot for technical assistance for saliva analysis.

References

1. Bojanowski, V.; Hummel, T. Retronasal perception of odors. *Physiol. Behav.*, *107* (4), 484–4873012.
2. Ruijschop, R. M. A. J.; Boelrijk, A. E. M.; Burgering, M.; de Graaf, C.; Westerterp-Plantenga, M. S. Acute Effects of Complexity in Aroma Composition on Satiation and Food Intake. *Chem. Senses* **2010**, *35* (2), 91–100.
3. Fernandez, P.; Bensafi, M.; Rouby, C.; Giboreau, A. Does olfactory specific satiety take place in a natural setting? *Appetite* **2013**, *60*, 1–4.
4. Nakao, S.; Ishihara, S.; Nakauma, M.; Funami, T. Inhomogeneous Spatial Distribution of Aroma Compounds in Food Gels for Enhancement of Perceived Aroma Intensity and Muscle Activity during Oral Processing. *J. Texture Stud.* **2013**, *44* (4), 289–300.
5. Hummel, T.; Heilmann, S. Olfactory event-related potentials in response to ortho- and retronasal stimulation with odors related or unrelated to foods. *Int. Dairy J.* **2008**, *18* (8), 874–878.
6. Welge-Lüssen, A.; Husner, A.; Wolfensberger, M.; Hummel, T. Influence of simultaneous gustatory stimuli on orthonasal and retronasal olfaction. *Neurosci. Lett.* **2009**, *454*, 124–128.
7. Roudnitzky, N.; Bult, J. H. F.; De Wijk, R. A.; Reden, J.; Schuster, B.; Hummel, T. Investigation of interactions between texture and ortho- and retronasal olfactory stimuli using psychophysical and electrophysiological approaches. *Behav. Brain Res.* **2011**, *216* (1), 109–115.
8. Halpern, B. P. Retronasal and orthonasal smelling. *ChemoSense* **2004**, *6* (3), 1–7.
9. Heilman, S.; Hummel, T. A new method for comparing orthonasal and retronasal olfaction. *Behav. Neurosci.* **2004**, *118*, 412–419.
10. Sun, B. C.; Halpern, B. P. Identification of air phase retronasal and orthonasal odorant pairs. *Chem. Senses* **2005**, *30* (8), 693–706.
11. Linforth, R. S. T.; Taylor, A. J. Persistence of Volatile Compounds in the Breath after Their Consumption in Aqueous Solutions. *J. Agric. Food Chem.* **2000**, *48* (11), 5419–5423.

12. Ferreira, V.; Petka, J.; Cacho, J. Intensity and Persistence Profiles of Flavor Compounds in Synthetic Solutions. Simple Model for Explaining the Intensity and Persistence of Their Aftersmell. *J. Agric. Food Chem.* **2006**, *54* (2), 489–496.
13. Tromelin, A.; Merabtine, Y.; Andriot, I.; Lubbers, S.; Guichard, E. Retention-release equilibrium of aroma compounds in polysaccharide gels: Study by quantitative structure-activity/property relationships approach. *Flavor Fragrance J.* **2010**, *25* (6), 431.
14. Rodriguez, O.; Teixeira, M. A.; Rodrigues, A. E. Prediction of odour detection thresholds using partition coefficients. *Flavor Fragrance J.* **2011**, *26* (6), 421–428.
15. Kurtz, D. B.; Zhao, K.; Hornung, D. E.; Scherer, P. Experimental and numerical determination of odorant solubility in nasal and olfactory mucosa. *Chem. Senses* **2004**, *29* (9), 763–773.
16. Buffo, R. A.; Rapp, J. A.; Krick, T.; Reineccius, G. A. Persistence of aroma compounds in human breath after consuming an aqueous model aroma mixture. *Food Chem.* **2005**, *89* (1), 103–108.
17. Ruijschop, R. M. A. J.; Burgering, M.; Jacobs, M. A.; Boelrijk, A. E. M. Retro-Nasal Aroma Release Depends on Both Subject and Product Differences: A Link to Food Intake Regulation? *Chem. Senses* **2009**, *34* (5), 395–403.
18. Déléris, I.; Saint Eve, A.; Dakowski, F.; Sémon, E.; Le Quéré, J. L.; Souchon, I. The dynamics of aroma release during the consumption of candies with different structures. Relationship with temporal perception. *Food Chem.* **2011**, *127*, 1615–1624.
19. Frank, D. C.; Eyres, G. T.; Piyasiri, U.; Delahunty, C. M. Effect of food matrix structure and composition on aroma release during oral processing using in vivo monitoring. *Flavor Fragrance J.* **2012**, *27* (6), 433–444.
20. Buettner, A. Prolonged retronasal aroma perception - a phenomenon influenced by physiological factors and food matrix composition. In *Flavour Research at the Dawn of the Twenty-first Century, Proceedings of the 10th Weurman Flavour Research Symposium*; Le Quere, J. L., Etievant, J. X., Eds.; Lavoisier: London, 2003; pp 188−193.
21. Repoux, M.; Sémon, E.; Féron, G.; Guichard, E.; Labouré, H. Inter-individual variability in aroma release during sweet mint consumption. *Flavor Fragrance J.* **2012**, *27*, 40–46.
22. Doyennette, M.; Déléris, I.; Féron, G.; Guichard, E.; I., S.; Tréléa, I. C. Main individual and product characteristics influencing in-mouth flavour release during eating masticated food products with different textures: mechanistic modelling and experimental validation. *J. Theor. Biol.* **2014**, *340*, 209–221.
23. Benjamin, O.; Silcock, P.; Beauchamp, J.; Buettner, A.; Everett, D. W. Tongue Pressure and Oral Conditions Affect Volatile Release from Liquid Systems in a Model Mouth. *J. Agric. Food Chem.* **2012**, *60* (39), 9918–9927.
24. Keyhani, K.; Scherer, P. W.; Mozell, M. M. A numerical model of nasal odorant transport for the analysis of human olfaction. *J. Theor. Biol.* **1997**, *186* (3), 279–301.

25. Moharamzadeh, K.; Colley, H.; Murdoch, C.; Hearnden, V.; Chai, W. L.; Brook, I. M.; Thornhill, M. H.; MacNeil, S. Tissue-engineered Oral Mucosa. *J. Dent. Res.* **2012**, *91* (7), 642–650.
26. Diaz del Consuelo, I.; Falson, F.; Guy, R. H.; Jacques, Y. Ex vivo evaluation of bioadhesive films for buccal delivery of fentanyl. *J. Controlled Release* **2007**, *122* (2), 135–140.
27. Samson, G.; de la Calera, A. G.; Dupuis-Girod, S.; Faure, F.; Decullier, E.; Paintaud, G.; Vignault, C.; Scoazec, J. Y.; Pivot, C.; Plauchu, H.; Pirot, F. Ex vivo study of bevacizumab transport through porcine nasal mucosa. *Eur. J. Pharm. Biopharm.* **2012**, *80* (2), 465–469.
28. Annaert, P.; Brouwers, J.; Bijnens, A.; Lammert, F.; Tack, J.; Augustijns, P. Ex vivo permeability experiments in excised rat intestinal tissue and in vitro solubility measurements in aspirated human intestinal fluids support age-dependent oral drug absorption. *Eur. J. Pharm. Sci.* **2010**, *39*, 15–22.
29. Deschl, U.; Vogel, J.; Aufderheide, M. Development of an in vitro exposure model for investigating the biological effects of therapeutic aerosols on human cells from the respiratory tract. *Exp. Toxicol. Pathol.* **2011**, *63* (6), 593–598.
30. Kastner, P. E.; Le Calve, S.; Zheng, W. Y.; Casset, A.; Pons, F. A dynamic system for single and repeated exposure of airway epithelial cells to gaseous pollutants. *Toxicol. In Vitro* **2013**, *27* (2), 632–640.
31. Pariselli, F.; Sacco, M. G.; Rembges, D. An optimized method for in vitro exposure of human derived lung cells to volatile chemicals. *Exp. Toxicol. Pathol.* **2009**, *61* (1), 33–39.
32. Annaert, P.; Tukker, J. J.; van Gelder, J.; Naesens, L.; De Clercq, E.; van den Mooter, G.; Kinget, R.; Augustijns, P. In vitro, ex vivo and in situ intestinal absorption characteristics of the antiviral ester prodrug adefovir Dipivoxil. *J. Pharm. Sci.* **2000**, *89* (8), 1054–1062.
33. Kataoka, M.; Sugano, K.; Mathews, C. D.; Wong, J. W.; Jones, K. L.; Masaoka, Y.; Sakuma, S.; Yamashita, S. Application of Dissolution/Permeation System for Evaluation of Formulation Effect on Oral Absorption of Poorly Water-Soluble Drugs in Drug Development. *Pharm. Res.* **2012**, *29* (6), 1485–1494.
34. Howie, N. M.; Trigkas, T. K.; Cruchley, A. T.; Wertz, P. W.; Squier, C. A.; Williams, D. M. Short-term exposure to alcohol increases the permeability of human oral mucosa. *Oral Dis.* **2001**, *7* (6), 349–354.
35. Squier, C. A.; Mantz, M. J.; Wertz, P. W. Effect of menthol on the penetration of tobacco carcinogens and nicotine across porcine oral mucosa ex vivo. *Nicotine Tob. Res.* **2010**, *12* (7), 763–767.
36. Wise, P. M.; Preti, G.; Eades, J.; Wysocki, C. J. The effect of menthol vapor on nasal sensitivity to chemical irritation. *Nicotine Tob. Res.* **2011**, *13* (10), 989–997.
37. Aufderheide, M.; Knebel, J. W.; Ritter, D. An improved in vitro model for testing the pulmonary toxicity of complex mixtures such as cigarette smoke. *Exp. Toxicol. Pathol.* **2003**, *55* (1), 51–57.

38. Morales, J. O.; McConville, J. T. Manufacture and characterization of mucoadhesive buccal films. *Eur. J. Pharm. Biopharm.* **2011**, *77* (2), 187–199.
39. Shakya, P.; Madhav, N. V. S.; Shakya, A. K.; Singh, K. Palatal mucosa as a route for systemic drug delivery: a review. *J. Controlled Release* **2011**, *151* (1), 2–9.
40. Laffleur, F.; Bernkop-Schnurch, A. Strategies for improving mucosal drug delivery. *Nanomedicine* **2013**, *8* (12), 2061–2075.
41. Patel, V. F.; Liu, F.; Brown, M. B. Modeling the oral cavity: In vitro and in vivo evaluations of buccal drug delivery systems. *J. Controlled Release* **2012**, *161* (3), 746–756.
42. Linforth, R. S. T.; Martin, F.; Carey, M.; Davidson, J.; Taylor, A. J. Retronasal Transport of Aroma Compounds. *J. Agric. Food Chem.* **2002**, *50*, 1111–1117.
43. Hodgson, M.; Parker, A.; Linforth, R. S. T.; Taylor, A. J. In vivo studies on the long term persistence of volatiles in the breath. *Flavor Fragrance J.* **2004**, *19* (6), 470–475.
44. Normand, V.; Avison, S.; Parker, A. Modeling the Kinetics of Flavour Release during Drinking. *Chem. Senses* **2004**, *29* (3), 235–245.
45. Buettner, A.; Beauchamp, J. Chemical input – Sensory output: Diverse modes of physiology–flavour interaction. *Food Qual. Pref.* **2010**, *21* (8), 915–924.
46. Buettner, A. Influence of Human Saliva on Odorant Concentrations. 2. Aldehydes, Alcohols, 3-Alkyl-2-methoxypyrazines, Methoxyphenols, and 3-Hydroxy-4,5-dimethyl-2(5H)-furanone. *J. Agric. Food Chem.* **2002**, *50* (24), 7105–7110.
47. Buettner, A. Influence of Human Salivary Enzymes on Odorant Concentration Changes Occurring in Vivo. 1. Esters and Thiols. *J. Agric. Food Chem.* **2002**, *50* (11), 3283–3289.
48. Neyraud, E.; Palicki, O.; Schwartz, C.; Nicklaus, S.; Peron, G. Variability of human saliva composition: Possible relationships with fat perception and liking. *Arch. Oral Biol.* **2012**, *57* (5), 556–566.
49. Ettre, L. S.; Welter, C.; Kolb, B. Determination of gas-liquid partition coefficients by automatic equilibrium headspace-gas chromatography utilizing the phase ratio variation method. *Chromatographia* **1993**, *35,* (1/2), 73–84.
50. Atlan, S.; Tréléa, I. C.; Saint-Eve, A.; Souchon, I.; Latrille, E. Processing gas chromatographic data and confidence interval calculation for partition coefficients determined by the phase ratio variation method. *J. Chromatogr. A* **2006**, *1110* (1-2), 146–155.
51. Doyennette, M.; De Loubens, C.; Déléris, I.; Souchon, I.; Tréléa, I. C. Mechanisms explaining the role of viscosity and post-deglutitive pharyngeal residue on in vivo aroma release: A combined experimental and modeling study. *Food Chem.* **2011**, *128* (2), 380–390.
52. Repoux, M.; Laboure, H.; Courcoux, P.; Andriot, I.; Semon, E.; Yven, C.; Feron, G.; Guichard, E. Combined effect of cheese characteristics and food

oral processing on in vivo aroma release. *Flavor Fragrance J.* **2012**, *27* (6), 414–423.
53. Buettner, A.; Beer, A.; Hanning, C.; Settles, M. Observation of the Swallowing Process by Application of Videofluoroscopy and Real-time Magnetic Resonance Imaging—Consequences for Retronasal Aroma Stimulation. *Chem. Senses* **2001**, *26* (9), 1211–1219.
54. Smart, J. D. The basics and underlying mechanisms of mucoadhesion. *Adv. Drug Delivery Rev.* **2001**, *48* (2-3), 173–193.

Chapter 13

Determination of Thresholds for Capsaicin in Aqueous and Oil-Based Solutions

Désirée Schneider,[*,1] Ingrid Seuß-Baum,[1] and Elmar Schlich[2]

[1]Department of Food Technology, University of Applied Sciences Fulda, Marquardstr. 35, 36039 Fulda, Germany
[2]Department of Process Engineering in Food and Servicing Business, Justus Liebig University Giessen, Stephanstr. 24, 35390 Giessen, Germany
[*]E-mail: desiree.schneider@hs-fulda.de.

The pungency detection thresholds for capsaicin in an aqueous and an oil-based solution were determined and compared. Thresholds were measured for 21 students (12 chili users and 9 non-users) using a 3-Alternative Forced Choice sensory test with ascending concentrations of capsaicin in either an aqueous (0.0225 ppm, 0.045 ppm, 0.090 ppm, 0.180 ppm and 0.360 ppm) or an oil solution (0.150 ppm, 0.450 ppm, 1.350 ppm, 4.050 ppm). In addition, the panelists were asked to identify the site of irritation (throat, tongue or both). The Best Estimate Thresholds for capsaicin were determined to be 0.080 ppm in the aqueous solution and 0.826 ppm in the oil solution. Chili users generally showed lower thresholds compared to non-users. Differences between users and non-users were not significant, suggesting that no desensitization effects occur; however, differences were more pronounced in the oil-based solution. In most cases, the primary irritation was experienced in the throat.

Introduction

Chili peppers are very popular and are often used as spices throughout the world. Capsaicinoids are the pungent components in plants of the genus Capsicum, and capsaicin is the most important active sensory component and provides the oral sensation of heat (*1*). The primary sensory responses to this component include burning, tingling, and pain (*2*).

Capsaicin and dihydrocapsaicin, the two major capsaicinoids in chili peppers, are responsible for approximately 90 % of the total pungency (*3*).

The most common method for the sensory determination of pungency attributed to capsaicinoids was defined by Wilbur Scoville in 1912 (*4*). According to this method, an alcoholic extract of the dried pepper is produced. The alcoholic solution is then added drop by drop to sweetened water until a distinct but weak pungency is perceptible on the tongue. The degree of dilution represents a value on the Scoville scale and is then expressed in Scoville Heat Units (SHU). According to this method, pure capsaicin shows a Scoville rating of approximately 16.000.000 SHU. Additional examples of Scoville ratings are listed in Table 1:

Table 1. Examples for Capsicum Fruits and Preparations and Their Corresponding Scoville Rating (Modified from BfR (*5*))

Product	*Scoville rating (in SHU)*
Paprika powder (piquant)	100 -500
Tabasco sauce	1.600-5.000
Green Jalapeno chili, fresh	2.500-8.000
Chili powder	30.000-50.000

However, the accuracy of this test is often criticized, and modified versions have been developed.

Problems with Heat Determination in Foods

Although multiple methods to analyze the pungency of chilies exist, there are only a few studies (*6–10*) on the pungency thresholds for capsaicin, and these studies display inconsistent results.

Another problem is that the intensity and duration of the heat are dependent on the food system (*11–13*). Amongst other ingredients, the fat content varies in food. Capsaicin is lipophilic, which could be a potential reason for the difference in the perception of pungency in products with a high fat content (*12*). Therefore, a prediction of capsaicin's sensory impact based on an aqueous system may overestimate the perceived intensity in fat-based foods.

Furthermore, individual differences in the perception of the pungency of chili consumption (*14*) and burn localization were determined (*8*).

The objectives of this study were to determine and compare the absolute thresholds for capsaicin in aqueous versus oil-based solutions, to determine whether chili users and non-users differ with respect to the perception of capsaicin, and to analyze the localization of the first sensation triggered by capsaicin.

Materials and Methods

Capsaicin Threshold Determination

Panel

The capsaicin threshold study included 21 healthy subjects (4 males, 17 females) between the ages of 20 and 25. The subjects were all students of the Department of Food Technology at the University of Applied Sciences Fulda. The selection criteria included availability, basic experience in sensory methods and evaluation, and an interest in participation. Chocolate and book vouchers were offered as incentives.

The subjects filled out a simple questionnaire on the frequency of consumption of pungent food and preferred intensity. The subjects were subsequently designated as either chili pepper users or non-users. Those who consumed chili at least once a week and who scored 5 or higher on a scale from 0-10 for preferred intensity of pungency were classified as users, and those with an intensity score lower than 5 and a chili consumption of less than once a week were classified as non-users. In this study, 12 subjects were classified as users and 9 as non-users.

Stimuli/Sample Preparation

The irritant stimulus was capsaicin, the main pungent component of the chili pepper. For the threshold determination in water, the test samples consisted of capsaicin dissolved in water. Capsaicin is not soluble in water, and therefore, the food-grade emulsifier polysorbate 80 was used to dissolve the stimulus in water. Because polysorbate 80 has a slightly bitter taste, the polysorbate 80-water mixture served as a reference sample. For the threshold determination in oil, the test samples consisted of sunflower oil containing capsaicin. Pure sunflower oil served as a reference sample. The concentrations of capsaicin in the test samples are shown in Figure 1.

Experimental Procedure

The detection threshold values of capsaicin were determined through the use of the 3-Alternative Forced Choice (3-AFC) procedure according to ISO 13301:2002 (*15*). The subjects were asked to identify the odd sample in the triangle sets at each concentration.

The experimental design of the 3-AFC sensory test is illustrated in Figure 1.

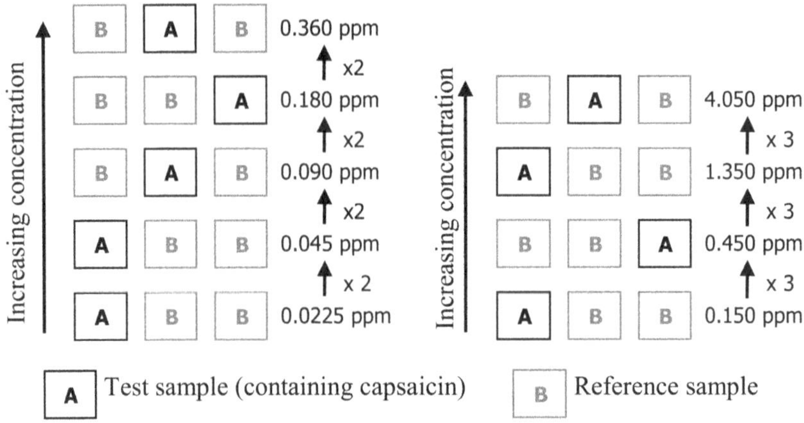

Figure 1. Experimental design of the 3-AFC sensory test: determination of thresholds of capsaicin in water (left) and in sunflower oil (right).

Each 5 ml sample was given a 3-digit code and presented to the panelists in a laboratory glass at room temperature. Five rows (in water) or four rows (in oil) of three samples in ascending concentrations, one test sample and two reference samples were presented to the panelists on a tray. To avoid positional bias, three combinations of orders of presentation (ABB, BAB, BBA) were balanced across the tray and among the panelists. The panelists were instructed to test each sample in the presented sequence, from left to right. The panelists were requested to put the whole sample into the mouth, swirl it around and then swallow it completely. The subjects were asked to identify the odd sample in the set of three at each concentration. If the panelists were uncertain, they were instructed to guess according to the forced choice principle.

The threshold concentrations of the sample solutions were 0.0225 ppm, 0.045 ppm, 0.090 ppm, 0.180 ppm and 0.360 ppm capsaicin; the concentrations of the series increased in 2-unit steps and were chosen based on pilot work. The concentrations for threshold testing in sunflower oil were 0.150 ppm, 0.450 ppm, 1.350 ppm and 4.050 ppm and increased in 3-unit steps.

The panelists also identified the localization of the burning sensation (tongue, throat or both) by placing a cross in the corresponding box at each concentration (Table 2). Furthermore, they were instructed not to give an answer if the exact localization of the burn was unknown. If they were uncertain and guessed which sample was odd, they were instructed to make a note.

The corresponding scoresheet of the 3-AFC test is shown in Table 2.

Table 2. Scoresheet 3-AFC Test (Threshold Determination of Capsaicin in Oil)

Task:	***Taste each sample from left to right. Start with row 1. Mark the odd sample in each row:***			***Localize the burn by placing a cross in the corresponding box:***
Row 4	950	547	063	O Throat O Tongue O Throat and tongue
	Note:			
Row 3	370	117	654	O Throat O Tongue O Throat and tongue
	Note:			
Row 2	630	241	852	O Throat O Tongue O Throat and tongue
	Note:			
Row 1	290	171	822	O Throat O Tongue O Throat and tongue
	Note:			

Data Analysis

The Best Estimate Threshold (BET) for each panelist was determined by calculating the geometric mean of the last missed concentration and the first concentration with a correct response followed by correct answers in the remaining sequence. The BET of the group (group threshold) was determined as the geometric mean of the individual BETs (*15*). An analysis of variance (ANOVA) was performed to test the significance of differences in thresholds of capsaicin between water and oil and between chili users and non-users.

Results

Threshold Comparison of Capsaicin in Water versus Oil

The individual thresholds of the panel in water and in oil are illustrated in Figure 2.

The x-axis represents the concentrations of the individual thresholds, and the y-axis represents the percentage of panelists with the corresponding individual thresholds.

Most of the panelists recognized the pungency at a concentration of 0.0636 ppm in water and 0.7794 ppm in oil. The group BETs for capsaicin were 0.080 ppm in aqueous and 0.826 ppm in oil solutions as shown in Figure 3. ANOVA showed that the threshold in water was significantly lower ($p < 0.01$) than in oil-based solutions.

The thresholds in oil were higher than those in water by a factor of approximately 10. The differences between users and non-users were not statistically significant in either water or oil. Users of pungent food, however, tended to show lower thresholds in water and oil than non-users. This observation was particularly notable in the oil-based solution, where users showed lower thresholds than non-users, suggesting that sensitization effects may play a role. However, because of the small number of panelists, there was no statistical significance.

Figure 4 illustrates the capsaicin concentrations plotted against the percentage of correct answers. The percentage of correct answers increased at the higher concentrations of capsaicin in water and oil. No clear differences between users and non-users can be observed in water (data not shown). However, compared to non-users, more chili users perceived the pungency at lower concentrations in the oil-based solution. For example, 50% of the chili users perceived the pungency at a concentration of approximately 0.5 ppm, but 50% of the non-users did not perceive the pungency until a concentration of approximately 1.3 ppm.

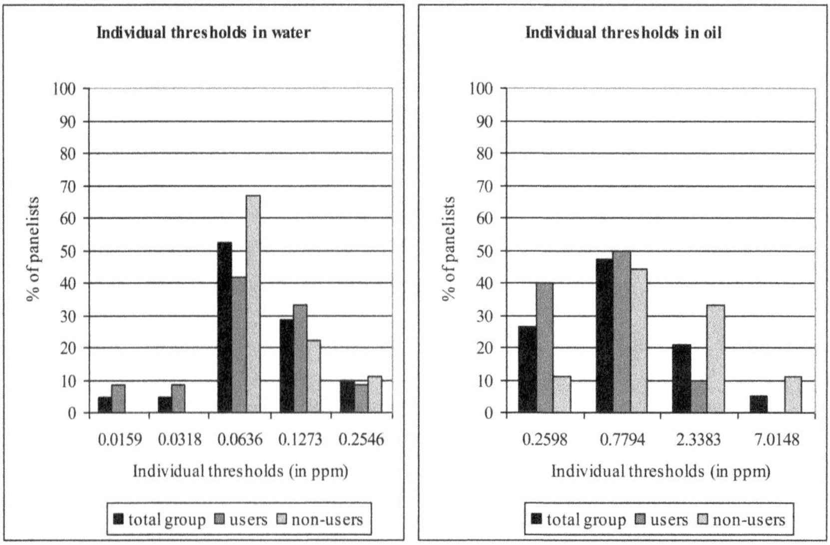

Figure 2. Individual thresholds for capsaicin in water (left) and in oil (right).

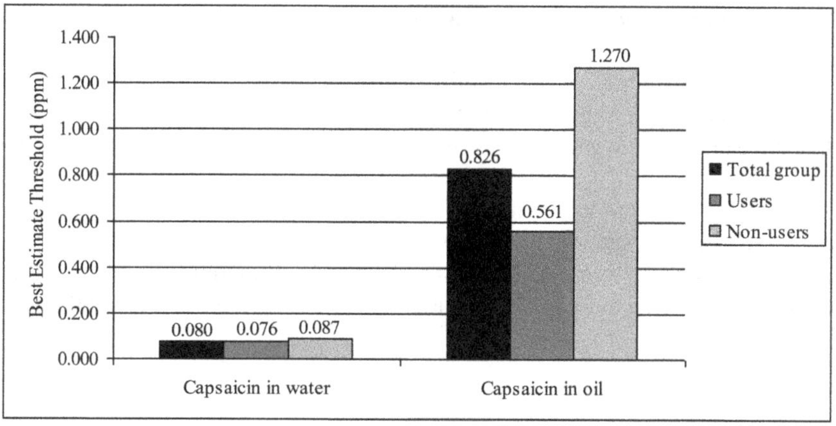

Figure 3. Comparison of Best Estimate Thresholds (group BET, geometric mean in ppm) for capsaicin in water and oil for users and non-users of chili peppers.

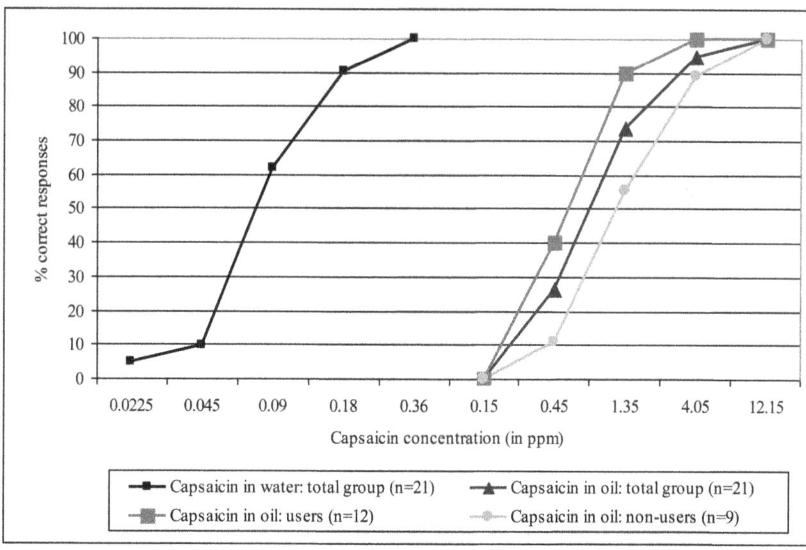

Figure 4. Capsaicin concentrations plotted against the percentage of correct answers (3AFC-test).

Determination of the Burn Localization

At each concentration, the panelists had to determine the localization of the first sensation triggered by capsaicin: either in the throat, on the tongue, or both. Only the panelists who gave correct answers in the triangles were accepted for the analysis.

The results for the localization of the burning sensation caused by capsaicin in water and oil are shown in Figure 5. Most of the panelists perceived the burn predominantly in the throat with both the water and oil solutions. However, other panelists perceived a burn on the tongue or both on the tongue and in the throat.

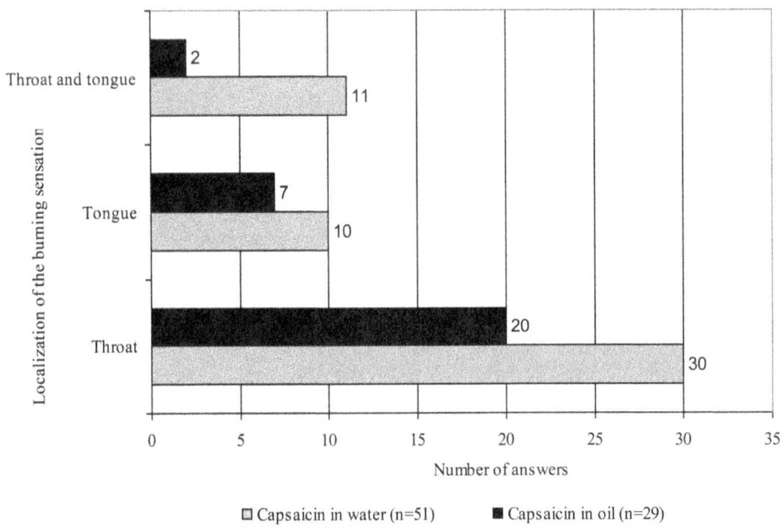

Figure 5. Localization of the burning sensation (number of answers).

Discussion

Thresholds in Oil and Water

As shown in Table 3, the results of the present study are consistent with the results of other groups (*6*) that reported lower thresholds for capsaicin in water compared to oil. Lawless and colleagues reported a threshold in water of approximately 0.31 ppm and in oil of approximately 11.75 ppm (*6*).

Table 3. Comparison of Thresholds for Capsaicin with Results from the Literature

Threshold	*Lawless and Hartono (2000) (6)*	*Present study*
In oil	11.75 ppm	0.83 ppm
In water	0.31 ppm	0.08 ppm

The clear difference between the thresholds in Table 3 was most likely caused by the use of different methods. In this study, the panelists were instructed to swallow the sample, but in the study from Lawless and Hartono (*6*), the panelists were presumably instructed to expectorate the sample after "swirling" the sample around in the mouth (*6*).

These results support previous findings (*8*), which indicated that the oral cavity is not uniformly sensitive to chemical irritants and that the throat plays an important role in the perception of chemesthetic stimuli during ingestion.

Other studies have shown an inverse relationship between response to heat intensity and fat content (*11*, *12*, *16*). Threshold differences in oil versus water can be explained by the idea that the lipophilic capsaicin dissolves better in an oily carrier than in water, and thus less capsaicin interacts with the trigeminal receptors in the oral cavity (*11*, *13*).

However, other groups could not confirm that a higher fat content leads to a reduction in the burning sensation evoked by capsaicin (*17*, *18*). One possible reason for these findings may be that the oral irritant is suppressed in an oily carrier rather than a pre-triggered stimulus because of the capsaicin-receptor interaction.

Differences between Users and Nonusers

Differences between users and non-users were not statistically significant ($p > 0.05$) in either water or oil. Users of pungent food, however, showed a lower threshold in water and oil compared to non-users. A trend could be observed, especially in oil, that users showed lower thresholds compared to non-users, suggesting that sensitization effects occur (*6*, *19*). In addition, Lawless and Hartono observed that higher chili consumption is associated with lower thresholds but that these differences were less pronounced in the water-based stimuli (*6*).

Orellana-Escobedo et al. (2012) (*9*) reported an absolute threshold concentration of 0.050 ppm for the total amount of capsaicinoids, defining the total capsaicinoid content as the sum of capsaicin and dihydrocapsaicin in the tested samples (*9*). Thus, Mexican participants, who regularly consume chili products, showed marginally lower thresholds compared to the current German participants.

This outcome begs the question: why is the difference between chili users and non-users, more pronounced in oil than in water? It is possible that the group size in the present study was too small to achieve statistical power, especially in the aqueous media group.

Conversely, other authors have reported that chili users show a lower response to irritants because of desensitization effects evoked by regular chili consumption, which led to lower intensity scores of pungency compared to non-users (*14*, *20*). It is possible that chili users perceive pungency at lower concentrations than non-users, but rate higher concentrations with a lower intensity.

Differences in Study Design

As illustrated in Table 4, the study design and choice of emulsifier used to dissolve capsaicin in water may have influenced the results.

Table 4. Comparison of Thresholds for Capsaicin in Water from Different Studies

Reference	*Threshold in water (ppm)*	*Emulsifier*	*Sugar added*	*Methods of testing*
(6)	0.310	ethanol or polysorbate 80	no	expectorate
(7)	0.039 - 0.078	ethanol	yes	swallow
(8)	0.098 (throat) 0.299 (tongue)	ethanol and polysorbate 80	yes	swallow
(9)	0.050	polysorbate 80	no	swallow
current study	0.080	polysorbate 80	no	swallow

Emulsifiers were used because capsaicin is not soluble in water. In some studies (*6–8*), ethanol was used as an emulsifier. However, ethanol is also an activator of the capsaicin receptor and potentiates the response of the receptor to capsaicin (*21*). Accordingly, polysorbate 80 was used as an emulsifier in the current study, although Lawless and Hartono (2000) could not find a clear effect in threshold determination when comparing the use of ethanol versus polysorbate 80 as a solvent (*6*).

Other authors, such as Krajewska and Powers (*7*) and Rentmeister-Bryant and Green (*8*), used water that contained sugar as a carrier. The latter group determined a higher threshold trend, indicating the masking effects of sugar. Despite these findings, other research groups (*7*) showed no clear evidence of higher thresholds when using sugar-based water solutions as carriers.

Furthermore, the threshold variation may be explained by a difference in study design. In one study (*6*), the panelists were presumably instructed to expectorate the sample after "swirling" the sample around in the mouth. In other studies (*7–9*), the panelists were instructed to swallow the sample after a few minutes.

Burn Localization

The results support previous findings that indicated that the oral cavity was not uniformly sensitive to chemical irritants, and further suggest that the throat, which is innervated by two nerves (the glossopharyngeal and vagus nerve), plays an important role in the perception of chemesthetic stimuli during ingestion (*8*). Rentmeister-Bryant and Green also demonstrated that the thresholds determined for the throat were lower than the thresholds on the front or back of the tongue in aqueous solutions, and reported threshold concentrations of 0.098 ppm for the throat and 0.299 ppm for the tongue (*8*). In this study, it was observed that at low concentrations near the threshold, pungency was predominantly perceived in the throat, and at higher concentrations, the pungency sensation increased on the tongue and in the oral cavity.

Conclusion

Lower thresholds for capsaicin were exhibited in water (0.080 ppm) compared to oil (0.826 ppm). Furthermore, participants with high chili consumption (users) exhibited lower thresholds compared to those with low chili consumption (non-users). The majority of both the users and non-users perceived the burn predominantly in the throat.

References

1. Contreras-Padilla, M.; Yahia, E. M. Changes in Capsaicinoids during Development, Maturation, and Senescence of Chile Peppers and Relation with Peroxidise Activity. *J. Agric. Food Chem.* **1998**, *46*, 2075–2079.
2. Cliff, M. A.; Heymann, H. Descriptive analysis of oral pungency. *J. Sens. Stud.* **1992**, *7*, 279–290.
3. Barbero, G. F.; Liazid, A.; Palma, M.; Barroso, C. G. Fast determination of Capsaicinoids from peppers by high-performance liquid chromatography using a reversed phase monolithic column. *Food Chem.* **2008**, *107*, 1276–1282.
4. Scoville, W. L. Note on Capsicum. *J. Am. Pharm. Assoc*. **1912**, *1*, 453–454.
5. Federal Institute for Risk Assessment (BfR). Too Hot Isn't Healthy - Foods with very high capsaicin concentrations can damage health, *BfR Opinion* No. 053/2011, October 18, 2011.
6. Lawless, H.; Hartono, C. Thresholds and suprathreshold intensity functions for capsaicin in oil and aqueous based carriers. *J. Sens. Stud.* **2000**, *15*, 437–447.
7. Krajewska, A. M.; Powers, J. J. Sensory properties of naturally occurring capsaicinoids. *J. Food Sci.* **1988**, *53*, 902–905.
8. Rentmeister-Bryant, H.; Green, B. Perceived irritation during ingestion of capsaicin or piperine: comparison of trigeminal and non-trigeminal areas. *Chem. Senses* **1997**, *22*, 257–266.
9. Orellana-Escobedo, L.; Ornelas-Paz, J. J.; Olivas, G. I.; Guerrero-Beltran, J. A.; Jimenez-Castro, J.; Sepulveda, D. R. Determination of absolute threshold

and just noticeable difference in the sensory perception of pungency. *J. Food Sci.* **2012**, *77*, 135–139.
10. Schneider, D. J.; Seuß-Baum, I.; Schlich, E. Comparison between chemical senses thresholds for capsaicin and dihydrocapsaicin in aqueous solutions and identification of the area of burning sensation. *Adv. J. Food Sci. Technol.* **2014**, *6*, 36–41.
11. Baron, R. F.; Penfield, M. P. Capsaicin heat intensity – concentration, carrier, fat level, and serving temperature effects. *J. Sens. Stud.* **1996**, *11*, 295–316.
12. Schneider, D. J.; Seuß-Baum, I.; Schlich, E. Relationship between pungency and food components – A comparison of chemical and sensory evaluations. *Food Qual. Pref.* **2014**, *38*, 98–106.
13. Kostyra, E.; Barylko-Pikielna, N.; Dabrowska, U. Relationship of pungency and leading flavour attributes in model food matrices – temporal aspects. *Food Qual. Pref.* **2010**, *21*, 197–206.
14. Prescott, J.; Stevenson, R. J. Effects of oral chemical irritation on tastes and flavors in frequent and infrequent users of chili. *Physiol. Behav.* **1995**, *58*, 1117–1127.
15. ISO 13301: Sensory analysis -- Methodology -- General guidance for measuring odour, flavour and taste detection thresholds by a three-alternative forced-choice (3-AFC) procedure; 2002.
16. Carden, L. A.; Penfield, M. P.; Saxton, A. M. Perception of heat in cheese sauces as affected by capsaicin concentration, fat level, fat mimetic and time. *J. Food Sci.* **1998**, *64*, 175–179.
17. Hutchinson, S. E.; Trantow, L. A.; Vickers, Z. M. The effectiveness of common foods for reduction of capsaicin burn. *J. Sens. Stud.* **1990**, *4*, 157–164.
18. Nasrawi, C. W.; Pangborn, R. M. Temporal effectiveness of mouth-rinsing on capsaicin mouth-burn. *Physiol. Behav.* **1990**, *47*, 617–623.
19. Ludy, M. J.; Mattes, R. D. Comparison of sensory, physiological, personality, and cultural attributes in regular spicy food users and non-users. *Appetite* **2012**, *58*, 19–27.
20. Reinbach, H. C.; Meinert, L.; Ballabio, D.; Aaslyng, M. D.; Bredie, W. L. P.; Olsen, K.; Moller, P. Interactions between oral burn, meat flavour and texture in chilli spiced pork patties evaluated by time-intensity. *Food Qual. Pref.* **2007**, *18*, 909–919.
21. Trevisani, M.; Smart, D.; Gunthorpe, M. J.; Tognetto, M.; Barbieri, M.; Campi, B.; Amadesi, S.; Gray, J.; Jerman, J. C.; Brough, S. J.; Owen, D.; Smith, G. D.; Randall, A. D.; Harrison, S.; Bianchi, A.; Davis, J. B.; Geppetti, P. Ethanol elicts and potentiates nociceptor responses via the vanilloid receptor-1. *Nat. Neurosci.* **2002**, *5*, 546–551.

Chapter 14

Do Salivary Proteins Play a Role in Tasting Bitter Substances?

Guy H. Carpenter*

Salivary Research, Flr 17, Tower wing, Kings College London Dental Institute, London, U.K.
***E-mail: guy.carpenter@kcl.ac.uk.**

The detection of bitterness occurs via well characterized receptors located in taste buds on the tongue surface. However, genetic variations in these receptors do not account for all the variations between subjects in their perception of bitter substances, which suggests there may be a role for salivary proteins or other factors in their detection. Several groups have shown associations of bitterness perception with carbonic anhydrase 6, proline-rich proteins and cystatins either by GWAS studies or proteomic studies, suggesting they potentially play a role. Some preliminary data is shown by the authors which provides further evidence that salivary proteins may play a role but the responsible protein is not determined, which may reflect the multifunctionality of salivary proteins.

Introduction

Generally saliva is considered to have a rather inert role in taste. In contrast to the anti-microbial properties of salivary proteins, in which nearly every salivary protein has some anti-microbial efficacy, very few proteins have been implicated in mediating taste (*1*). Probably the best known is gustin, later identified as carbonic anhydrase 6, for which there is some evidence by association. The reasons why so few proteins appear to have a role in taste are probably manifold, but a major reason could be the multifunctionality of salivary proteins. This term denotes that one protein can perform many roles but also that other proteins can substitute for another protein if absent. This makes it rather difficult to identify functions if certain subjects lack, or have low levels, of one particular protein, which commonly occurs through alternative splicing as salivary proteins are well known for being highly polymorphic (*2*). Amylase illustrates this rather well. It is the single most abundant protein in saliva and has well-known starch hydrolytic abilities (*3*) which relates to its function in the perception of sugar/starch. One study has related the gene-copy number of amylase to different populations of humans (agricultural vs arctic) based on the amount of starch in their diet (*4*) and furthermore to levels of obesity (*5*). However, these studies examine gene-copy number rather than actual protein levels (for which the correlation is not as strong), which suggests these are inherited traits rather than causal effects. There are many subjects who lack amylase and yet have normal taste detection thresholds (a commonly used indicator of taste). A more causal effect of salivary amylase has been implicated in the modification of the physical properties of food by hydrolyzing starch polymers in potato fries (chips) (*6*) and starch-thickened drinks (*7*) during in-mouth processing. However, most dietary starch is hydrolysed by pancreatic amylase in the gastro-intestinal tract. Thus perhaps it is more useful to consider the broader context of tasting foods in terms of their mastication and breakdown as well as the transport of tastants to taste receptors located at the taste buds. Although the receptors on taste cells within taste buds on the tongue are now well characterized not all the variation in subject to subject perception of bitterness is fully understood. Some preliminary data is presented in this chapter to suggest that salivary proteins facilitate the transport of bitter substances to the taste bud-located taste receptors.

Saliva Composition

Saliva is secreted by the major salivary glands including the parotid, submandibular and sublingual as well as the minor glands. The major glands are located remotely from the mouth and therefore require long ducts to deliver the saliva into the oral cavity. In contrast, the minor salivary glands are located within the oral mucosa and have short ducts to the surface. Nearly all salivary secretion is regulated by autonomic nerves (*8*), as is evidenced by the cessation of secretion when subjects are unconscious. Salivary secretion is described as having two rates, resting (or unstimulated) and a faster stimulated rate. The faster rate is stimulated by taste, chewing and olfactory stimuli being released from food in the mouth. The resting flow rate is dependent on nerve signals from the higher centres of the brain and thus follows a diurnal variation in flow rate and is lowest when we are asleep. Although Pavlov's work on dogs implied that humans can become conditioned to food cues to salivate, this is not unequivocally supported by other studies (*9*). Or at least thinking about food cannot induce a sustained secretion above resting flow rate, whereas taste, chewing and smell can. Since salivary secretion is a reflex most salivary secretion is not influenced by the type of stimuli that elicits it, which appears to be true for secretion from a single gland such as the parotid. However some studies have shown there are differences in certain proteins, such as secretory IgA due to differences in the flow rate (*10*). Larger differences in protein concentration between types of stimuli (chewing, taste and smell) have been detected when whole mouth saliva was analyzed, most likely due to differential activation of each salivary gland (*9*). More detailed analysis using proteomic methods has shown other differences in levels of salivary proteins following stimulation by different tastes (*11*).

Salivary Proteins

Many of the proteins in saliva are highly specialized proteins with unusual amino acid compositions, which has driven their initial naming. Hence the most abundant class proteins are the proline-rich proteins (PRPs). Although only encoded by six genes (*2*) there are over 30 different isoforms with varying size, overall charge and degree and type (N- or O- linked) of glycosylation. Many of the PRPs have O-linked glycans (*12*) and some also have N-linked glycans (*13*), whilst the majority are also sialylated. These glycans are important for their binding and agglutination of bacteria in the mouth (*14*). A cysteine-rich protein, named cystatin, forms another large group of proteins, whilst histidine-rich histatins are also present in large amounts, and both have anti-microbial activites (*15*). Another abundant group of proteins are the mucins. In saliva Mucin-5b and Mucin-7 are the two main species whilst Mucin 1 is present on epithelial cells that regularly slough (desqamate) into saliva. The emergence of proteomics has revealed that there are over a thousand proteins in saliva (*16*), although most are at lower amounts than those just mentioned. Other proteins which are of particular relevance to this chapter are Carbonic anhydrase 6, which reversibly catalyses the conversion of carbonic acid to water and carbon dioxide and is a major buffer of pH within saliva, Statherin, a surface active protein which maintains calcium homeostasis in the mouth, and Albumin as the blood-derived protein. Although most protein in saliva is the product of salivary gland synthesis, a small portion of salivary proteins is derived from the blood circulation. In addition to leakage via gingival crevicular fluid through the periodontium surrounding teeth, there is also a small leakage of albumin within the glands and into saliva (as shown by the presence of albumin in parotid saliva).

Salivary Peptides, Free Amino Acids, and Other Metabolites

Due to the high bacterial load in saliva, the breakdown of proteins is probably to be expected but it is not as random as might be anticipated and indeed shows some consistency-suggesting order (*17*). Furthermore some proteins are cleaved even before they are excreted into the oral cavity (*18*); that is, the protein precursor is cleaved already in the salivary gland cell, or at the cell membrane to yield two new proteins. This process seems particularly prevalent for PRPs but is also a feature of transport of IgA across the epithelial cell by the polymeric immunoglobulin receptor (*19*). In addition to the peptides in saliva there are also free amino acids, presumably either derived from the complete breakdown of proteins by proteases or released from serum (*20*). The scientific progress of metabolomics has also contributed to its impact on taste detection (*21*). A largely under-investigated area, though, is the effect of bacteria and other microflora on taste. At the two extremes are the effect of reduced bacteria following antibiotic use (*22*) and bacterial overgrowth (halitosis) where there are clear effects on taste. How bacterial biofilms growing on the fissured tongue affects taste is largely unknown. Saliva contains a large number and variety of bacteria, over half of which cannot be cultured (*23*), which makes studying them problematic.

The Structure of the Tongue

The taste buds are mostly located on the tongue in three main areas: the front, sides and back of the tongue. These are associated with visible bumps on the tongue known as the fungiform, foliate and the circumvallate papillae. In addition there are the filiform papillae – these are tough keratinized ridges on the tongue, particularly noticeable in domestic cats, that provide the roughness of the tongue but have no taste buds associated with them. The most obvious papillae are the red dotted fungiform papillae on the tip of the tongue. In animals each of these may have one or more taste buds (*24*) but this is quite variable between human subjects (*25*). The taste maps of the tongue often reproduced in textbooks are now largely discounted – there is abundant evidence to show most areas of the tongue are able to detect most tastes (*26*). By far the most abundant taste buds are located on the posterior part of the tongue around the foliate and circumvallate papillae. These taste buds are located within crypts on the tongue (see Figure 1) that are filled with saliva produced by serous minor glands also known as the von Ebner's glands. These glands secrete very small amounts of saliva but govern the environment around taste buds within the crypts. These glands are not well studied due to the difficulties of collecting the tiny amounts of saliva but do have some interesting taste-related molecules (*27*). Lingual lipase is one such product of the von Ebner's glands. Although it has been suggested that this enzyme may digest fat in the mouth and facilitate its detection, the very small amounts secreted from von Ebner's glands in the mouth makes it highly unlikely that it can digest fat from foods in time to detect their taste (*28*). Instead it probably functions to clear the tongue of any deposited fat layer, as happens following the ingestion of an oil (*29*).

Much progress has been made in characterizing the different channels responsible for the detection of the basic tastes by taste bud cells (*30*). Salt taste is transmitted by sodium and possibly potassium channels located on the apical surface of taste bud cells and signal to afferent nerves via ATP (adenosine triphosphate) molecules whereas sour taste (which are protons) is detected by a separate channel (*31*). Receptors for bitter tastes (*32*) and glutamate have also been determined (*33*). Now that specific receptors have been cloned more studies are examining the confounding factors of taste receptors, such as age (*34*). The detection of fat by taste cells is still debated. In rodent models a likely receptor was found (CD36), although its functional expression in humans has yet to be shown and emerging evidence points towards long chain fatty acids as being the key tastants – whereas CD36 detects the short chain fatty acids (*35*). Although free fatty acids may be the tastants within fatty foods, they do not seem to evoke a fatty taste but instead induce a bitter or rough (astringent) sensation (*35*). It is widely assumed that the preference for fatty foods relates to their improved mouthfeel as a result of increased lubrication (*36*).

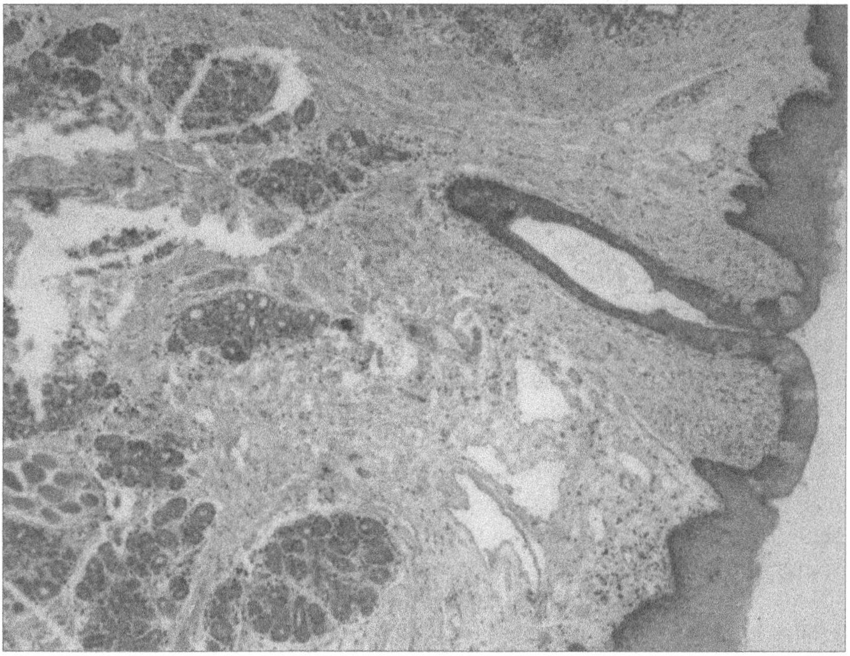

Figure 1. Haematoxylin & Eosin stained section of human tongue dorsum showing taste buds in the epithelium and von Ebner's salivary glands in the deeper dermis layers. Courtesy of Professor Peter Morgan, King's college London, U.K.

Perception of Bitterness and Its Relation to Salivary Proteins

Although most of the 25 or so bitterness receptors have been cloned the perception of bitterness still seems to be complex, implicating the role of other (unknown) factors. PROP (propylthiouracil) is a useful example which has previously been linked to the human tatse receptor TAS2R38 (*37*). More recent studies have used a GWAS (genome wide association study) approach. These studies are impressive for the immense size, with over 4000 participants in one such study which demonstrated that TAS2R38 variations did not explain all the perceived variation in PROP tasting (*38*). Other GWAS studies have contrasted PROP with caffeine, another well known bitter molecule, to show they act through different receptors (*39*) and may be associated with some salivary protein polymorphisms probably the same basic proline-rich proteins described previously (*40*). A commonly described candidate salivary protein is carbonic anhydrase 6, previously known as gustin (*41*). This was originally implicated in taste as it was deficient in some patients with altered taste (*42*). This enzyme mostly functions in the mouth to moderate or neutralize acids by catalyzing the addition of protons with bicarbonate ions to form carbon dioxide and water via carbonic acid (*43*). Although levels of carbonic anhydrase 6 are associated with supertasters of bitter compounds (*44*) and it has been suggested that carbonic anhydrase 6 acts as a trophic factor for taste buds on fungiform papillae (*45*), it is unlikely these relationships are causal. The best evidence comes from the carbonic anhydrase 6 knockout mouse (*46*), which showed no changes in taste bud number on fungiform papillae and only a very slight decrease in liking of bitter substances. However, even this study is not definitive as the multifunctionality of salivary proteins, mentioned earlier, may overcome for a loss of carbonic anhydrase 6 by some other salivary proteins.

The association of carbonic anhydrase 6 with taste sensitivities may stem from its variable upregulation by neural signals. Certainly it is well documented that most salivary protein expression and secretion are tightly regulated by neural signals (*8*). Principally these signals come from sensory nerves in the mouth via the brain stem, as described earlier. It might even be possible that salivary gland cells have bitter receptors and upregulate salivary proteins in response to serum levels of caffeine as shown *in vivo* in rats (*47*) and in human immortalized salivary gland cells (*48*). However, the *in vitro* study using immortalized salivary gland cells upregulated cystatin in response to caffeine (*48*). Work from the same groups have shown changes in bitter perception linked to amylase and albumin (*49*), zinc-alpha2 glycoprotein (*50*) and calgranulin (*11*) in addition to carbonic anhydrase 6. Certainly there is a clear role for carbonic anhydrase in the detection of carbon dioxide in fizzy drinks although this identifies four (membrane bound) rather than six (secreted form) as the main candidate (*51*). Whether a single protein is involved seems unlikely as other proteins such as albumin (*52*) or even amino acids such as arginine (*53*) also interact with bitter substances and are both present in saliva. Hence it seems likely that changes in the salivary protein profile can influence tasting of bitterness, but differentiating which protein is important will be difficult. It is most likely that the heterogeneous nature of bitter tastants (*54*) has resulted in a large selection of taste receptors and thus it is possible that several salivary proteins may exert an influence in transporting the bitter substance to the receptors.

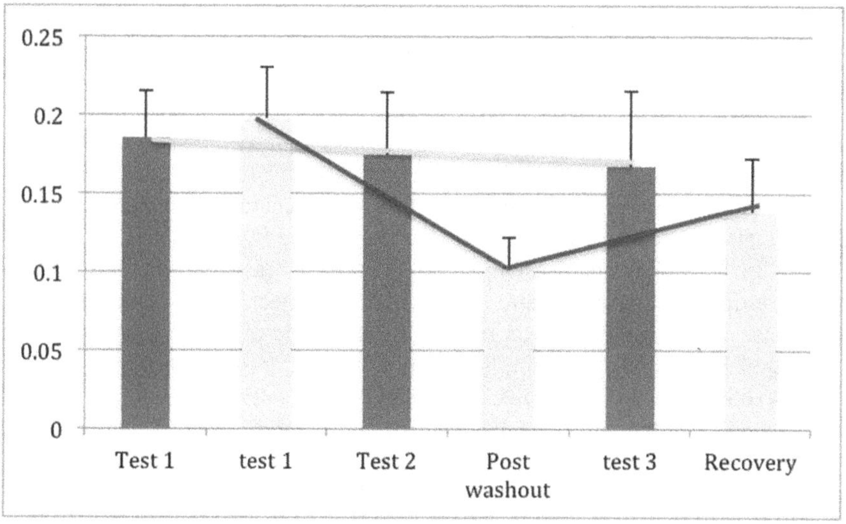

Figure 2. Mean ± SD (n=3, single subject) parotid salivary flows (ml/min) in response to the same solution of caffeine (50 mM in water) given repeatedly with 5 min rests between each 1 min period of tastant (dark green bars with yellow trend line). In a second series (light green bars) the mouth was washed out with water before test 2 (Post washout). The third test (Recovery) was the same as the first series. The post washout flow rate was significantly lower than without a water washout ($p < 0.05$, student's t-test) suggesting that the water washout has reduced the tasting of the caffeine solution.

Preliminary Evidence for the Role of Salivary Proteins

Previously we used a simple method to show that salivary proteins attached to the mucosa (termed the mucosal pellicle) were also involved in polyphenol-induced astringency (*55*). The method used was to wash the mouth out with copious quantities of water which removed most of saliva which forms a thin film on all surfaces of the mouth (*56*). Instead of reducing astringency this procedure increased the apparent astringency of a solution of black tea. Previously the theory of how astringency occurred was via the interaction and subsequent precipitation of salivary proline-rich proteins and histatins by the polyphenols, destroying the lubricating layer of the mouth either by disrupting PRPs attached to the mucosa or by creating nano-particulates (*57*). These particulates were thought to act like sand creating roughness by their presence. However, several pieces of evidence point to another mechanism. Firstly, in a detailed study we could not find any proline-rich proteins or histatins attached to oral epithelial cells (*58*). In fact the mucosal pellicle is composed mostly of salivary mucins (Mucin- 5B and Mucin-7), Secretory IgA and carbonic anhydrase 6. The second piece of evidence is that polyphenols also interact with mucins (*59*). Our ideas of how astringency occurs have been modified to include the mucosal pellicle (*60*) by the use of a simple washout experiment. Similar methods have been used in rats to show the role of saliva in taste (*61*). We therefore used the same technique (water washout) to determine if salivary proteins mediate the perception of bitterness and obtained some consistently variable results. This sounds like an oxymoron but in one subject a water washout between tests of caffeine consistently caused a significant ($p<0.05$) decrease in the parotid salivary flow (see Figure 2). We used parotid salivary flow as it is easier to measure more accurately and controlled for the adaptation to tastants, as shown by the yellow line in Figure 2. This subject repeatedly showed this decrease in all nine repeats of the same test. In contrast, when we did the same procedure on seven other subjects, five subjects showed a decrease, two subjects showed an increase, but none of these changes in flow rate were statistically significant, hence our consistent (within an individual) but variable (between subjects) results. Further analysis of salivary proteins from these subjects indicated that the first subject which consistently showed a decrease with washout had the lowest total salivary protein (Bicinchoninic acid assay), including Carbonic anhydrase 6 (specific immunoblot). We infer from this data that salivary proteins probably do influence or aid the detection of bitter substances but that the effect is difficult to detect due to the multifunctionality of salivary proteins and the role of absorbed salivary proteins, such as Carbonic anhydrase 6, in the mucosal pellicle.

Conclusions

There appears to be substantial evidence, from a variety of sources and techniques, to suggest that salivary proteins do mediate the detection of bitter substances. The most likely candidate is carbonic anhydrase 6 although functional inhibitor studies, by blocking the enzymatic activity with acetazolamide, seem to be lacking. If a causal link is found between salivary proteins and bitter tastant perception the prospects for new bitter-masking strategies seems a fruitful area of research.

Acknowledgments

The preliminary data shown in this chapter was generated by Richard Fitzgerald as part of his intercalated BSc project under the supervision of the author.

References

1. Nieuw Amerongen, A. V.; Veerman, E. C. I. Saliva - the defender of the oral cavity. *Oral Dis.* **2002**, *8* (1), 12–22.
2. Azen, E. A. Genetics of Salivary Protein Polymorphisms. *Crit. Rev. Oral Biol. Med.* **1993**, *4* (3-4), 479–485.
3. Butterworth, P. J.; Warren, F. J.; Ellis, P. R. Human alpha-amylase and starch digestion: An interesting marriage. *Starch/Staerke* **2011**, *63* (7), 395–405.

4. Perry, G. H.; Dominy, N. J.; Claw, K. G.; Lee, A. S.; Fiegler, H.; Redon, R.; Werner, J.; Villanea, F. A.; Mountain, J. L.; Misra, R.; Carter, N. P.; Lee, C.; Stone, A. C. Diet and the evolution of human amylase gene copy number variation. *Nat. Genet.* **2007**, *39* (10), 1256–1260.
5. Falchi, M.; El-Sayed Moustafa, J. S.; Takousis, P.; Pesce, F.; Bonnefond, A.; Andersson-Assarsson, J. C.; Sudmant, P. H.; Dorajoo, R.; Al-Shafai, M. N.; Bottolo, L.; Ozdemir, E.; So, H. C.; Davies, R. W.; Patrice, A.; Dent, R.; Mangino, M.; Hysi, P. G.; Dechaume, A.; Huyvaert, M.; Skinner, J.; Pigeyre, M.; Caiazzo, R.; Raverdy, V.; Vaillant, E.; Field, S.; Balkau, B.; Marre, M.; Visvikis-Siest, S.; Weill, J.; Poulain-Godefroy, O.; Jacobson, P.; Sjostrom, L.; Hammond, C. J.; Deloukas, P.; Sham, P. C.; McPherson, R.; Lee, J.; Tai, E. S.; Sladek, R.; Carlsson, L. M. S.; Walley, A.; Eichler, E. E.; Pattou, F.; Spector, T. D.; Froguel, P. Low copy number of the salivary amylase gene predisposes to obesity. *Nature Genetics* **2014**, *46* (5), 492–497.
6. Boehm, M. W.; Warren, F. J.; Moore, J. E.; Baier, S. K.; Gidley, M. J.; Stokes, J. R. Influence of hydration and starch digestion on the transient rheology of an aqueous suspension of comminuted potato snack food. *Food Funct.* **2014**, *5* (11), 2775–2782.
7. Hanson, B.; Cox, B.; Kaliviotis, E.; Smith, C. H. Effects of Saliva on Starch-thickened Drinks with Acidic and Neutral pH. *Dysphagia* **2012**, *27* (3), 427–435.
8. Proctor, G. B.; Carpenter, G. H. Regulation of salivary gland function by autonomic nerves. *Autonomic Neurosci. Basic Clin* **2007**, *133* (1), 3–18.
9. Ilangakoon, Y.; Carpenter, G. H. Is the mouthwatering sensation a true salivary reflex? *J. Texture Stud.* **2011**, *42* (3), 212–216.
10. Proctor, G. B.; Carpenter, G. H. Chewing stimulates secretion of human salivary secretory immunoglobulin A. *J. Dent. Res.* **2001**, *80* (3), 909–913.
11. Neyraud, E.; Sayd, T.; Morzel, M.; Dransfield, E. Proteomic analysis of human whole and parotid salivas following stimulation by different tastes. *J. Proteome Res.* **2006**, *5* (9), 2474–2480.
12. Carpenter, G. H.; Proctor, G. B. O-Linked glycosylation occurs on basic parotid salivary proline-rich proteins. *Oral Microbiol. Immunol.* **1999**, *14* (5), 309–315.
13. Ramachandran, P.; Boontheung, P.; Xie, Y. M.; Sondej, M.; Wong, D. T.; Loo, J. A. Identification of N-linked glycoproteins in human saliva by glycoprotein capture and mass spectrometry. *J. Proteome Res.* **2006**, *5* (6), 1493–1503.
14. Levine, M. J.; Herzberg, M. C.; Levine, M. S.; Ellison, S. A.; Stinson, M. W.; Li, H. C.; Vandyke, T. Specificity of salivary-bacterial interactions - role of terminal sialic-acid residues in interaction of salivary glycoproteins with streptococcus-sanguis and streptococcus-mutans. *Infect. Immun.* **1978**, *19* (1), 107–115.
15. De Smet, K.; Contreras, R. Human antimicrobial peptides: defensins, cathelicidins and histatins. *Biotechnol. Lett.* **2005**, *27* (18), 1337–1347.
16. Denny, P.; Hagen, F. K.; Hardt, M.; Liao, L.; Yan, W.; Arellanno, M.; Bassilian, S.; Bedi, G. S.; Boontheung, P.; Cociorva, D.; Delahunty, C. M.; Denny, T.; Dunsmore, J.; Faull, K. F.; Gilligan, J.; Gonzalez-Begne, M.;

Halgand, F.; Hall, S. C.; Han, X.; Henson, B.; Hewel, J.; Hu, S.; Jeffrey, S.; Jiang, J.; Loo, J. A.; Loo, R. R. O.; Malamud, D.; Melvin, J. E.; Miroshnychenko, O.; Navazesh, M.; Niles, R.; Park, S. K.; Prakobphol, A.; Ramachandran, P.; Richert, M.; Robinson, S.; Sondej, M.; Souda, P.; Sullivan, M. A.; Takashima, J.; Than, S.; Wang, J.; Whitelegge, J. P.; Witkowska, H. E.; Wolinsky, L.; Xie, Y.; Xu, T.; Yu, W.; Ytterberg, J.; Wong, D. T.; Yates, J. R., III; Fisher, S. J. The proteomes of human parotid and submandibular/sublingual gland salivas collected as the ductal secretions. *J. Proteome Res.* **2008**, *7* (5), 1994–2006.

17. Vitorino, R.; Barros, A.; Caseiro, A.; Domingues, P.; Duarte, J.; Amado, F. Towards defining the whole salivary peptidome. *Proteomics: Clin. Appl.* **2009**, *3* (5), 528–540.
18. Chan, M.; Bennick, A. Proteolytic processing of a human salivary proline-rich protein precursor by proprotein convertases. *Eur. J. Biochem.* **2001**, *268* (12), 3423–3431.
19. Song, W. X.; Apodaca, G.; Mostov, K. Transcytosis of the polymeric immunoglobulin receptor is regulated in multiple intracellular compartments. *J. Biol. Chem.* **1994**, *269* (47), 29474–29480.
20. Barnes, V. M.; Kennedy, A. D.; Panagakos, F.; Devizio, W.; Trivedi, H. M.; Jonsson, T.; Guo, L. N.; Cervi, S.; Scannapieco, F. A. Global Metabolomic Analysis of Human Saliva and Plasma from Healthy and Diabetic Subjects, with and without Periodontal Disease. *PLoS One* **2014**, *9* (8), 8.
21. Mounayar, R.; Morzel, M.; Brignot, H.; Tremblay-Franco, M.; Canlet, C.; Lucchi, G.; Ducoroy, P.; Feron, G.; Neyraud, E. Nutri-metabolomics Applied to Taste Perception Phenotype: Human Subjects with High and Low Sensitivity to Taste of Fat Differ in Salivary Response to Oleic Acid. *OMICS* **2014**, *18* (11), 666–672.
22. Logan, R. P. H.; Gummett, P. A.; Schaufelberger, H. D.; Greaves, R.; Mendelson, G. M.; Walker, M. M.; Thomas, P. H.; Baron, J. H.; Misiewicz, J. J. Eradication of helicobacter-pylori with clarithromycin and omeprazole. *Gut* **1994**, *35* (3), 323–326.
23. Wade, W. G. The oral microbiome in health and disease. *Pharmacol. Res.* **2013**, *69* (1), 137–143.
24. Ichikawa, H.; Terayama, R.; Yamaai, T.; De Repentigny, Y.; Kothary, R.; Sugimoto, T. Dystonin deficiency reduces taste buds and fungiform papillae in the anterior part of the tongue. *Brain Re.* **2007**, *1129* (1), 142–146.
25. Arvidson, K. Location and variation in number of taste-buds in human fungiform papillae. *Scand. J. Dent. Res.* **1979**, *87* (6), 435–442.
26. Arvidson, K.; Friberg, U. Human taste - response and taste bud number in fungiform papillae. *Science* **1980**, *209* (4458), 807–808.
27. Eliasson, L.; Carlen, A. An update on minor salivary gland secretions. *Eur. J. Oral Sci.* **2010**, *118* (5), 435–442.
28. Voigt, N.; Stein, J.; Galindo, M. M.; Dunkel, A.; Raguse, J. D.; Meyerhof, W.; Hofmann, T.; Behrens, M. The role of lipolysis in human orosensory fat perception. *J. Lipid Res.* **2014**, *55* (5), 870–882.

29. Adams, S.; Singleton, S.; Juskaitis, R.; Wilson, T. In-vivo visualisation of mouth-material interactions by video rate endoscopy. *Food Hydrocolloids* **2007**, *21* (5-6), 986–995.
30. Chandrashekar, J.; Hoon, M. A.; Ryba, N. J. P.; Zuker, C. S. The receptors and cells for mammalian taste. *Nature* **2006**, *444* (7117), 288–294.
31. Huang, A. L.; Chen, X. K.; Hoon, M. A.; Chandrashekar, J.; Guo, W.; Trankner, D.; Ryba, N. J. P.; Zuker, C. S. The cells and logic for mammalian sour taste detection. *Nature* **2006**, *442* (7105), 934–938.
32. Behrens, M.; Meyerhof, W. Bitter taste receptor research comes of age: From characterization to modulation of TAS2Rs. *Semin. Cell Dev. Biol.* **2013**, *24* (3), 215–221.
33. Iwatsuki, K.; Ichikawa, R.; Uematsu, A.; Kitamura, A.; Uneyama, H.; Torii, K. Detecting sweet and umami tastes in the gastrointestinal tract. *Acta Physiol.* **2012**, *204* (2), 169–177.
34. Mennella, J. A.; Pepino, M. Y.; Duke, F. F.; Reed, D. R. Age modifies the genotype-phenotype relationship for the bitter receptor TAS2R38. *BMC Genet* **2010**, *11*, 60.
35. Running, C. A.; Mattes, R. D. Different oral sensitivities to and sensations of short-, medium-, and long-chain fatty acids in humans. *Am. J. Physiol.* **2014**, *307* (3), G381–G389.
36. Stokes, J. R.; Boehm, M. W.; Baier, S. K. Oral processing, texture and mouthfeel: From rheology to tribology and beyond. *Curr. Opin. Colloid Interface Sci.* **2013**, *18* (4), 349–359.
37. Bartoshuk, L. M.; Duffy, V. B.; Miller, I. J. PTC/PROP tasting - anatomy, psychophysics, and sex effects. *Physiol. Behav.* **1994**, *56* (6), 1165–1171.
38. Hayes, J. E.; Bartoshuk, L. M.; Kidd, J. R.; Duffy, V. B. Supertasting and PROP bitterness depends on more than the TAS2R38 gene. *Chem. Senses* **2008**, *33* (3), 255–265.
39. Reed, D. R.; Zhu, G.; Breslin, P. A. S.; Duke, F. F.; Henders, A. K.; Campbell, M. J.; Montgomery, G. W.; Medland, S. E.; Martin, N. G.; Wright, M. J. The perception of quinine taste intensity is associated with common genetic variants in a bitter receptor cluster on chromosome 12. *Human Mol. Genet.* **2010**, *19* (21), 4278–4285.
40. Cabras, T.; Melis, M.; Castagnola, M.; Padiglia, A.; Tepper, B. J.; Messana, I.; Barbarossa, I. T. Responsiveness to 6-n-Propylthiouracil (PROP) Is Associated with Salivary Levels of Two Specific Basic Proline-Rich Proteins in Humans. *Plos One* **2012**, *7* (2).
41. Thatcher, B. J.; Doherty, A. E.; Orvisky, E.; Martin, B. M.; Henkin, R. I. Gustin from human parotid saliva is carbonic anhydrase VI. *Biochem. Biophys. Res. Commun.* **1998**, *250* (3), 635–641.
42. Henkin, R. I.; Martin, B. M.; Agarwal, R. P. Decreased parotid saliva gustin/carbonic anhydrase VI secretion: An enzyme disorder manifested by gustatory and olfactory dysfunction. *Am. J. Med. Sci.* **1999**, *318* (6), 380–391.
43. Pedersen, A. M.; Bardow, A.; Jensen, S. B.; Nauntofte, B. Saliva and gastrointestinal functions of taste, mastication, swallowing and digestion. *Oral Dis.* **2002**, *8* (3), 117–129.

44. Calo, C.; Padiglia, A.; Zonza, A.; Corrias, L.; Contu, P.; Tepper, B. J.; Barbarossa, I. T. Polymorphisrns in TAS2R38 and the taste bud trophic factor, gustin gene co-operate in modulating PROP taste phenotype. *Physiol. Behav.* **2011**, *104* (5), 1065–1071.
45. Melis, M.; Atzori, E.; Cabras, S.; Zonza, A.; Calo, C.; Muroni, P.; Nieddu, M.; Padiglia, A.; Sogos, V.; Tepper, B. J.; Barbarossa, I. T. The Gustin (CA6) Gene Polymorphism, rs2274333 (A/G), as a Mechanistic Link between PROP Tasting and Fungiform Taste Papilla Density and Maintenance. *PLoS One* **2013**, *8* (9).
46. Patrikainen, M.; Pan, P.; Kulesskaya, N.; Voikar, V.; Parkkila, S. The role of carbonic anhydrase VI in bitter taste perception: evidence from the Car6(-/-) mouse model. *J. Biomed. Sci.* **2014**, *21*.
47. Turner, S.; Manuele, M. G.; Davicino, R.; Ferraro, G.; Filip, R.; Anesini, C. Effect of caffeine on the secretion of peroxidase in rat submandibular gland: A study of its mechanism of action. *Arch. Oral Biol.* **2009**, *54* (2), 179–184.
48. Dsamou, M.; Morzel, M.; Le Corre, L.; Severin, I.; Chagnon, M.-C. Caffeine increases the expression of cystatin SN in human submandibular acinar-like HSG cells. *Arch. Oral Biol.* **2013**, *58* (10), 1511–1516.
49. Dsamou, M.; Palicki, O.; Septier, C.; Chabanet, C.; Lucchi, G.; Ducoroy, P.; Chagnon, M.-C.; Morzel, M. Salivary Protein Profiles and Sensitivity to the Bitter Taste of Caffeine. *Chem. Senses* **2012**, *37* (1), 87–95.
50. Morzel, M.; Chabanet, C.; Schwartz, C.; Lucchi, G.; Ducoroy, P.; Nicklaus, S. Salivary protein profiles are linked to bitter taste acceptance in infants. *Eur. J. Pediatr.* **2014**, *173* (5), 575–582.
51. Chandrashekar, J.; Yarmolinsky, D.; von Buchholtz, L.; Oka, Y.; Sly, W.; Ryba, N. J. P.; Zuker, C. S. The Taste of Carbonation. *Science* **2009**, *326* (5951), 443–445.
52. Wang, W. N.; Zhang, W.; Duan, Y. K.; Jiang, Y.; Zhang, L. R.; Zhao, B.; Tu, P. F. Investigation of the binding sites and orientation of caffeine on human serum albumin by surface-enhanced Raman scattering and molecular docking. *Spectrochim. Acta, Part A* **2013**, *115*, 57–63.
53. Melis, M.; Aragoni, M. C.; Arca, M.; Cabras, T.; Caltagirone, C.; Castagnola, M.; Crnjar, R.; Messana, I.; Tepper, B. J.; Barbarossa, I. T. Marked Increase in PROP Taste Responsiveness Following Oral Supplementation with Selected Salivary Proteins or Their Related Free Amino Acids. *PLoS One* **2013**, *8* (3), 8.
54. Wiener, A.; Shudler, M.; Levit, A.; Niv, M. Y. BitterDB: a database of bitter compounds. *Nucl. Acids Res.* **2012**, *40* (D1), D413–D419.
55. Nayak, A.; Carpenter, G. H. A physiological model of astringency. *Physiol. Behav.* **2008**, *95* (3), 290–294.
56. Pramanik, R.; Osailan, S. M.; Challacombe, S. J.; Urquhart, D.; Proctor, G. B. Protein and mucin retention on oral mucosal surfaces in dry mouth patients. *Eur. J. Oral Sci.* **2010**, *118* (3), 245–253.
57. Jobstl, E.; O'Connell, J.; Fairclough, J. P. A.; Williamson, M. P. Molecular model for astringency produced by polyphenol/protein interactions. *Biomacromolecules* **2004**, *5* (3), 942–949.

58. Gibbins, H. L.; P., G.; Yakubov, G. E.; Wilson, S.; Carpenter, G. H. Concentration of salivary protective proteins in the bound oral mucosal pellicle. *Oral Dis.* **2013** DOI:10.1111/odi.12194.
59. Davies, H. S.; Pudney, P. D. A.; Georgiades, P.; Waigh, T. A.; Hodson, N. W.; Ridley, C. E.; Blanch, E. W.; Thornton, D. J. Reorganisation of the salivary mucin network by dietary components: insights from green tea polyphenols. *PLoS One* **2014**, *9* (9), e108372.
60. Gibbins, H. L.; Carpenter, G. H. Alternative mechanisms of astringency-what is the role of saliva? *J. Texture Stud.* **2013**, *44* (5), 364–375.
61. Matsuo, R.; Yamauchi, Y.; Morimoto, T. Role of submandibular and sublingual saliva in maintenance of taste sensitivity recorded in the chorda tympani of rats. *J. Physiol.* **1997**, *498* (3), 797–807.

Chapter 15

Molecular Determinants of the Bittersweet Janus Head of Steviol Glycosides from *Stevia rebaudiana* (Bert.) Bertoni

C. Dawid,[1] C. Well,[1] A. Brockhoff,[2] F. Stähler,[2] W. Meyerhof,[2] and T. Hofmann[*,1]

[1]Chair of Food Chemistry and Molecular Sensory Science, Technische Universität München, Lise-Meitner-Strasse 34, 85354 Freising, Germany
[2]Department of Molecular Genetics, German Institute of Human Nutrition Potsdam-Rehbruecke (DIfE), Arthur-Scheunert-Allee 114-116, 14558 Nuthetal, Germany
[*]E-mail: thomas.hofmann@tum.de.

To identify the structural requirements for the organoleptic properties of steviol glycosides from *Stevia rebaudiana*, we correlated *in vivo* data obtained from human psychophysical experiments with *in vitro* data from cell-based taste receptor assays. While sensory evaluation demonstrated the structural features causing the sweet and bitter taste of these sweeteners, screening experiments with the 25 human bitter taste receptors revealed hTAS2R4 and hTAS2R14 to be the general sensors for bitter taste elicited by steviol glycosides. These results help to navigate breeding of *Stevia rebaudiana* and improve postharvest downstream processing toward the production of preferentially sweet and least bitter tasting *Stevia* extracts.

Introduction

Due to a high number of undesirable health effects such as obesity, dental caries, type-2 diabetes or cardiovascular diseases and its risk factors which were associated with increasing sucrose consumptions (*1–7*), a new field of food research has emerged since the last decades - namely the investigation of low-calorie sweeteners. In this context, artificial sweeteners such as saccharine, aspartame, acesulfame K, sucralose or neotame found widespread use in food industry in the U.S.A. as well as in the European Union (*8*, *9*). Since consumers were more and more aware of what ingredients go into their foods, special emphasis was given by producers to find taste active compounds of natural sources featuring functional, nutritional, dietary and tasty properties.

Particularly, following the worldwide consumer demand for both non-nutritive high-potency and natural sweeteners, with no organoleptic drawbacks compared to sucrose, steviol glycosides the sweet principle of *Stevia rebaudiana* (Bert.) Bertoni, have recently approved as food additive in the EU (*10*). The leaves of *Stevia rebaudiana*, the so-called "sweet herb", which are very popular, have a very long history as natural sweetener as they were already used by the native population in South America to sweeten and to mask the bitter off-taste of herbal teas (*11*, *12*). Flavor research performed within the last years has shown that diterpenic *ent*-kaurene glycosides, all of which share steviol (**1**) as the common aglycone, are the sweet principle of Stevia (cf. Figure 1). Although, stevioside (**2**) was reported to be 210 to 300 times sweetener than sucrose, depending on the sensory protocol, especially, rebaudioside A (**3**) has the reputation for being the most potent sweetener with the most pleasant taste profile (*13*, *14*). However, the data published on systematic and comparative sensory analysis of the purified individual steviol glycosides are rather fragmentary.

Apart from their attractive sweetness the taste profile of steviol glycosides is hindered by a bitter off-taste and an unpleasant lingering aftertaste which is often the reason for consumer reactions and therefore a major problem for food makers (*13*, *15*). Although several studies showed evidence that steviol glycosides play an important role in inducing the bitter taste of *Stevia rebaudiana*, it is still unclear which key structural requirements of the molecules do contribute to the overall bitterness of this non-nutritive sweetener.

In general, sweet and bitter belong to the five basic taste modalities and are mediated by G protein-coupled receptors (GPCRs) expressed by taste receptor cells (bitter: hT2Rs and sweet: hT1R2/hT1R3) (*16*). In the last decade, functional expression studies have successfully enabled the identification of a broad range of cognate agonists for most of the 25 hT2R bitter taste receptors as well as for the heteromeric sweet taste receptor hTAS1R2/hTAS1R3 [e.g.: (*17–23*)]. Although, at atomic resolution the whole structure has still not been resolved, it is assumed that both sweet taste receptor subunits possess a large amino-terminal ectodomain, including a venus-flytrap binding domain that likely contains the orthosteric binding side for several sweet tasting activators like stevioside (*23*, *24*).

The objectives of the present work were, therefore, to investigate the structural requirements for sweet and bitter taste activities of the most important key steviol glycosides. Next to the characterisation of their psychophysical functions by means of different sensory experiments, the responses of the sweet taste receptor (hTAS1R2/hTAS1R3) to the most abundant steviol glycosides should be verified by functional expression studies in human embryonic kidney (HEK)-cells. In addition, to compare human psychophysical data with those obtained from cell-based taste receptor assays, the hTAS2 bitter taste receptors responding to the non-nutritive sweeteners ought to be identified by means of a similar functional receptor assay.

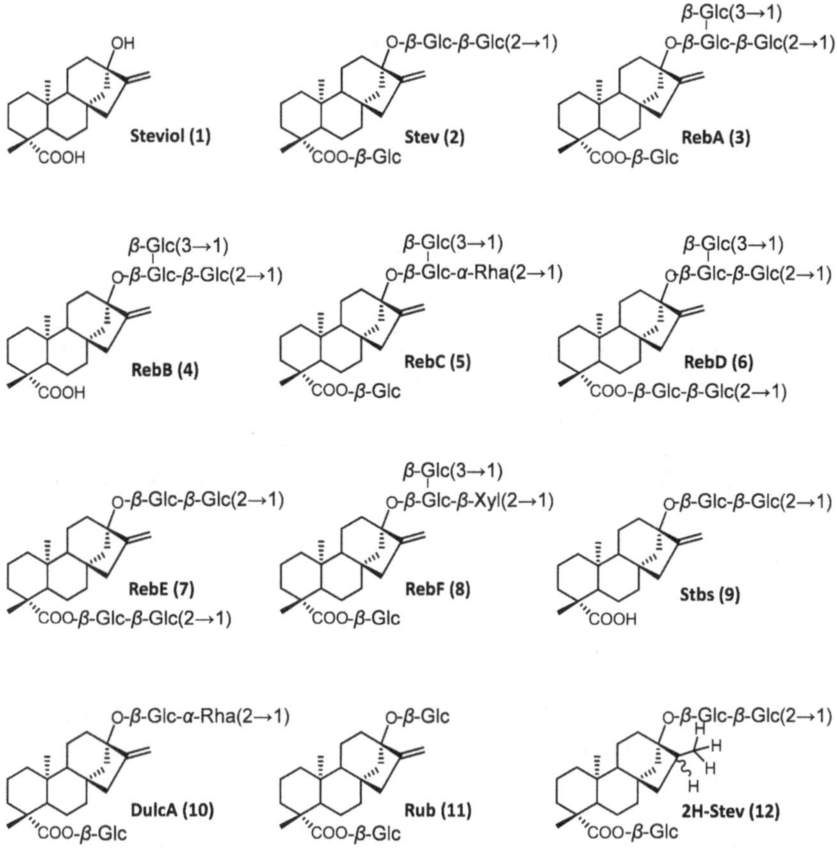

*Figure 1. Chemical structure of the aglycone steviol (**1**), steviol glycosides **2-11** and 16,17-dihydrostevioside (**12**).*

Experimental

The following compounds were obtained commercially: sucralose (Merck, Darmstadt, Germany), colchicine, and aristolochic acid (Sigma-Aldrich, Steinheim, Germany). Lactisole was provided by Cargill (Minneapolis, U.S.A.). Dihydrostevioside (2H-Stev, **13**), rebaudioside B (**4**) and steviolbioside (**9**) were synthesized as described before (*25–27*). While stevioside (**2**), rebaudioside A (**3**), rebaudioside C (**5**), rebaudioside D (**6**), and dulcoside A (**10**) were isolated and purified from commercial Stevia extracts (Cargill, Minneapolis, U.S.A.), rubusoside (**11**) was generated from a commercial extract of *Rubus suavissimus* (MedHerbs, Wiesbaden, Germany) following literature procedures (*26–30*). Prior to the psychophysical experiments and cell-culture assays, spectroscopic data and the purity (>98%) of each individual steviol derivatives **2–6** and **9–12** were checked by means of $^1H/^{13}C$ NMR, LC-MS/MS, and LC-TOF-MS experiments. Thereby, spectroscopic data were in good agreement with those published in the literature.

All psychophysical and functional expression experiments were performed as reported previously (*27*).

Results and Discussion

Aimed at characterizing the alluring sweetness next to the undesirable bitter off-taste of steviol glycosides from *Stevia rebaudiana* data from human psychophysical experiments were combined with functional expressions of TAS1 sweet taste and TAS2 bitter taste receptors. Therefore, first, the individual steviol glycosides **2–6** and **9–11** and **12** were isolated or synthesized and their purities were confirmed by means of NMR, LC-MS and LC-TOF-MS experiments.

Human Sensory Studies on Steviol Glycosides

To characterize the sensory activity of the selected chemosensates **2–6** and **9–12** their human sweet and bitter recognition thresholds were determined by means of triangle tests. While the human threshold concentrations determined for sweetness ranged from 5.3 to 32.9 μmol/L, the oral threshold concentration for bitterness ranged from 23 to 194 μmol/L and, was always above the recognition threshold determined for sweetness as given in Figure 2.

Moreover, among the steviol glycosides, the lowest threshold concentration for sweetness was found for rebaudioside D (**6**), bearing the most β-glucose residues (5 β-glucose residues), followed by rebaudioside A (**3**, 4 residues) and stevioside (**2**, 3 residues), the highest thresholds were observed for the two rhamnose bearing glycosides no. **5** and **10**, followed by the least hydrophilic glycosides no. **11** and **9**. In general, the amount of glucose moieties influences the sweet recognition values significantly. For example, while **2** and **4**, both bearing three β-glucose moieties, showed no significant differences in their threshold concentrations, the threshold of **4** significantly differed from that of compound **3**, bearing four glucose moieties. In addition, the chemosensates decorated with rhamnose moieties induced higher sweet taste threshold concentrations.

Besides the glycone chain length and pyranose substitution, additionally the exocyclic double bond plays an essential role for the orosensory impression of the highly appreciated low-calorie sweeteners. Hydrogenation of the double bond at position C(16), as found in **12**, resulted in a significant increase of the threshold from 11.1 (**1**) to 28.1 μmol/L (**12**).

Unlike their sweet recognition values, the bitter threshold concentrations of **2–6** and **9–12** could not be correlated to the amounts of β-glucose moieties, linked to the aglycone. But, interestingly, steviol glycosides evaluated with the highest sweet threshold values, exhibited the lowest bitter thresholds and, depending on their chemical structure, these showed low recognition thresholds between 49 and

84 µmol/L (**5**, **9-11**). Thereby, the threshold values of Dulc A (**10**) and Reb C (**5**), both containing rhamnose residues, are only ~1.5 times higher compared to their sweet thresholds. Moreover, the lowest bitter threshold concentration was determined for compound **12**, the hydrogenated analogon of stevioside (**1**).

In conclusion, saturation of the double bond in compound **12** compared to **1** induced an archaic increase in bitterness and is a prerequisite for its bitterness.

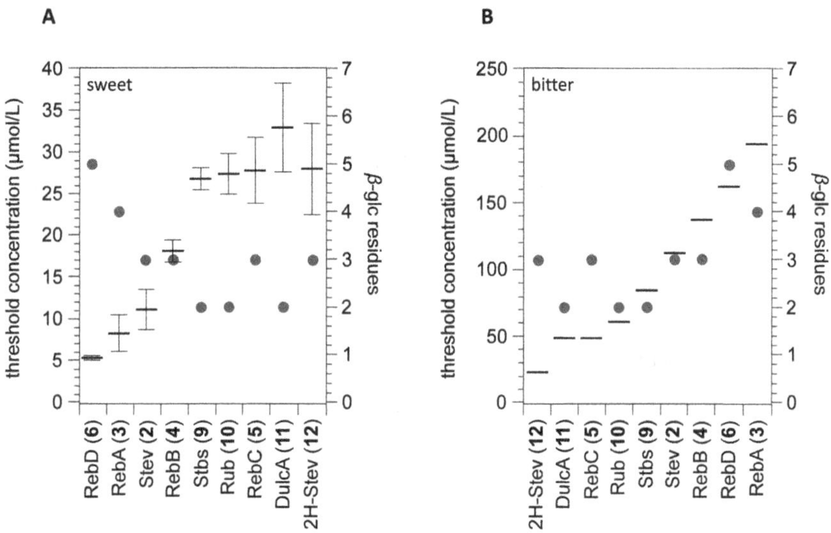

Figure 2. Correlation of taste threshold concentrations (bars) measured for sweetness (A) (error bars indicate standard deviation) and bitterness (B) and the amount of β-glycosylic residues (○) of different steviol glycosides and of 12. The structures of the individual compounds are given in Figure 1.

Sweet Taste Receptor Responses to Steviol Glycosides

In order to assess the structure/activity relationships of steviol glycosides on the human sweet taste receptor, functional experiments were carried out using the human embryonic kidney cell line HEK293 expressing the human sweet taste receptor subunits hTAS1R2 and hTAS1R3 and the chimeric G protein subunit Gα15Gαi3 following the protocol reported recently (*23*, *27*). Thereby, the functional sweet taste receptor heteromer is implemented by stable expression of the subunit hTAS1R2, and inducible expression of the second subunit, hTAS1R3, through a tetracycline-responsive element (*27*, *31*, *32*). Especially, the G protein subunit couples the sweet taste receptor to the release of calcium from intracellular stores that can be monitored by means of a calcium-indicator fluorescent dye (cf. Figure 3).

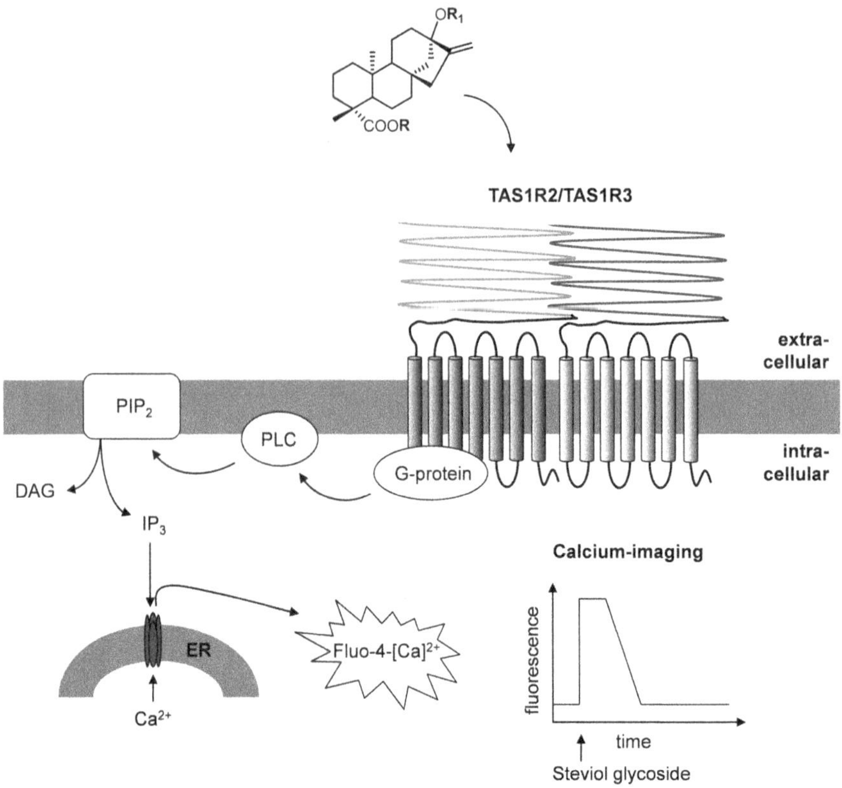

Figure 3. Scheme of the functionally expressed human sweet taste receptor hTAS1R2/hTAS1R3 assay according to Behrens et al., 2011.

Using this strategy, we observed that cells expressing hTAS1R2/hTAS1R3 responded with a transient increase of calcium fluorescence to application of all tested steviol glycosides (**2**–**6** and **9**–**12**). Being well in agreement with the findings of the psychophysical experiments, the onset of responses from the sweet receptor-expressing cells were in the same range as the sensory data observed in vivo (cf. Figure 2 and Table 1). For example, substance no. **6** was found to be the most sweet potent steviol glycoside *in vitro* as well as *in vivo* (threshold concentration *in vivo*: 5.3 µmol/L; *in vitro*: 2.2 µmol/L). Intriguingly, also in the cell assay steviol glycosides, exhibiting a high number of β-glycosyl residues, such as Reb A (**3**) and Reb D (**6**), revealed the lowest threshold concentrations, while high threshold values could be observed for stevia compounds which contain rhamnose residues (cf. Figure 2 and Table 1). Therefore, the recently identified Rebaudiside M, bearing six β-glucose moieties, could be a highly promising stevia sweetener (*33*).

Table 1. Threshold Concentration for Activation of Bitter Receptors TAS2R4 and TAS2R14 and the TAS1R2/TAS1R3 Sweet Taste Receptor by Steviol Glycosides

	Threshold concentration[1] (µmol/L) in cells expressing		
Compound[2]	*TAS1R2/TAS1R3*	*TAS2R4*	*TAS2R14*
Stev (**2**)	4.3	200	600
RebA (**3**)	4.3	200	600
RebB (**4**)	12.9	200	1000
RebC (**5**)	38.8	400	400
RebD (**6**)	2.2	n.s.[3]	n.s.[3]
Stbs (**9**)	12.9	400	n.s.[4]
Dulc A (**10**)	38.8	200	50
Rub (**11**)	25.9	50	400
2H-Stev (**12**)	38.8	n.d.	n.d.

[1] The structures of the individual compounds are given in Figure 1. [2] Threshold concentration is defined as the lowest concentration which was used and led to a cellular response which is significant higher than that obtained by applying buffer solutions to the cell. [3] No response to the test compound up to the maximal soluble concentration of 400 µmol/L. [4] No response to the test compound up to the maximal soluble concentration of 800 µmol/L. n.d. Not determined due to receptor-independent fluorescence signal in control cells.

Although most of the sweet receptor responses to the steviol glycosides were rather similar to the data observed in the psycophysical experiments, there are noteworthy differences in potency between compound no. **2** and its hydrated derivative **12**, which are much more pronounced *in vitro* than *in vivo* and which are apparent by the 9-fold increased threshold value of **12** compared to **2**.

Identification of the hTAS2 Bitter Taste Receptors Responding to Bitter Steviol Glycosides

To analyze the bitter off-taste of steviol glycosides in more detail, we selected stevioside (**2**) as a representative to identify the responding bitter taste receptors (cf. Figure 4). Therefore, we used HEK293T Gα16gust44 cells, which transiently expressed each of the 25 hTAS2Rs individually. As already described for the sweet receptor assay also the activation of hTAS2R receptors was coupled to the release of Ca^{2+} from intracellular stores, which could be measured using a calcium-sensitive fluorescence dye (*27*).

Intriguingly, two of the 25 bitter taste receptors were activated by compound no. **2**, namely hTAS2R4 and hTAS2R14 (cf. Figure 4).

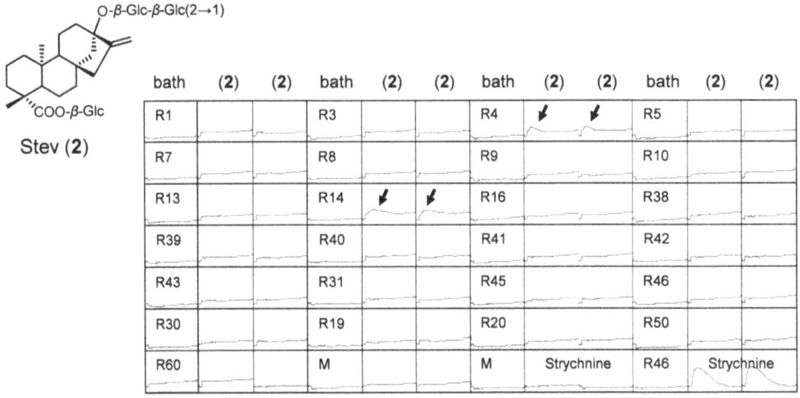

Figure 4. Calcium responses of HEK-293T-Gα16gust44 cells expressing one of the 25 hTAS2Rs or empty vector (mock, M) elicited by bath application of stevioside (2) (1mM). Arrows point to positive fluorescence signals in TAS2R4- and TAS2R14-expressing cells. As positive control calcium responses of hTAS2R46 to 10 μmol/L strychnine were recorded.

Due to their structural similarities, we conclude that these two bitter receptors are selective sensors for all steviol glycosides. In order to compare the potency of the steviol glycosides, the cells expressing hTAS2R4 and hTAS2R14, respectively, were challenged with the tastants **2**–**6** and **9**–**12** and their threshold concentration were determined. Therefore, increasing concentrations (up to 1.2 mmol/L) of each compound were tested in our cell-assay and the lowest concentration leading to a significant fluorescence signal was determined. All steviol glycosides that have been used in this assay were capable of activating hTAS2R4 and/or the hTAS2R14 (cf. Table 1), except of rebaudioside D (**6**). Due to its limited solubility, rebaudioside D was measured at a lower concentration compared to other steviol glycosides.

Comparison of the threshold data revealed, that the two bitter receptors showed neither the same rank order of potency nor were equally sensitive to the steviol glycosides. But among the test compounds, lower threshold concentrations were found for the activation of hTAS2R4, than for that of hTAS2R14. Interestingly, hTAS2R4 seems to be more sensitive to short short-chained and rhamnose-containing steviol glycosides since comparatively low threshold concentrations were observed for **10**, **5** and **11** (cf. Table 1). In addition, the high threshold values for Reb A and Reb B indicated that hTAS2R14 is less sensitive to steviol glycosides with a high number of β-glycosyl moieties. In sum, the bitter taste thresholds determined *in vitro* were well in line with those observed *in vivo*.

Conclusion

Based on the findings of the present study, we conclude that, glycone chain length, pyranose substitution, and the C16 double bond play an essential role for the orosensory impression of the highly appreciated low-calorie steviol glycosides (**2-11**). Sensory and cell-based studies revealed and characterized the sweet profiles of individual steviol glycosides. For the first time, comprehensive screening experiments with the 25 human bitter taste receptors demonstrated that two members, hTAS2R4 and hTAS2R14, act as general sensors for the bitter off-taste elicited by steviol glycosides. These results might help to develop commercially available preferentially sweet and least bitter tasting *Stevia* extracts.

References

1. Newbrun, E. Sucrose, the arch criminal of dental caries. *ASDC J. Dent. Child.* **1969**, *36*, 239–24.
2. Nizel, A. E. Dental caries: protein, fats and carbohydrates. A literature review. *N. Y. State Dent. J.* **1969**, *35*, 71–81.
3. Walker, A. R. P. S. Sugar intake and diabetes mellitus. *Afr. Med. J.* **1977**, *51*, 842–851.
4. Wolfram, G. What is the etiologic role of sugar in cardiovascular disease. *Z. Ernaehrungswiss.* **1990**, *29*, 35–38.
5. Grenby, T. H. Prospects for sugar substitutes. *Chem. Br.* **1991**, *27*, 342–345.
6. Howard, B. V.; Wylie-Rosett, J. Sugar and cardiovascular disease: A statement for healthcare professionals from the committee on nutrition of the council on nutrition, physical activity, and metabolism of the American heart association. *Circulation* **2002**, *106*, 523–527.
7. Anderson, C. A.; Curzon, M. E. J.; van Loveren, C.; Tatsi, C.; Duggal, M. S. Sucrose and dental caries: a review of the evidence. *Obesity Rev.* **2009**, *10*, 41–54.
8. Duffy, V. B.; Anderson, G. H. Position of the American Dietetic Association: Use of nutritive and nonnutritive sweeteners. *J. Am. Diet. Assoc.* **1998**, *98*, 580–587.
9. *Official Journal of the European Union*, Commission Directive, 2009/163/ EU of 22.12.2009.
10. Commision E., 201. No. 1131/2011, L 295/205, 12.11.2011.
11. Brandle, J. E.; Starratt, A. N.; Gijzen, M. *Stevia rebaudiana*: Its agricultural, biological, and chemical properties. *Can. J. Plant Sci.* **1998**, *78*, 527–536.
12. Geuns, J. M. C. Stevioside. *Phytochemistry* **2003**, *64*, 913–921.
13. Kinghorn, A. D.; Soejarto, D. D. Sweetening agents of plant origin. *Crit. Rev. Plant Sci.* **1986**, *4*, 79–120.
14. Crammer, B.; Ikan, R. Progress in the Chemistry and Properties of the Rebaudiosides. In *Developments in Sweeteners*; Grenby, T. H., Ed.; Elsevier Applied Science: London, 1987; pp 45–64.
15. Kinghorn, A. D.; Soejarto, D. D. Intensely sweet compounds of natural origin. *Med. Res. Rev.* **1989**, *9*, 91–115.
16. Lindemann, B. Receptors and transduction in taste. *Nature* **2001**, *413*, 219–225.
17. Xu, H.; Staszewski, L.; Tang, H.; Adler, E.; Zoller, M.; Li, X. Different functional roles of T1R subunits in the heteromeric taste receptors. *Proc. Natl. Acad. Sci. U.S.A.* **2004**, *101* (39), 14258–14263.

18. Jiang, P.; Cui, M.; Zhao, B.; Liu, Z.; Snyder, L. A.; Benard, L. M.; Osman, R.; Margolskee, R. F.; Max, M. Lactisole interacts with the transmembrane domains of human T1R3 to inhibit sweet taste. *J. Biol. Chem.* **2005**, *280*, 15238–15246.
19. Sandell, M. A.; Breslin, P. A. S. Variability in a taste-receptor gene determines whether we taste toxins in food. *Curr. Biol.* **2006**, *16*, 792–794.
20. Brockhoff, A.; Behrens, M.; Massarotti, A.; Appendino, G.; Meyerhof, W. Broad Tuning of the Human Bitter Taste Receptor hTAS2R46 to Various Sesquiterpene Lactones, Clerodane and Labdane Diterpenoids, Strychnine, and Denatonium. *J. Agric. Food Chem.* **2007**, *55*, 6236–6243.
21. Ide, N.; Sato, E.; Ohta, K.; Masuda, T.; Kitabatake, N. Interactions of the Sweet-Tasting Proteins Thaumatin and Lysozyme with the Human Sweet-Taste Receptor. *J. Agric. Food Chem.* **2009**, *57*, 5884–5890.
22. Intelmann, D.; Batram, C.; Kuhn, C.; Haseleu, G.; Meyerhof, W.; Hofmann, T. Three TAS2R Bitter Taste Receptors Mediate the Psychophysical Response to Bitter Compounds of Hops (*Humulus lupulus* L.) and Beer. *Chem. Percept.* **2009**, *2*, 118–132.
23. Behrens, M.; Meyerhof, W.; Hellfritsch, C.; Hofmann, T. Sweet and umami taste: natural products, their chemosensory targets, and beyond. *Angew. Chem., Int. Ed.* **2011**, *50*, 2220–2242.
24. Zhang, F.; Klebansky, B.; Fine, R. M.; Liu, H.; Xu, H.; Servant, G.; Zoller, M.; Tachdjian, C.; Li, X. Molecular mechanism of the sweet taste enhancers. *Proc. Natl. Acad. Sci. U.S.A.* **2010**, *107* (10), 4752–4757.
25. Wood, H. B.; Allerton, R.; Diehl, H. W.; Fletcher, H. G.; Stevioside, I. The Structure of the Glucose Moieties. *J. Org. Chem.* **1955**, *20*, 875–883.
26. Kohda, H.; Kasai, R.; Yamasaki, K.; Murakami, K.; Tanaka, O. New sweet diterpene glucosides from Stevia rebaudiana. *Phytochemistry* **1976**, *15*, 981–983.
27. Hellfritsch, C.; Brockhoff, A.; Stähler, F.; Meyerhof, W.; Hofmann, T. Human Psychometric and Taste Receptor Responses to Steviol Glycosides. *J. Agric. Food Chem.* **2012**, *60*, 6782–6793.
28. Kobayashi, M.; Horikawa, S.; Degrandi, I. H.; Ueno, J.; Mitsuhashi, H. Dulcosides, A and B, new diterpene glycosides from *Stevia rebaudiana. Phytochemistry* **1977**, *16*, 1405–1408.
29. Sakamoto, I.; Yamasaki, K.; Tanaka, O. Application of ^{13}C NMR spectroscopy to chemistry of plant glycosides: Rebaudioside D and rebaudioside E, new sweet diterpene-glucosides of *Stevia rebaudiana* Bertoni. *Chem. Pharm. Bull.* **1977**, *25*, 3437–3439.
30. Sakamoto, I.; Yamasaki, K.; Tanaka, O. Application of ^{13}C NMR spectroscopy to chemistry of natural glycosides: Rebaudioside C, a new sweet diterpene glycoside of *Stevia rebaudiana. Chem. Pharm. Bull.* **1977**, *24*, 844–846.
31. Galindo-Cuspinera, V.; Waeber, T.; Antille, N.; Hartmann, C.; Stead, N.; Martin, N. Reliability of threshold and suprathreshold methods for taste phenotyping: characterization with PROP and sodium chloride. *Chemosens. Percept.* **2009**, *2*, 214–228.

32. Hennings, J. K.; Burhenne, N.; Stähler, F.; Winnig, M.; Walter, B.; Meyerhof, W.; Schmale, H. Sweet taste receptor interacting protein CIB1 is a general inhibitor of InsP(3)-dependent $Ca^{(2+)}$-release *in vivo*. *J. Neurochem.* **2008**, *106*, 2249–2262.
33. Prakash, I.; Markosyan, A.; Bunders, C. Development of next generation stevia sweetener: Ruberoside M. *Foods* **2014**, *3*, 162–175.

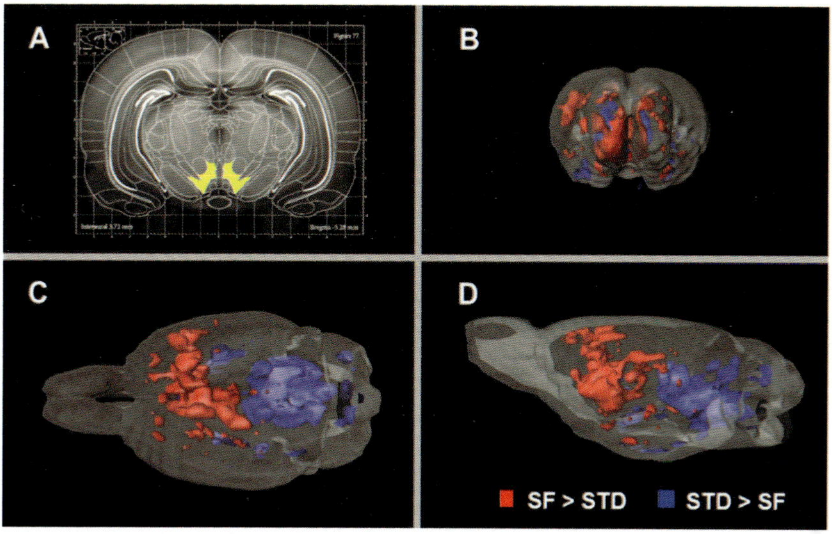

Figure 4. In (A) the overlay of a slice of the reconstructed dataset with the corresponding atlas slice (Bregma -5.28 mm) from the rat brain atlas by Paxinos and Watson is shown with one of the smallest analyzed regions (VTA) marked in yellow. In (B), (C) and (D), significantly differently activated brain areas due to the intake of snack food (SF, potato chips) and standard chow (STD) are displayed in a rat brain surface. Areas colored in red are significantly (p<0.05) activated by the snack food whereas areas colored in blue are significantly (p<0.05) deactivated by the snack food. The results are displayed in axial (B), horizontal (C) and sagittal (D) view. Reproduced with permission from reference (13). (Chapter 10)

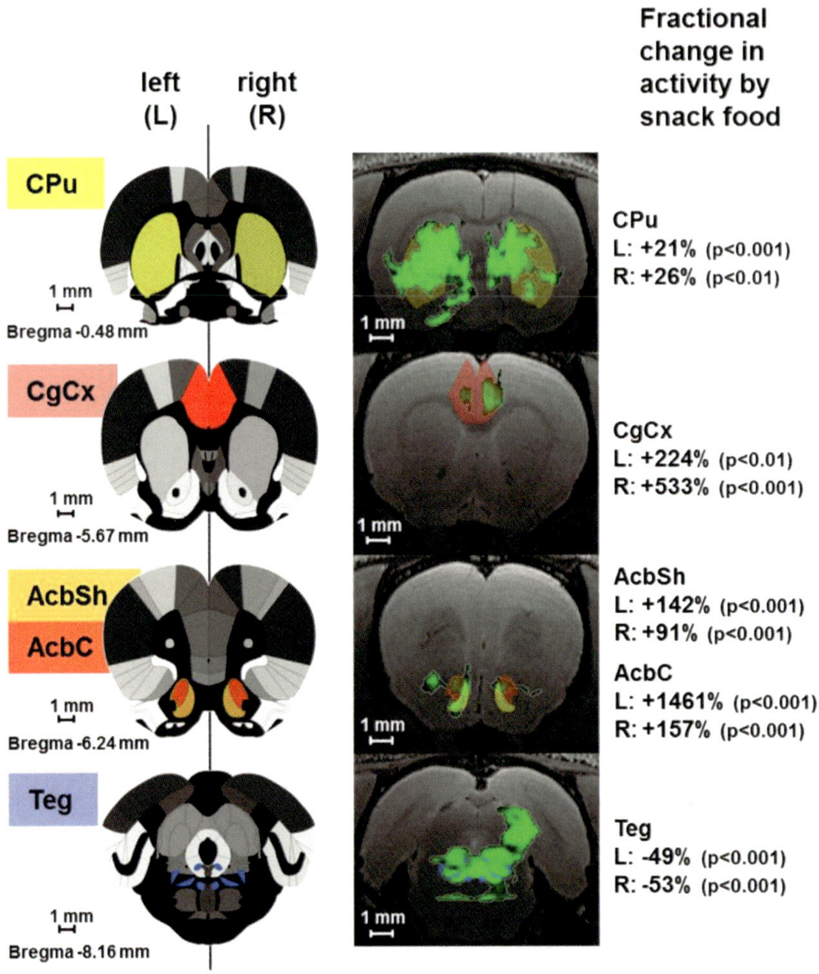

Figure 5. In the left column, the localization of five brain areas is shown on slices of the rat brain atlas by Paxinos and Watson . The middle column shows the significantly differently activated areas (VBM) in an overlay with the corresponding standard T2 weighted MRI anatomy and atlas labels. In the right column, the respective fractional change in activity by snack food intake is shown. CPu = Caudate Putamen, CgCx = Cingulate Cortex, AcbSh = Shell region of the Nucleus Accumbens, AcbC = Core region of the Nucleus Accumbens, Teg = Tegmental Nuclei. Reproduced with permission from reference (13). (Chapter 10)

Chapter 16

Nanotechnological Methods of Antioxidant Characterization

Reşat Apak,*,[1] Esra Çapanoğlu,[2] and Ayşem Üzer Arda[1]

[1]Department of Chemistry, Faculty of Engineering, Istanbul University, Avcilar 34320, Istanbul-Turkey
[2]Department of Food Engineering, Faculty of Chemical and Metallurgical Engineering, Istanbul Technical University, Maslak 34469, Istanbul-Turkey
***E-mail: rapak@istanbul.edu.tr. Fax: +90 212 4737180.**

Nanoparticle (NP)-based analytical methods have displayed a rapid development at the interface of analytical chemistry, food chemistry, biochemistry, and nanotechnology, together with their related industries. For the design of novel antioxidant assays, NPs can be used as colorimetric or electrochemical probes, components in chemical and biological detectors, and radical generation systems. Most applications of NPs used as probes for food chemicals and biochemicals are associated with the use of Au, Ag, magnetite (Fe_3O_4) or titania (TiO_2) nanoparticles and quantum dots. Chemical reduction-based nanotechnological colorimetric assays of antioxidant capacity make use of the formation or enlargement of noble metal nanoparticles (AuNPs, AgNPs, *etc.*) upon reaction of Au(III) or Ag(I) salts with antioxidant compounds acting as chemical reductants. In this chapter, NP-based methods for the measurement of total antioxidant capacity involving chemical reduction together with the methods for the detection of reactive oxygen and nitrogen species (ROS/RNS) and determination of their scavenging activity have been reviewed. Within this scope, spectroscopic methods associated with electron transfer and noble metal nanoparticles as well as electroanalytical biosensor-originated antioxidant activity/capacity methods using nanostructures have been evaluated. On the other hand, methods with different mechanisms for reactive species estimation, NP-based methods for the detection of hydrogen peroxide and its scavengers, and limitations of NP-based antioxidant assays have also been discussed.

Introduction

The growing field of nanotechnology has transformed many sectors of industry with its breakthrough applications in the areas of biotechnology, electronics, medicinal drug delivery, cosmetics, material science, aerospace engineering, and biosensors (*1*, *2*). The applications of nanotechnology in food and agricultural systems are also growing rapidly, with the number of related publications increasing exponentially since 1990. One important application of nanotechnology in food and nutrition is to design and develop novel functional food ingredients with improved water solubility, thermal stability, oral bioavailability, sensory attributes, and physiological performance (*3*). Some others include the development of improved food packaging, *e.g.*, 'smart sensors' embedded in labels (*4*), development of nanosensors aimed at ensuring food safety, and use of nanoparticles, such as micelles, liposomes, nanoemulsions, biopolymeric nanoparticles and cubosomes (*5*). Recent developments in nanotechnology also offer opportunities to develop new antioxidant assays and improve the sensitivity or analytical performance of the existing ones. Nanoparticle-based analytical methods have undergone a rapid development at the interface of analytical chemistry, food chemistry, biochemistry, and nanotechnology, together with their related industries. When applied to the design of novel antioxidant assays, nanoparticles (NPs) can be used as colorimetric or electrochemical probes, components in chemical and biological detectors, and radical generation systems; most applications of NPs used as probes for food chemicals and biochemicals are associated with the use of Au, Ag, magnetite (Fe_3O_4) or titania (TiO_2) nanoparticles and quantum dots (*6*).

Due to their small size, high surface-to-volume ratio and electron configurations, NPs have unique opto-electronic and catalytic properties, making them ideal analytical probes for various substances bearing biochemical and environmental significance. When noble metal NPs are dispersed in liquid media, they exhibit a strong UV-Vis absorption band not present in the spectrum of the bulk metal. This strong surface plasmon resonance (SPR) absorption is attributed to the collective oscillation of electrons in the conduction band of these particles in resonance with the wavelength of incoming light, with a periodic change in electron density at the surface. The SPR absorption of nano-sized particles having near-zero dielectric constant gives rise to a localized surface plasmon resonance (LSPR) band. Known LSPR sensors typically monitor shifts in the peak position or absorption in response to local refractive index changes in the close vicinity of the NP surface. The shape of NPs is also important, as the optical absorption band of spherical colloidal Au-NPs can be seen in the visible spectrum at ~520 nm (*7*), as opposed to the bathochromic (red-shifted) bands of other anisotropic shapes like nanorods and nanoshells having different aspect ratios (*i.e.* relative ratio of major optical axis to minor one).

Nanoparticle-Based Methods of Total Antioxidant Capacity Measurement Involving Chemical Reduction

Spectroscopic Methods Associated with Electron Transfer and Noble Metal Nanoparticles

Antioxidants are health-beneficial compounds capable of scavenging reactive oxygen/nitrogen species (ROS/RNS), thereby retarding or preventing undesired oxidative degradation reactions of biological macromolecules like lipid, protein, and DNA which otherwise eventually leads to tissue damage. Antioxidants may act as hydrogen-atom or electron donators toward reactive species, and this function is simulated in chemical reduction-based assays of total antioxidant capacity (TAC) measurement. The primary reduction-based assays of TAC assessment include Folin-Ciocalteu (*8*), FRAP (*9*), ferricyanide (*10*), and CUPRAC (*11*) tests, based on Mo(VI)-Mo(V), Fe(III)-Fe(II) or Cu(II)-Cu(I) reduction in the presence of suitable ligands; the widely used ABTS (*12*) and DPPH (*13*) radical tests also work on a mixed-mode hydrogen- and electron-transfer mechanism.

Chemical reduction-based nanotechnological methods of colorimetric TAC assay make use of the formation or enlargement of noble metal nanoparticles (AuNPs, AgNPs, etc.) upon reaction of Au(III) or Ag(I) salts with antioxidant compounds acting as chemical reductants. The standard reduction potentials for Au(III)-Au(0) and Ag(I)-Ag(0) redox couples are 1.5 and 0.8 V, respectively, enabling the oxidation of most common phenolic antioxidants by these salts, thereby producing the nanoparticles of the corresponding noble metal. The strong visible light absorption at a characteristic wavelength arises from the LSPR absorption of metal nanoparticles. The intensity of the resulting NP plasmon absorption bands was found to correlate well with the redox characteristics of some selected phenolic acids estimated from cyclic voltammetry experiments, where the highest ability of reducing Au(III) to AuNPs corresponded to the highest TAC, consistent with the tendency of phenolic acids to donate electrons (*14*).

The redox potentials of most phenolic antioxidants (i.e. E^o values of the corresponding phenoxy radical/phenol redox couple) are thermodynamically favorable for producing noble metal NPs via chemical reduction of the corresponding noble metal salts, however kinetic restrictions may not enable quantitative conversion at room temperature within a reasonable time. If the formation (rather than growth) of AuNPs in the presence of antioxidants was targeted, sigmoidal {absorbance versus concentration} curves of low linearity ($r^2 \geq 0.990$) and low reproducibility (on the basis of KAuNP values, corresponding to AuNPs production per polyphenol concentration unit) were obtained, with RSD $\leq$ 11% for polyphenol standards and RSD $\leq$ 15% for food samples (*15*). Thus, when a linear absorbance response versus antioxidant concentration is strictly required, coating of preformed noble metal NP seeds may work better than initial seeding reactions where a higher linear correlation between absorbance and concentration can be obtained, because, for core-shell NPs, it is the outermost layer, and not the core, that dominates the interaction with incoming light (*16*). For example, citrate-stabilized Ag-NPs were formed first in solution, and later the polyphenols-containing sample was added to enable seed-mediated particle growth with the purpose of developing a AgNPs-based TAC assay for phenolic antioxidants (*17*), because the SPR absorption of citrate-stabilized Ag-NPs was greatly intensified upon the addition of polyphenols (Figure 1). The substoichiometric character of the metal salt-to-zerovalent metal reduction reactions by antioxidants leading to NP formation may be the cause of less frequent use of NP-based TAC assays, compared to classical absorptimetric or fluorimetric methods (like ABTS, DPPH, ORAC, CUPRAC, FRAP) for antioxidant capacity assessment. An additional reason may be the inaccessibility of ideal sensitivity and selectivity in NP-based TAC assays as a result of various factors affecting LSPR absorption such as reaction stoichiometry, particle size and shape, and dielectric constants of both the metal and the surrounding medium. Moreover, possible interferent compounds in real food samples (other than dietary antioxidants) may also give rise to NP formation, thereby decreasing selectivity, because –unlike conventional reduction-based TAC assays– even certain surfactants may cause noble metal NPs formation.

Szydłowska-Czerniak *et al.* (*18*) used a silver nanoparticles-based method for determining the antioxidant capacity of rapeseed varieties by comparing the results with those found for sinapic, gallic, caffeic and ascorbic acids and quercetin. Szydłowska-Czerniak and Tułodziecka (*19*) compared the findings of this method for rapeseed varieties with those of previously established spectrophotometric methods, *i.e.* Folin-Ciocalteu, FRAP and DPPH tests, and found significant correlations (r: 0.59-0.91) among the proposed and reference methods. This SPR absorbance measurement method for the reduced Ag-NPs was reported not to require specialized equipment or special reagents. The main advantages of the method were summarized as the relative stability of the $AgNO_3$ reagent when kept in a dark bottle, and the working pH (*i.e.* pH 8.4, achieved with an ammonium buffer) close to physiological pH, as opposed to those of FRAP (pH 3.6) and Folin-Ciocalteu phenolic (pH 10.0) assays. A possible disadvantage may be the appearance of the maxium absorption wavelength (λ_{max}) of yellow-orange colored colloidal silver solutions at 405 nm, significantly shorter (and closer to the UV region) than the SPR λ of Ag-NPs generated with conventional methods (*17*), because Ag-NPs were deposited from a partly stable $Ag(NH_3)_2^+$ complex in ammonium buffer.

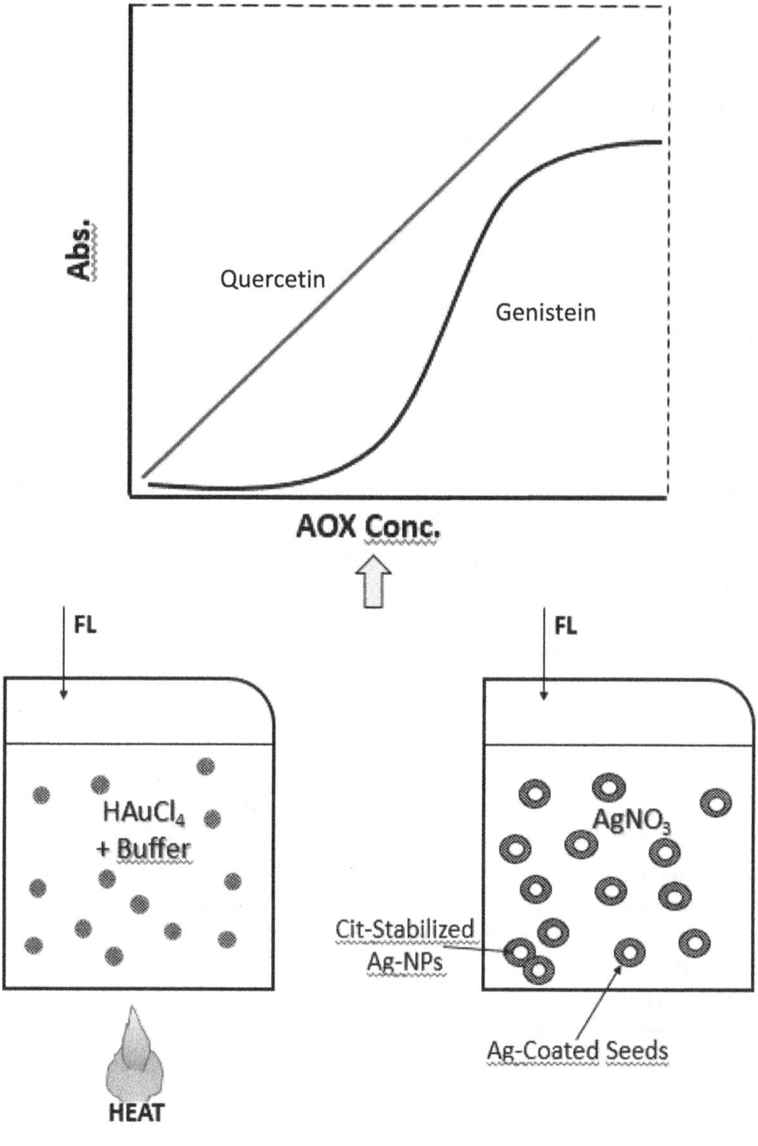

Figure 1. Direct seeding of AuNPs from $HAuCl_4$ solution with flavonoid (FL) addition gives sigmoidal curves (15), while controlled coating of preformed and citrate (Cit)-stabilized AgNPs seeds with antioxidant addition to $AgNO_3$ solution yields linear curves of absorbance versus concentration (17).

In another enlargement study, $NaBH_4$-seeded and citrate-stabilized AuNPs were enlarged with the addition of CTAC and flavonoids in pH 7 phosphate buffer heated to 45° C, but the increase in the peak intensity of SPR bands was not proportional to flavonoid concentration, probably due to differential kinetics of AuNPs formation in the presence of two reducing agents, flavonoid and surfactant (*20*). Using the same technique in electroanalysis, a cysteamine-modified Au-electrode was seeded with AuNPs and immersed in $HAuCl_4$ solution containing flavonoids to enlarge the AuNP seeds on the electrode surface; as a result, the peak currents in the corresponding cyclic voltammograms decreased and peak separation increased. However, neither the SPR band heights nor peak currents provided a linear concentration-dependent quantification of antioxidants, but the sequence of antioxidant power of flavonoids was similar to that of other well-documented antioxidant assays (*20*). Roy *et al.* (*21*) determined the antioxidant activity of aqueous and methanolic extracts of the stem bark of 'Indian Rosewood' plant having gold nanoparticles formation potential by simplifying the procedures developed by Scampicchio *et al.* (*14*) and Wang *et al.* (*20*) for Au(III)-Au(0) reduction, producing Au-NPs having the characteristic SPR band for the colorimetric assay of antioxidant activity.

In an attempt at antioxidant-surfactant combination by making use of light scattering rather than absorbance measurements, the capability of antioxidant compounds to enable Au(III)-AuNPs reduction was kinetically studied (in the presence of CTAB) with the stopped-flow mixing technique and a resonance light scattering (RLS) detection system, where the initial reaction rate for the formation of AuNPs was measured (*22*). When the initial rate of the system (as RLS signal per unit time) in the presence and absence of antioxidant was recorded against antioxidant concentration, a linear calibration curve was empirically obtained with relatively low limit of detection (LOD) values around 10 nmol L^{-1} for some food antioxidants.

Since biothiols are important scavengers of reactive species (ROS/RNS) by hydrogen atom and electron donation and the redox potential of the oxidized/reduced forms of glutathione (GSSG/2GSH) is a basic indicator of the redox environment within a cell, selective thiol quantification is a major challenge in bioanalytical chemistry (*23*). An optical sensor for quantifying biologically important thiols was designed with the use of Ellman's reagent: 5,5'-Dithio-bis(2-nitrobenzoic acid) (DTNB) for functionalizing Au-NPs through non-covalent interactions, and the absorbance changes associated with the formation of the yellow-colored 5-thio-2-nitrobenzoate (TNB^{2-}) anion as a result of reaction with biothiols was measured at 410 nm (*24*), because thiols were preferentially adsorbed on AuNPs releasing the pre-adsorbed disulfide (*25*). This AuNP-based thiol sensor gave a linear response over a wide concentration range of standard biothiols comprising cysteine, glutathione, homocysteine, cysteamine, dihydrolipoic acid and 1,4-dithioerythritol (*24*), and its linearity was better than that of Nile Red dye-derivatized AuNPs sensor for thiol determination (*26*).

Electroanalytical Biosensor-Based Methods Using Nanostructures (Including Biosensors Measuring DNA Damage)

Electroanalytical biosensor-originated antioxidant activity/capacity measurements based on the attenuation of oxidative damage caused by superoxide anion radicals ($O_2^{\bullet-}$) generally involve the use of cytochrome *c* (Cyt *c*) heme protein or DNA (as summarized below). The oxidizibility of DNA bases during the use of DNA-based bioelectrosensors in the absence and presence of antioxidants can be taken as a measure of antioxidant activity.

(a) $O_2^{\bullet-}$ was produced by the xanthine-XOD enzyme system, using a Cyt *c*-modified electrode, where Cyt *c* was reported to communicate with the nano-Au electrode through self-assembled monolayers of short chain alkanethiols. The immobilised Cyt *c* was reduced by $O_2^{\bullet-}$ and rapidly regenerated at the surface of the electrode polarized at the oxidation potential, where the current generated by electron transfer from the radical to the electrode through Cyt *c* was proportional to the radical concentration (*27*). In the presence of antioxidants, the oxidation current was decreased, concomitant with a decrease in the concentration of scavenged radicals. Using this bioelectrosensor, the order of antioxidant activity established for flavonoids was: flavanols > flavonols > flavones > flavonones > isoflavonones (*28*).

(b) DNA-based bioelectrosensors worked on the principle of immobilizing (usually calf thymus–originated) double stranded (ds)-DNA on screen-printed carbon electrodes (SPE), followed by the detection of the guanine oxidation peak between +0.8 and +1.0 V (*vs.* Ag/AgCl) by square wave voltammetry. Since the peak current intensity was proportional to the concentration of DNA base, guanine, the immersion of the DNA-modified electrode into a Fenton solution (such as $Fe(II)+H_2O_2$) produced a signal decrease in the peak current intensity, whereas the presence of antioxidants restored the signal (demonstrating DNA integrity) due to hydroxyl radical quenching (*29*). In order to measure this signal alteration more effectively, screen-printed carbon was doped with TiO_2 nanoparticles, creating a porous surface structure on which ds-DNA adsorbed better because of specific DNA phosphate-TiO_2 interactions (*30*). A redox mediator, namely *tris*-2,2'-bipyridine (bipy) ruthenium(II), was electrooxidized on the electrode surface to subsequently oxidize both the adsorbed ds-DNA and the antioxidants in solution. Divalent and trivalent ruthenium-bipy species, *i.e.* $Ru_{DNA}^{(II)}$ and $Ru_{DNA}^{(III)}$, represented those redox mediators that were in the vicinity of TiO_2 nanoclusters. Here, the oxidative damage was produced by $[Ru(bipy)_3]^{3+}$, an efficient oxidant of guanine and adenine (*i.e.* those DNA bases that are most sensitive to oxidation). Thus, ds-DNA oxidation damage was monitored by measuring the catalytic oxidation of $[Ru(bipy)_3]^{2+}$ where electro-generated $[Ru(bipy)_3]^{3+}$ was used as a ds-DNA redox oxidant in the absence and presence of antioxidant compounds, enabling antioxidant activity evaluation (*30*).

Wang, Jiao and Yu (*31*) recently fabricated an electrochemical DNA damage biosensor for measuring antioxidant activity of orange juice by constituting a hybridized membrane of poly-L-glutamic acid and nano-Ag with an outside layer of chitosan (CS)/double stranded DNA (ds-DNA). A Fenton solution (Fe^{2+}/H_2O_2) was used to generate hydroxyl radicals (•OH) and induce ds-DNA damage, especially on the guanine moiety. When $Ru(NH_3)_6^{3+}$ was used as electroactive indicator, it would oxidize DNA (guanine) and form a catalytic redox cycle, where the linear sweep voltammetric response was recorded in the absence and presence of antioxidants, the latter protecting DNA from •OH damage. Other than sensor development in the qualitative sense, neither a concentration *versus* response curve and a table of interferences nor a comparison with the findings of other antioxidant activity assays were presented by the authors (*31*).

Nanoparticle-Based Methods for the Detection and Scavenging Activity Determination of Reactive Species (ROS/RNS)

During normal cellular oxidative metabolism, ROS and RNS are formed in small quantities and can be removed effectively by natural defense mechanisms. However, uncontrolled and excessive generation of ROS/RNS have been implicated in the pathogenesis of various diseases including cancer, cardiovascular and neurodegenerative diseases as well as aging (*32*, *33*). The main features of most antioxidant assays are based on an efficient generation system of free radicals, a suitable oxidation reaction induced by free radicals and an appropriate measurement of the suppressive signal (*34*). In this regard, one of the antioxidant detection methods with a nanotechnological approach is to use NPs to generate free radicals that were further trapped by various antioxidants and detected using different strategies or techniques (*6*). Biosensors prepared with NPs may be an attractive alternative for the determination of free radicals and antioxidants, characterized by selectivity, low cost, easy automation and simplicity (*35*). Functional NPs (electronic, optical and magnetic) bound to biological macromolecules (*e.g.*, peptides, proteins, nucleic acids) have been developed for use in biosensors to detect and amplify various signals. Some of the NP-based sensors include acoustic wave biosensors, optical biosensors, magnetic and electrochemical biosensors (*36*). On the other hand, chemically modified electrodes (CMEs) with nanomaterials have started to become popular due to certain advantageous properties of nanomaterials including large surface area, high catalytic activity, and, in some cases, excellent conductivity (*37–39*).

Nanoparticle-Based Methods Having Different Mechanisms for Reactive Species Estimation

A simple and sensitive electrochemical method for the determination of antioxidant capacity using NPs was reported by Wang *et al.* (*34*). In this study, photocatalytic oxidation of water by UV-illuminated TiO_2 nanoparticles was chosen as a clean and reproducible method for the generation of •OH radicals, and 4-hydroxybenzoic acid (4-HBA) was used as a trapping agent for the photogenerated ${}^{\bullet}OH$, leading to the formation of 3,4-dihydroxybenzoic acid (3,4-DHBA) subsequently measured by voltammetry. Addition of antioxidants induced the competition between 4-HBA and antioxidants toward ${}^{\bullet}OH$ elimination, so the antioxidants scavenging ability could be correlated with the decrease of 3,4-DHBA peak current. Antioxidant capacity of standard substances were measured with the proposed method and compared to the findings of a fluorimetric method also utilizing illuminated TiO_2 as the ${}^{\bullet}OH$ source and therephthalic acid as probe, producing the fluorescent 2-hydroxyterephthalic acid upon hydroxylation.

Recently, an interesting nanoparticle-based antioxidant assay was developed by Wang, Ma, Gan *et al.* (*40*) in which antioxidants directly reduced the trapped holes (h^+) photogenerated on the surface of a graphene-TiO_2 nanohybrid. The photoelectrochemical platform (PEC) established for this purpose consisted of an ITO (indium tin oxide) electrode, the surface of which was coated with sulfonated graphene-nano TiO_2 composite; PEC current was generated under visible light and open circuit conditions, as a result of phenol-phenoxyl radical oxidation mainly by h^+ separated from electrons (e^-) on the platform (Figure 2). For gallic and caffeic acids and tea catechins, the authors achieved nearly two orders-of-magnitude linear response of current intensity *versus* concentration. The main advantage of the proposed method was its potential applicability to colored solutions and miniaturized systems (*40*). However, the additivity of current response with respect to the concentrations of individual antioxidants in mixtures was not tested. Moreover, the possible oxidation of polyphenols by either hydroxyl radicals or h^+ generated on PEC cannot be precisely differentiated, with possible adverse consequences on the stoichiometry and reproducibility of conversion of phenolic antioxidants to phenoxyl radicals.

In another study performed by Liu *et al.* (*41*), the generation of ${}^{\bullet}OH$ at a palladium oxide NPs-modified electrode during the concomitant reduction of palladium oxide in the presence of hydrogen peroxide either added to the solution or produced *in situ* by oxygen reduction was investigated. The catalytic decomposition of H_2O_2 on the freshly exposed palladium metal generated hydroxyl radicals, confirmed by a terephthalic acid probe that produced the fluorescent 2-hydroxyterephthalic acid product upon hydroxylation. The electrochemical antioxidant sensor of this study was a palladium oxide (PdO) NPs-modified indium tin oxide (ITO) electrode, and cathodic scanning from a positive potential reduced PdO to Pd metal onto which dissolved oxygen was reduced to form H_2O_2. The catalytic decomposition of H_2O_2 on the freshly exposed Pd metal generated ${}^{\bullet}OH$, which in turn re-oxidized Pd to PdO, thereby increasing the cathodic current associated with the reduction of PdO. Antioxidants, when present, caused a decrease in this catalytic reduction current *via* scavenging ${}^{\bullet}OH$. Liu *et al.* (*41*) claim that their method may be more selective toward antioxidants than other similar methods exploiting electrochemical oxidation of antioxidants at anodic potentials, because PdO reduction occurred at negative potentials where antioxidants are usually not oxidized. However, a major disadvantage of the proposed method was that the intensity of the catalytic reduction current could only be correlated to antioxidant concentration with a non-linear (exponential) curve.

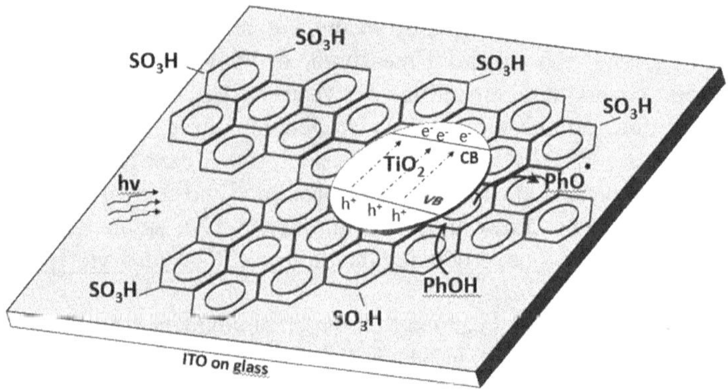

$$TiO_2/SG \rightarrow (+h\nu) \rightarrow TiO_2\,(h^+)\text{–}SG\,(e^-)$$

$$TiO_2\,(h^+) + Ph\text{-}OH \rightarrow TiO_2 + PhO^{\bullet}$$

Figure 2. Photoelectrochemical (PEC) platform of TiO_2 – Sulfonated Graphene (SG) on indium tin oxide (ITO) electrode on which phenolic antioxidants (PhOH) are oxidized to phenoxyl radicals ($PhO^{\bullet}$). PEC current was generated under visible light and open circuit conditions, as a result of phenol-phenoxyl radical oxidation mainly by h^+ separated from electrons (e^-) on the platform (31).

A portable nanoparticle based-assay for rapid and sensitive detection of food antioxidants was proposed by Sharpe *et al.* (*42*), based on the use of immobilized ceria (cerium oxide) nanoparticles. Due to the reversible oxidation state of cerium Ce(III)/Ce(IV) on the NP surface, nanoceria is capable of changing redox states and surface properties after interaction with antioxidants. Thus, two strategies for antioxidant assay with nano-ceria were reported: (i) antioxidants can reduce cerium on the NP surface from Ce(IV) to Ce(III) in a concentration-dependent manner; (ii) surface adsorbed ceria-peroxyl complexes are formed on ceria NPs after treatment with H_2O_2, accompanying a color change from white/yellow to dark brown (Ce-peroxyl complexes); antioxidants, when present, caused a concentration-dependent decrease in color intensity of the cerium complex. This sensor was tested for the rapid detection of standards of common antioxidant compounds as well as real samples (teas and medicinal mushrooms). The authors mention that the proposed assay could be appealing for remote sensing applications, where specialized equipment is not available, and also for high throughput analysis of large numbers of samples.

In continuation of their work with nano-ceria, Sharpe *et al.* (*43*) reported a novel chemical sensing array, based on metal oxide nanoparticles (*i.e.* cerium oxide, titanyl oxalate, TiO_2, Fe_2O_3, ZrO_2, ZnO and SiO_2) immobilized onto cellulose, as a portable and inexpensive paper-based colorimetric method for polyphenol detection and field characterization of antioxidant containing samples. Multiple metal oxide nanoparticles with various polyphenol charge-transfer complexation properties were used as active sensing materials to develop the colorimetric sensor array and establish a database of polyphenol standards. Since the charge-transfer complexes of the sensor metal oxides with each polyphenol may absorb light at different wavelengths, the selectivity of this method is questionable, because there is no unique wavelength of measurement for all antioxidant polyphenols as encountered in the standard colorimetric tests based on the formation of a single redox product with antioxidants. Another disadvantage may be the non-linear response of the proposed sensor array, *i.e.* linear increase of color intensity with the logarithm of antioxidant concentration. The antioxidant activity of each sample was calculated and validated against the oxygen radical absorbance capacity (ORAC) assay showing good comparability. Besides, Sharpe and Andreescu (*44*) suggested that the sensor they prepared for the detection of polyphenolic antioxidants is a good candidate for analyzing the antioxidant character of food, drink, botanical medicines, and physiological fluids.

For the analysis of phenolic compounds, amperometric biosensors based on tyrosinase have been shown to be simple and convenient due to their high sensitivity, effectiveness, and simplicity (*45*). However, immobilization of enzymes within these procedures could sometimes be complicated and could have low sensitivity, stability and bioactivity (*46*). On the other hand, biocompatible nanomaterials have several advantages in enzyme immobilization such as retaining enzyme activity in a favorable microenvironment, and enhancing the direct electron transfer between the enzyme's active sites and the electrode (*47*). In the mediator-free phenol biosensor using ZnO nanoparticles developed by Li *et al.* (*45*), the low-isoelectric point tyrosinase was adsorbed on the surface of high-isoelectric point ZnO nanoparticles (nano-ZnO) facilitated by electrostatic interactions, and then immobilized on a glassy carbon electrode *via* film formation by chitosan. It was reported that the nano-ZnO matrix provided an advantageous microenvironment in terms of its favorable isoelectric point for tyrosinase loading in which immobilized tyrosinase basically retained its activity. The researchers reported that phenolic compounds were determined with quantitative recovery by the direct reduction of biocatalytically generated quinone species at -200 mV (*vs* saturated calomel electrode), and that the proposed biosensor exhibited high sensitivity and fast response without a need for other electron mediators.

A novel electrochemical tyrosinase biosensor for determining phenolic compounds based on the use of a glassy carbon electrode (GCE) modified with tyrosinase-Fe_3O_4 magnetic nanoparticles-chitosan nanobiocomposite film was proposed by Wang *et al.* (*48*). This biosensor exhibited fast and sensitive amperometric responses to various phenolic compounds in the absence of other electron mediators. The nanobiocomposite film provided a suitable microenvironment, enabling high loading of the enzyme and preventing the leaching of the immobilized enzyme. In spite of the fact that only three phenolic compounds were tested, the authors foresee great promise for the rapid, simple, and cost effective analysis of phenolics with this biosensor.

In another study, a generic method for the immobilization of enzymes in chemically synthesized gold polypyrrole (Au–PPy) nanocomposite in combination with an amperometric biosensor using this composite was developed and optimized. Three enzyme systems including cytochrome *c*, glucose oxidase and polyphenol oxidase were used as model systems. The synthesis and deposition of the nanocomposite onto a GCE were optimized, followed by procedure development for enzyme immobilization and subsequent fabrication of glucose and phenol biosensors. According to the results, the presence of Au-NPs provided enhanced electrochemical activity, and the biosensors enabled a rapid, simple, and accurate measurement of glucose and phenol with high sensitivities, low detection limits and fast response times for rapid detection of enzyme substrates and inhibitors (*49*).

SOD has a net negative charge at physiological pH, which enables its electrostatic retention onto the positively-charged ammonium groups of a thiol-amino acid forming a self-assembled monolayer on a gold electrode through the -S-Au bond. SOD confined on the electrode normally retains its inherent enzymatic activity for dismutation of the superoxide ion. The peak reduction current required to convert the surface-immobilized SOD into the reduced state is significantly decreased in the presence of superoxide radical which can be scavenged by the enzyme. Enzyme immobilization techniques onto a modified electrode share some common patterns. For example, a composite fiber electrode may be fabricated by dispersing AuNPs onto poly(methyl methacrylate) (PMMA)–polyaniline (PANI) core–shell electrospun nanofibers (*50*). This architecture (containing the conducting polymer: PANI) proved to be a favorable environment for the immobilization of the enzyme to produce a third-generation biosensor without a redox mediator, where the direct electron transfer between the active site of SOD and the electrode surface was affected by the scavenging action of $O_2^{\bullet-}$ by SOD. Likewise, a superoxide dismutase biosensor can be fabricated (*51*) by immobilizing SOD on cysteine-derivatized Au/CFME (carbon fiber microelectrodes), and SOD scavenges $O_2^{\bullet-}$ with the formation of oxygen and hydrogen peroxide *via* a cyclic oxidation/reduction mechanism by the reactions:

$$SOD(Cu^{2+}) + O_2^{\bullet-} \rightarrow SOD(Cu^{+}) + O_2$$

$$SOD(Cu^{+}) + O_2^{\bullet-} + 2H^{+} \rightarrow SOD(Cu^{2+}) + H_2O_2$$

Immobilized SOD possesses bifunctional enzymatic catalytic activity for the detection of $O_2^{\bullet-}$ through redox cycling of the active site (SOD-$Cu^{2+/1+}$). Superoxide radical was produced through the addition of xanthine oxidase to oxygen-saturated PBS buffer containing xanthine, and subsequently a superoxide-dependent oxidation current was generated at the electrode, the intensity of which was proportional to $O_2^{\bullet-}$ concentration. The signal depression due to the decomposition of $O_2^{\bullet-}$ was proportional to SOD activity (*50*). Antioxidants, when present, may scavenge $O_2^{\bullet-}$, thereby giving rise to less H_2O_2 production and changing the current intensity (Figure 3). It should also be added that both reaction products, O_2 and H_2O_2, can be alternatively detected using amperometric transducers, though with some interference for H_2O_2 determinations in biological samples (*29*, *50*).

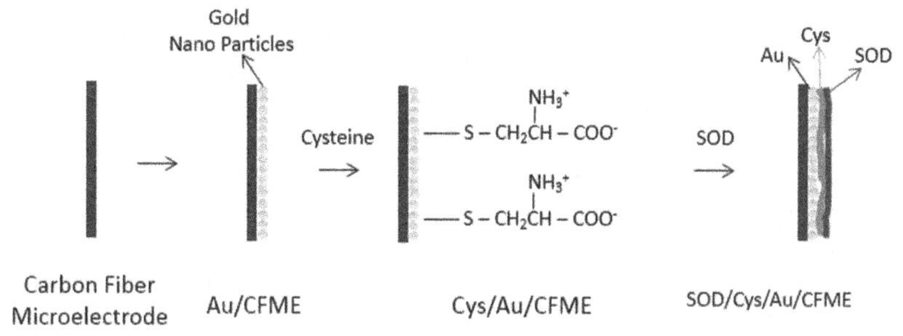

Figure 3. The process of carbon fiber microelectrode (CFME) modification for immobilizing superoxide dismutase (SOD) on cysteine-derivatized Au/CFME (29, 51).

Li, Zhou, Wu *et al.* (*52*) reported that nano-sized iron has been widely applied in the degradation of halogenated organic compounds and other persistent toxic substances due to its ability to catalyze redox processes. Catalytic amounts of iron are sufficient to yield hydroxyl radicals (•OH) from superoxide ($O_2^{\bullet-}$) and hydrogen peroxide (H_2O_2) (*53, 54*). On the other hand, for investigating the electrocatalytic reduction of H_2O_2, AuNPs-modified electrodes were also investigated (*55, 56*). A gold electrode, modified with an amino and a thiol compound, AuNPs, and Prussian blue, showed a wider pH adaptive range, better electrochemical stability and larger response current to the reduction of H_2O_2 (*55*).

Due to the large specific surface area and good biocompatibility of AuNPs, horseradish peroxidase (HRP) can be adsorbed onto an AuNP layer for the detection of H_2O_2 without loss of biological activity (*57*). Shi *et al.* (*58*) confirmed that this kind of HRP-AuNPs biosensor exhibited long-term stability and good reproducibility.

Du *et al.* (*59*) synthesized Au-SnO_2/graphenes–single-walled carbon nanotubes (SWCNs) nanocomposite and fabricated a highly sensitive electrochemical sensor based on this nanocomposite-modified glassy carbon (GC) electrode. Researchers reported that Au/SnO_2 provides a catalytically active center for reactions between electrons on the electrode and redox species in solution, leading to the detection of *tert*-butylhydroquinone (TBHQ) which is an important antioxidant additive in the food industry. The TBHQ oxidation peak observed in the modified electrode shifted to a more negative value than that of the bare GC electrode due to faster reaction kinetics, and additionally, the peak current was seven times intensified. Du *et al.* (*59*) concluded that this nanocomposite-modified electrode was deliberately designed using green chemistry to combine the virtues of metal, semiconductor, and carbon to build an electroanalytical sensor showing promise for TBHQ detection in the food industry.

Karunakaran *et al.* (*60*) worked on nano-and micro-ZrO_2 and TiO_2 particles, and with the use of the DPPH antioxidant activity assay, they observed that the radical scavenging activity of these particles is dose as well as size dependent, nano-sized particles exhibiting higher radical scavenging ability than the corresponding micro-sized particles for ZrO_2 and TiO_2. This finding may be important in the development of new antioxidant assays utilizing these nano-sized oxides on testing platforms so as to avoid possible interferences in antioxidant activity measurement. On the other hand, enhanced anti-radical (DPPH scavenging) activity with decreasing particle size for iron oxide (Fe_2O_3) nanoparticles was first shown by Paul *et al.* (*61*), followed by the demonstration of a similar DPPH-radical quenching effect by nickel oxide (*62*).

Nanoparticle-Based Methods for the Detection of Hydrogen Peroxide and Its Scavengers

Hydrogen peroxide is an interesting compound which may act both as an oxidant and reductant depending on the chemical environment, and its oxidizing power is greatly enhanced by the catalytic effect of traces of transition metal ions. At the same time, H_2O_2 is an important ROS that may cause Fenton-type oxidations in cells and tissues, giving rise to lipid peroxidation, protein carbonyl formation, enzyme inactivation, and DNA strand breaks (*63*). Therefore, the detection of hydrogen peroxide as well as of its scavengers, using both optical and electrochemical methods, are important.

The optical determination of hydrogen peroxide scavenging activity with the use of NP-based assays differs in principle from chemical reduction-based assays, because neither seeding nor growth of Au/Ag-NPs by antioxidants (acting as reductants) is targeted, and an alternative strategy relying on the inhibition of H_2O_2-induced growth of gold nanoshells (GNSs) by hydrogen peroxide scavenging compounds is adapted (*64*). H_2O_2 enlarged the AuNPs on the surface of GNSs precursor nanocomposites (SiO_2/AuNPs), and the pre-adsorbed AuNPs served as nucleation sites for Au deposition. The mechanism of H_2O_2-mediated enlargement of AuNPs was described by Zayats *et al.* (*65*), where the formation of nanocrystalline clusters at the intersection of the faces in addition to the deposition of gold on the NP faces was observed, according to the reaction equation (*66*):

$$2\ HAuCl_4 + 3\ H_2O_2 \ \rightarrow(\text{AuNP seeds})\rightarrow\ 2\ Au^0 + 8\ HCl + 3\ O_2$$

Increasing concentrations of H_2O_2 caused the enlargement of AuNPs on the SiO_2 cores, accompanying the spectral changes in plasmon absorption bands of AuNPs, and H_2O_2-mediated formation of GNSs were limited in the presence of antioxidant compounds, enabling the measurement of H_2O_2 quenching activity (*67*). In another example, H_2O_2-induced growth of GNSs was inhibited by the addition of phenolic acids such as the potent scavenger caffeic acid, which affected the wavelength of surface plasmon absorption (*68*). In these sensing efforts, phenolic acids are presumably oxidized with hydrogen peroxide to phenoxyl radicals, and structure-activity relationships dictate that the number of phenolic –OH groups as well as their position in the aromatic ring, the presence of other electron-donating subsituents in suitable positions of the ring, and possible stabilization of the 1-e oxidation products (*i.e.* phenoxyl radicals) by intra- and inter-molecular hydrogen bonding all play a part in the determined H_2O_2 scavenging action.

A problem with the spectroscopic determination of hydrogen peroxide *via* AuNPs enlargement methods is that the measurement wavelength does not remain constant during gradual increase of analyte concentration. For example, inspection of the spectral changes of the AuNP seeds upon treatment with H_2O_2 revealed that at low H_2O_2 concentrations, a *ca.* 15 nm red shift in the absorbance maximum was observed, while at higher H_2O_2 concentrations, concomitant to the absorbance growth, a *ca.* 10 nm blue shift in the absorbance maxima occurred (*65*). As a result, although the absorbance of the SPR band of SiO_2/AuNPs increased with H_2O_2 concentration, the λ_{max} showed a bathochromic rise (red shift) of ≈ 200 nm accompanying an increase in H_2O_2 concentration from 20 to 200 μM (*67*), invalidating a fixed-wavelength calibration curve. A similar strategy based on the inhibition of H_2O_2-mediated growth of Au nanoshells on 3D ordered SiO_2/ AuNPs precursor nanocomposites as active SERS substrates was also used in other assays, where the hydrogen peroxide quenching ability of biochemically important compounds could be measured (*69*).

Most H_2O_2 sensors work on electroanalytical principles, as reviewed in the current literature (*70–72*). Enhancement in conductivity, bio-immobilization and bio sensors are at the forefront in electrochemical methods. In general, the detection methods of H_2O_2 scavenging activity utilize the peroxidase enzyme (*63, 73*). Unfortunately, the peroxidase-based H_2O_2 scavenging activity assays suffer from two possible interferences: (i) the tested antioxidant should not be a substrate for peroxidase, in competition with H_2O_2; (ii) peroxidase is also inhibited in a ROS generation system involving superoxide, weakening the specificity for H_2O_2 (*73*).

Nanoparticles of noble metals have been used to fabricate biosensors for the detection of hydrogen peroxide (*e.g.*, in conjunction with glucose sensing) by making use of the oxidase- and peroxidase-like activity of NPs (*74–81*). For example, AuNPs with different coatings and bovine serum albumin (BSA)-stabilized Au clusters exhibited peroxidase-like activity along with good biocompatibility, which may make AuNPs promising candidates for use in biosensors (*82–84*). He *et al.* (*85*) have reported that AuNPs can catalyze the rapid decomposition of H_2O_2, and that this decomposition is accompanied by the formation of hydroxyl radicals at lower pH and oxygen at higher pH.

Silver nanoparticles have the ability to reduce ROS with concomitant silver oxidation, enabling the development of a number of H_2O_2 sensors. Since the standard reduction potential for O_2/H_2O_2 redox couple is 0.70 V while that of H_2O_2/H_2O couple is 1.76 V, it may be understood why the development of most NPs-based probes take advantage of the oxidizing ability of H_2O_2, such as the oxidative decomposition of AgNPs by H_2O_2. Therefore, methods weakening or annihilating the intensity of the SPR absorption band of AgNPs are much more common in the literature. For example, the highly clustered nanoparticles in a polymer (polyvinylpyrrolidone) coating were decomposed with their oxidation, and consequently, the LSPR absorption band intensity decreased in a concentration-dependent manner, enabling the quantitation of H_2O_2 (*86*). Filippo *et al.* (*87*) synthesized a colorimetric H_2O_2 sensor (using the principle of its catalytic degradation) based on LSPR of poly(vinyl alcohol) capped AgNPs; the yellow color of the AgNPs–polymer solution was gradually changed to transparent upon NP degradation, and regained by thermal treatment, showing reversibility of the sensor. In a similar study by Vasileva *et al.* (*88*), stable and uniform sized starch-stabilized AgNPs were synthesized using ultrasound mediated reduction of $AgNO_3$ by glucose. The degradation of AgNPs, induced by the catalytic decomposition of hydrogen peroxide, caused a considerable change in the absorbance strength of LSPR band depending on the H_2O_2 concentration, enabling its determination over a concentration interval covering five orders-of-magnitude (*87*). Related to the same phenomenon of catalytic degradation of AgNPs with hydrogen peroxide, a fiber optic sensor utilizing a combination of SPR and LSPR for the detection of hydrogen peroxide was fabricated and characterized using polyvinyl alcohol (PVA) embedded silver nanoparticles over the silver coated core of the optical fiber (*89*). The decomposition of H_2O_2 resulted in the degradation of AgNPs which in turn changed the refractive index of the nanocomposite layer and caused a shift in resonance absorption wavelength enabling the quantitation of H_2O_2 over a wide concentration range, and therefore, Bhatia *et al.* (*89*) suggested potential use of this sensor in remote sensing applications. The basic challenge with sensors exploiting the decrease (or annihilation) of LSPR band intensity of AgNPs accompanying the decomposition of H_2O_2 is the lack of selectivity, because other oxidizing agents may also degrade NPs. On the contrary, by considering the comparable reduction potentials of O_2/H_2O_2 and Ag^+/Ag^0 redox couples, it is rather difficult if not impossible to manufacture LSPR enhancement sensors of H_2O_2 detection based AgNPs formation/enlargement. One of the rare examples of this case where the LSPR band intensity of AgNPs was strengthened rather than attenuated in the presence of hydrogen peroxide was developed by Zhang *et al.* (*90*), in which the superoxide anion radicals presumably produced from H_2O_2 by UV light acted as a reducing agent toward $AgNO_3$, resulting in the intensification of SPR absorption of AgNPs. This method suffered from relatively slow reaction kinetics and low sensitivity.

Another problem with H_2O_2 electrochemical sensors is that the direct reduction or oxidation of H_2O_2at bare electrodes is not suited for analytical applications due to the slow electrode kinetics and high overpotentials required for redox reactions of H_2O_2 on many electrode materials. To overcome these obstacles, redox mediators have been widely used in order to decrease the overpotential and accelerate the electron transfer kinetics (*91*). For example, HRP immobilized on nano-Au displayed good electrocatalytical activity for the reduction of H_2O_2, the latter being subsequently determined in the presence of hydroquinone as a mediator to transfer electrons between the electrode and HRP (*92*). Direct electrochemistry of hemoglobin was observed at carbon nanotube interface for H_2O_2 detection, where the adsorbing hemoglobin −acting as mediator− could transfer electrons directly at carbon nanotube interface compared with common carbon material (*93*). The negatively-charged nano-structured

WO_3 surface at neutral pH adsorbed Cyt *c* with positive charge which facilitated direct and fast electron transfer; the proposed electrochemical method for H_2O_2 detection was claimed to be free from not only common anodic interferences like ascorbic acid, uric acid, and dopamine metabolite: 3,4-dihydroxyphenylacetic acid, *etc.*, but also cathodic interference of oxygen (*94*). In addition to the use of Pt, Au and Pd as catalysts for electrochemical detection of H_2O_2 (*92, 95, 96*), other nano-sized catalytic materials such as MnO_2 (*97*) and CuO (*98–101*) were also investigated as alternatives, with good electrochemical response to H_2O_2. Multi-walled carbon nanotubes (MWCNs) were also incorporated into H_2O_2 sensors to enhance the electrode response; *e.g.*, a novel hydrogen peroxide sensor was developed based on MWCN/AgNPs nanohybrids modified gold electrode; this sensor had a favorable catalytic ability for the reduction of H_2O_2 (*102*). Carbon nanotubes can also be combined with ionic liquids to increase the conductivity, selectivity and stability, *e.g.*, a film consisting of polyaniline and single-walled carbon nanotubes (SWCNs) was electro-polymerized on a platinum electrode in a room temperature ionic liquid, resulting in a selective and non-enzymatic electrode for sensing H_2O_2 (*103*). Another ionic liquid, *n*-octylpyridinium hexafluorophosphate, was used to fabricate a new carbon composite electrode with attractive electrochemical behavior; the carbon ionic liquid electrode (CILE) could accelerate the electron transfer rates at the interface and reduce the overpotentials for several electroactive species (*104*).

The presence of nanoparticles in electrochemical sensors can decrease the overpotentials of many analytes that occur at unmodified electrodes. The combination of metal NPs and miniaturized detection systems may offer a satisfactory platform for the detection of H_2O_2 with high performance. On the other hand, the rather expensive enzymatic sensors for H_2O_2 detection may be advantageous in terms of sensitivity (which may go down to several nanomolar concentrations) rather than selectivity, because the electrode immobilization and stabilization protocols of enzymes are very complicated, and the activity of oxidases can be easily affected by temperature, pH, humidity, toxic chemicals, *etc.*, leading to poor stability and low reproducibility. Naturally, when H_2O_2 is to be determined in the presence of scavengers, one should always be careful whether the method chosen for hydrogen peroxide detection is interfered by antioxidants. For example, if H_2O_2 acts as a reducing agent (such as in the case of noble metal NPs formation/enlargement), its reducing effect should be distinguished from that of the tested antioxidant compounds. In enzymatic electro-sensors, one should clearly understand whether the measured quantity corresponds to the inhibition of H_2O_2 or the enzyme itself, *etc.*

Limitations of Nanoparticle-Based Antioxidant Assays

In spite of major advantages of metal NPs over bulk structures of the same material such as small size resulting in large surface-to-volume ratio, chemically tailorable physical properties, unusual target binding properties, and structural robustness (*105*), NP-based antioxidant activity assays have rarely been used (until the last decade) in food science (*14*) probably because of problems in NP stability, method reproducibility and selectivity. In terms of short-term stability, control of agglomeration, usually with surfactants, is an important issue. For example, in the colorimetric detection of β-agonists (phenylethanolamines) acting as reducing agents toward $HAuCl_4$ solution to produce AuNPs, agglomeration may occur in the absence of the cationic surfactant (CTAC) bringing about Au-colloid precipitation and absorbance decrease at the characteristic SPR band (*106*), but surfactants alone may increase AuNPs formation which is nonspecific for the analyte. Noble metal NPs also undergo long-term storage stability problems in physiological buffers (*107*). NP-based antioxidant activity assays additionally suffer from reproducibility problems, because shifts may be observed in the maximum absorbance wavelength (λ_{max}) with respect to concentration (*14*), and direct seeding techniques (i.e. formation of noble metal NPs from direct antioxidant addition to the corresponding noble metal ion solution) having differential kinetics with respect to a number of antioxidants usually give nonlinear responses (*17*) adversely affecting analytical precision. A third drawback is selectivity, as antioxidants are not merely reducing agents but may also act with hydrogen atom transfer and metal chelating mechanisms, while noble metal NPs are formed from antioxidants mainly by chemical reduction. This raises selectivity problems for true antioxidant compounds, because other biological reducing agents such as dopamine, neurotransmitters, and sugars may additionally act as reducing agents toward $AgNO_3$ and $HAuCl_4$ solutions (*106*) to generate the corresponding metal NPs.

Oil-soluble antioxidants can only be tested in hydrophobic organic solvents, which tend to mediate the self-assembly of Au/AgNPs at the water-oil interface because of the reduction in Gibbs free energy (*108*). Difficulties may be encountered in the simultaneous determination of hydrophilic and lipophilic antioxidants in the same solution because hydrophobic solvents that should be added to the system to solubilize lipophilic antioxidants may decrease analytical selectivity by considerably changing the dielectric constant of the medium whereas addition of surfactants to keep both types of antioxidants in solution may also enhance new NPs generation. The measurement of reactive species (ROS/RNS) scavenging with the use of NP-based techniques is not as straightforward as chemical reduction, producing several products with complex mechanisms, and therefore frequently yielding nonlinear responses. Another inherent problem with hydroxyl radical scavenging assays is that $^{\bullet}OH$ has so great reactivity that, once formed in living systems, will react immediately with whatever biological molecule is in its vicinity (*i.e.* not necessarily antioxidants), producing secondary radicals of variable reactivity (*109*). NP-based light scattering techniques have certain disadvantages, because the measured signal not only depends on the size/shape of each particle but also on the orientation of particles on the surface and their interactions with other particles, the overall effect of which makes calibration difficult (*105*). NP-based techniques generally show a significant dependence on the variations of pH and solvent. Photo-bleaching of nanoprobes may be a serious problem in fluorescence measurements. Finally, challenges of enzyme-responsive NP systems comprise: control over ligand:NP stoichiometry, avoidance of non-specific binding and uncontrolled aggregation, and long-term stability (*107*).

References

1. Rao, C. N. R.; Muller, A.; Cheetham, A. K. *The Chemistry of Nanomaterials; Synthesis, Properties and Applications*; Wiley-VCH Verlag GmbH & Co.: Fairfor, U.K., 2004; Vol. 2, pp 450–560.
2. Manke, A.; Wang, L.; Rojanasakul, Y. Mechanisms of Nanoparticle-Induced Oxidative Stress and Toxicity. *BioMed Res. Int.* **2013**, *2013*, 1–15.
3. Huang, Q.; Yu, H.; Ru, Q. Bioavailability and delivery of nutraceuticals using nanotechnology. *J. Food Sci.* **2010**, *75* (1), 50–57.
4. Blasco, A. J.; González Crevillén, A.; González, M. C.; Escarpa, A. Direct electrochemical sensing and detection of natural antioxidants and antioxidant capacity *in vitro* systems. *Electroanalysis* **2007**, *19* (22), 2275–2286.
5. Sozer, N.; Kokini, J. L. Nanotechnology and its applications in the food sector. *Trends Biotechnol.* **2009**, *27* (2), 82–89.
6. Vasilescu, A.; Sharpe, E.; Andreescu, S. Nanoparticle-based technologies for the detection of food antioxidants. *Curr. Anal. Chem.* **2012**, *8*, 1–11.
7. Link, S.; Mohamed, M. B.; El-Sayed, M. A. Simulation of the Optical Absorption Spectra of Gold Nanorods as a Function of Their Aspect Ratio and the Effect of the Medium Dielectric Constant. *J. Phys. Chem. B.* **1999**, *103*, 3073–3077.
8. Singleton, V. L.; Orthofer, R.; Lamuela-Raventos, R. M. Analysis of Total Phenols and Other Oxidation Substrates and Antioxidants by Means of Folin-Ciocalteu Reagent. *Methods Enzymol.* **1999**, *299*, 152–178.
9. Benzie, I. F. F.; Strain, J. J. The Ferric Reducing Ability of Plasma (FRAP) as a Measure of 'Antioxidant Power': The FRAP Assay. *Anal. Biochem.* **1996**, *239*, 70–76.
10. Oyaizu, M. Studies on products of browning reaction: antioxidative activity of products of browning reaction prepared from glucosamine. *Jpn. J. Nutr.* **1986**, *44*, 307–315.
11. Apak, R.; Güçlü, K.; Özyürek, M.; Karademir, S. E. A Novel Total Antioxidant Capacity Index for Dietary Polyphenols, Vitamins C and E, Using Their Cupric Ion Reducing Capability in the Presence of Neocuproine: CUPRAC Method. *J. Agric. Food Chem.* **2004**, *52*, 7970–7981.
12. Re, R.; Pellegrini, N.; Proteggente, A.; Pannala, A.; Yang, M.; Rice-Evans, C. Antioxidant activity applying an improved ABTS radical cation decolorization assay. *Free Radical Biol. Med.* **1999**, *26*, 1231–1237.
13. Sanchez-Moreno, C.; Larrauri, J. A.; Saura-Calixto, F. A. A Procedure to Measure the Antiradical Efficiency of Polyphenols. *J. Sci. Food Agric.* **1998**, *76*, 270–276.
14. Scampicchio, M.; Wang, J.; Blasco, A. J.; Arribas, A. S.; Mannino, S.; Escarpa, A. Nanoparticle-Based Assays of Antioxidant Activity. *Anal. Chem.* **2006**, *78*, 2060–2063.
15. Vilela, D.; Gonzales, M. C.; Escarpa, A. Gold-nanosphere formation using food sample endogenous polyphenols for *in-vitro* assessment of antioxidant capacity. *Anal. Bioanal. Chem.* **2012**, *404*, 341–349.

16. Jiang, Z.-J.; Liu, C.-Y. Seed-Mediated Growth Technique for the Preparation of a Silver Nanoshell on a Silica Sphere. *J. Phys. Chem. B.* **2003**, *107*, 12411–12415.
17. Özyürek, M.; Güngör, N.; Baki, S.; Güçlü, K.; Apak, R. Development of a Silver Nanoparticle-Based Method for the Antioxidant Capacity Measurement of Polyphenols. *Anal. Chem.* **2012**, *84*, 8052–8059.
18. Szydłowska-Czerniak, A.; Tułodziecka, A.; Szłyk, E. A silver nanoparticlebased method for determination of antioxidant capacity of rapeseed and its products. *Analyst* **2012**, *137*, 3750–3759.
19. Szydłowska-Czerniak, A.; Tułodziecka, A. Comparison of a silver nanoparticle-based method and the modified spectrophotometric methods for assessing antioxidant capacity of rapeseed varieties. *Food Chem.* **2013**, *141*, 1865–1871.
20. Wang, J.; Zhou, N.; Zhu, Z.; Huang, J.; Li, G. Detection of flavonoids and assay for their antioxidant activity based on enlargement of gold nanoparticles. *Anal. Bioanal. Chem.* **2007**, *388*, 1199–1205.
21. Roy, N.; Laskar, R. A.; Sk, I.; Kumari, D.; Ghosh, T.; Begum, N. A. A detailed study on the antioxidant activity of the stem bark of *Dalbergia sissoo* Roxb., an Indian medicinal plant. *Food Chem.* **2011**, *126*, 1115–1121.
22. Andreu-Navarro, A.; Fernández-Romero, J. M.; Gómez-Hens, A. Determination of antioxidant additives in foodstuffs by direct measurement of gold nanoparticle formation using resonance light scattering detection. *Anal. Chim. Acta* **2011**, *695*, 11–17.
23. Schafer, F. Q.; Buettner, G. R. Redox environment of the cell as viewed through the redox state of the glutathione disulfide/glutathione couple. *Free Radical Biol. Med.* **2001**, *30*, 1191–1212.
24. Güçlü, K.; Güngör, N.; Özyürek, M.; Baki, S.; Apak, R. Selective optical sensing of biothiols with Ellman's reagent: 5,5'-dithio-bis(2-nitrobenzoic acid)-modified gold nanoparticles. *Anal. Chim. Acta* **2013**, *794*, 90–98.
25. Bain, C. D.; Biebuyck, H. A.; Whitesides, G. M. Comparison of Self-Assembled Monolayers on Gold: Coadsorption of Thiols and Disulfides. *Langmuir* **1989**, *5*, 723–727.
26. Chen, S.-J.; Chang, H.-T. Nile Red-Adsorbed Gold Nanoparticles for Selective Determination of Thiols Based on Energy Transfer and Aggregation. *Anal. Chem.* **2004**, *76*, 3727–3734.
27. Tammeveski, K.; Tenno, T. T.; Mashirin, A. A.; Hillhouse, E. W.; Manning, P.; McNeil, C. J. Superoxide electrode based on covalently immobilized cyt c: modelling studies. *Free Radical Biol. Med.* **1998**, *25*, 973–978.
28. Ignatov, S.; Shishniashvili, D.; Ge, B.; Scheller, F. W.; Lisdat, F. Amperometric biosensor based on a functionalized gold electrode for the detection of antioxidants. *Biosens. Bioelectron.* **2002**, *17*, 191–199.
29. Prieto-Simon, B.; Cortina, M.; Campas, M.; Calas-Blanchard, C. Electrochemical biosensors as a tool for antioxidant capacity assessment. *Sens. Actuators, B* **2008**, *129*, 459–466.

30. Liu, J.; Su, B.; Lagger, G.; Tacchini, P.; Girault, H. H. Antioxidant Redox Sensors Based on DNA Modified Carbon Screen-Printed Electrodes. *Anal. Chem.* **2006**, *78*, 6879–6.
31. Wang, X.; Jiao, C.; Yu, Z. Electrochemical biosensor for assessment of the total antioxidant capacity of orange juice beverage based on the immobilizing DNA on apoly l-glutamic acid doped silver hybridized membrane. *Sens. Actuators, B* **2014**, *192*, 628–633.
32. Bourdon, E.; Blache, D. The importance of proteins in defense against oxidation. *Antioxid. Redox Signaling* **2001**, *3* (2), 293–311.
33. Niki, E. Assessment of Antioxidant Capacity *in vitro* and *in vivo*. *Free Radical Biol. Med.* **2010**, *49*, 503–515.
34. Wang, Y.; Calas-Blanchard, C.; Cortina-Puig, M.; Baohong, L.; Marty, J.-L. An electrochemical method for sensitive determination of antioxidant capacity. *Electroanalysis* **2009**, *21*, 1395–1400.
35. Mello, L. D.; Kubota, L. T. Biosensors as a tool for the antioxidant status evaluation. *Talanta* **2007**, *72*, 335–348.
36. Jianrong, C.; Yuqing, M.; Nongyue, H.; Xiaohua, W.; Sijiao, L. Nanotechnology and biosensors. *Biotechnol. Adv.* **2004**, *22*, 505–518.
37. Guo, S.; Wen, D.; Zhai, Y.; Dong, S.; Wang, E. Platinum Nanoparticle Ensemble-on-Graphene Hybrid Nanosheet: One-Pot, Rapid Synthesis, and Used as New Electrode Material for Electrochemical Sensing. *ACS Nano* **2010**, *4*, 3959–3968.
38. Fatima Barroso, M.; de-los-Santos-Álvarez, N.; Delerue-Matos, C.; Oliveira, M. B. P. P. Towards a reliable technology for antioxidant capacity and oxidative damage evaluation: electrochemical (bio)sensors. *Biosens. Bioelectron.* **2011**, *30*, 1–12.
39. Niu, X.; Yang, W.; Guo, H.; Ren, J.; Gao, J. Highly sensitive and selective dopamine biosensor based on 3,4,9,10-perylene tetracarboxylic acid functionalized graphene sheets/multi-wall carbon nanotubes/ionic liquid composite film modified electrode. *Biosens. Bioelectron.* **2013**, *41*, 225–231.
40. Wang, L.; Ma, W.; Gan, S.; Han, D.; Zhang, Q.; Niu, L. Engineered Photoelectrochemical Platform for Rational Global Antioxidant Capacity Evaluation Based on Ultrasensitive Sulfonated Graphene-TiO_2 Hybrid. *Anal. Chem.* **2014**, *86*, 10171–10178.
41. Liu, J.; Lagger, G.; Tacchini, P.; Girault, H. H. Generation of OH radicals at palladium oxide nanoparticle modified electrodes, and scavenging by fluorescent probes and antioxidants. *J. Electroanal. Chem.* **2008**, *619–620*, 131–136.
42. Sharpe, E.; Frasco, T.; Andreescu, D.; Andreescu, S. Portable ceria nanoparticle-based assay for rapid detection of food antioxidants (NanoCerac). *Analyst* **2013**, *138*, 249–262.
43. Sharpe, E.; Bradley, R.; Frasco, T.; Jayathilaka, D.; Marsh, A.; Andreescu, S. Metal oxide based multisensor array and portable database for field analysis of antioxidants. *Sens. Actuators, B* **2014**, *193*, 552–562.
44. Sharpe, E.; Andreescu, S. Portable Nanoparticle Based Sensors for Antioxidant Analysis. *Methods Mol. Biol.* **2015**, *1208*, 221–231.

45. Li, Y. F.; Liu, Z. M.; Liu, Y. L.; Yang, Y. H.; Shen, G. L.; Yu, R. Q. A mediator-free phenol biosensor based on immobilizing tyrosinase to ZnO nanoparticles. *Anal. Biochem.* **2006**, *349* (1), 33–40.
46. Putzbach, W.; Ronkainen, N. J. Immobilization Techniques in the Fabrication of Nanomaterial-Based Electrochemical Biosensors: A Review. *Sensors* **2013**, *13*, 4811–4840.
47. Gorton, L.; Lindgren, A.; Larsson, T.; Munteanu, F. D.; Ruzgas, T.; Gazaryan, I. Direct electron transfer between heme-containing enzymes and electrodes as basis for third generation biosensors. *Anal. Chim. Acta* **1999**, *400*, 91–108.
48. Wang, S.; Tan, Y.; Zhao, D.; Liu, G. Amperometric tyrosinase biosensor based on Fe_3O_4 nanoparticles–chitosan nanocomposite. *Biosens Bioelectron.* **2008**, *23* (12), 1781–1787.
49. Njagi, J.; Andreescu, S. Stable enzyme biosensors based on chemically synthesized Au-polypyrrole nanocomposites. *Biosens. Bioelectron.* **2007**, *23*, 168–175.
50. Santhosh, P.; Manesh, K. M.; Lee, S. H.; Uthayakumar, S.; Gopalan, A. I.; Lee, K. P. Sensitive electrochemical detection of superoxide anion using gold nanoparticles distributed Poly(methyl methacrylate)-polyaniline core-shell electrospun composite electrode. *Analyst* **2011**, *136*, 1557–1561.
51. Tian, Y.; Mao, L.; Okajima, T.; Ohsaka, T. A carbon fiber microelectrode based third-generation biosensor for superoxide anion. *Biosens. Bioelectron.* **2005**, *21*, 557–564.
52. Li, H.; Zhou, Q.; Wu, Y.; Fu, J.; Wang, T.; Jiang, G. Effects of waterborne nano-iron on medaka (*Oryzias latipes*): Antioxidant enzymatic activity, lipid peroxidation and histopathology. *Ecotoxicol. Environ. Saf.* **2009**, *72*, 684–692.
53. Papanikolaou, G.; Pantopoulos, K. Iron metabolism and toxicity. *Toxicol. Appl. Pharmacol.* **2005**, *202*, 199–211.
54. Muller, K.; Skepper, J. N.; Posfai, M.; Trivedi, R.; Howarth, S.; Corot, C.; Lancelot, E.; Thompson, P. W.; Brown, A. P.; Gillard, J. H. Effect of ultra small super- paramagnetic ironoxide nanoparticles (Ferumoxtran-10)on human monocyte-macrophages in vitro. *Biomaterials* **2007**, *28*, 1629–1642.
55. Li, N. B.; Park, J. H.; Park, K.; Kwon, S. J.; Shin, H.; Kwak, J. Characterization and electrocatalytic properties of Prussian blue electrochemically deposited on nano-Au/PAMAM dendrimer-modified gold electrode. *Biosens. Bioelectron.* **2008**, *23*, 1519–1526.
56. Zhang, X.; Guo, Q.; Cui, D. Recent advances in nanotechnology applied to biosensors. *Sensors* **2009**, *9*, 1033–1053.
57. Gao, F.; Yuan, R.; Chai, Y.; Chen, S.; Cao, S.; Tang, M. Amperometric hydrogen peroxide biosensor based on the immobilization of HRP on nano-Au/Thi/poly (p-aminobenzene sulfonic acid)-modified glassy carbon electrode. *J. Biochem. Biophys. Methods* **2007**, *70*, 407–413.
58. Shi, A. W.; Qu, F. L.; Yang, M. H.; Shen, G. L.; Yu, R. Q. Amperometric H_2O_2 biosensor based on poly-thionine nanowire/HRP/nano-Au-modified glassy carbon electrode. *Sens. Actuators, B* **2008**, *129*, 779–783.

59. Du, Y.; Gao, X.; Ye, X.; Zheng, Z.; Feng, Q.; Wang, C.; Wu, K. Composition and architecture-engineered Au–SnO_2/GNs-SWCNTs nanocomposites as ultrasensitive and robust electrochemical sensor for antioxidant additives in foods. *Sens. Actuators, B* **2014**, *203*, 926–934.
60. Karunakaran, G.; Suriyaprabh, R.; Manivasakan, P.; Yuvakkumar, R.; Rajendran, V.; Kannan, N. Screening of *in vitro* cytotoxicity, antioxidant potential and bioactivity of nano-and micro-ZrO_2 and TiO_2 particles. *Ecotoxicol. Environ. Saf.* **2013**, *93*, 191–197.
61. Paul, S.; Saikia, J. P.; Samdarshi, S. K.; Konwar, B. K. Investigation of antioxidant property of iron oxide particles by 1'-1'diphenylpicryl-hydrazyle (DPPH) method. *J. Magn. Magn. Mater.* **2009**, *321*, 3621–3623.
62. Saikia, J. P.; Paul, S.; Konwar, B. K.; Samdarshi, S. K. Nickel oxide nanoparticles: a novel antioxidant. *Colloids Surf. B* **2010**, *78*, 146–148.
63. Magalhaes, L. M.; Segundo, M. A.; Reis, S.; Lima, J. L. F. C. Methodological aspects about *in vitro* evaluation of antioxidant properties. *Anal. Chim. Acta* **2008**, *613*, 1–19.
64. Nezhad, H. M. R.; Alimohammadi, M.; Tashkhourian, J.; Razavian, S. M. Optical detection of phenolic compounds based on the surface plasmon resonance band of Au nanoparticles. *Spectrochim. Acta A.* **2008**, *71*, 199–203.
65. Zayats, M.; Baron, R.; Popov, I.; Willner, I. Biocatalytic Growth of Au Nanoparticles: From Mechanistic Aspects to Biosensors Design. *Nano Lett.* **2005**, *5*, 21–25.
66. Zhou, N.; Wang, J.; Chen, T.; Yu, Z.; Li, G. X. Enlargement of Gold Nanoparticles on the Surface of a Self-Assembled Monolayer Modified Electrode: A Mode in Biosensor Design. *Anal. Chem.* **2006**, *78*, 5227–5230.
67. Li, H.; Ma, X.; Dong, J.; Qian, W. Development of Methodology Based on the Formation Process of Gold Nanoshells for Detecting Hydrogen Peroxide Scavenging Activity. *Anal. Chem.* **2009**, *81*, 8916–8922.
68. Ma, X.; Li, H.; Dong, J.; Qian, W. Determination of hydrogen peroxide scavenging activity of phenolic acids by employing gold nanoshells precursor composites as nanoprobes. *Food Chem.* **2011**, *126*, 698–704.
69. Xu, B. B.; Ma, X. Y.; Rao, Y. Y.; Dong, J.; Qian, W. P. Plasmonic biosensors and nanoprobes based on gold nanoshells. *Chin. Sci. Bull.* **2011**, *56*, 3234–3241.
70. Rad, A. S.; Mirabi, A.; Binaian, E.; Tayebi, H. A review on glucose and hydrogen peroxide biosensor based on modified electrode included silver nanoparticles. *Int. J. Electrochem. Sci.* **2011**, *6*, 3671–3683.
71. Schäferling, M.; Grögel, D. B. M.; Schreml, S. Luminescent probes for detection and imaging of hydrogen peroxide. *Microchim. Acta* **2011**, *174*, 1–18.
72. Chen, S.; Yuan, R.; Chai, Y.; Hu, F. Electrochemical sensing of hydrogen peroxide using metal nanoparticles: a review. *Microchim. Acta* **2013**, *180*, 15–32.
73. Pazdzioch-Czochra, M.; Widenska, A. Spectrofluorimetric determination of hydrogen peroxide scavenging activity. *Anal. Chim. Acta* **2002**, *452*, 177–184.

74. Gao, L.; Zhuang, J.; Nie, L.; Zhang, J.; Zhuang, Y.; Gu, N.; Wang, T.; Feng, J.; Yang, D.; Perrett, S.; Yan, X. Intrinsic peroxidase-like activity of ferromagnetic nanoparticles. *Nat Nanotechnol.* **2007**, *2*, 577–583.
75. Asati, A.; Santra, S.; Kaittanis, C.; Nath, S.; Perez, J. M. Oxidase-like activity of polymer-coated cerium oxide nanoparticles. *Angew. Chem., Int Ed.* **2009**, *48*, 2308–2312.
76. Dai, Z.; Liu, S.; Bao, J.; Ju, H. Nanostructured FeS as a mimic peroxidase for biocatalysis and biosensing. *Chem.−Eur. J.* **2009**, *15*, 4321–4326.
77. He, W.; Wu, X.; Liu, J.; Hu, X.; Zhang, K.; Hou, S.; Zhou, W.; Xie, S. Design of AgM Bimetallic Alloy Nanostructures (M = Au, Pd, Pt) with Tunable Morphology and Peroxidase-Like Activity. *Chem Mater.* **2010**, *22*, 2988–2994.
78. He, W.; Liu, Y.; Yuan, J.; Yin, J.-J.; Wu, X.; Hu, X.; Zhang, K.; Liu, J.; Chen, C.; Ji, Y.; Guo, Y. Au@Pt nanostructures as oxidase and peroxidase mimetics for use in immunoassays. *Biomaterials.* **2011**, *32*, 1139–1147.
79. He, W.; Jia, H.; Li, X.; Lei, Y.; Li, J.; Zhao, H.; Mi, L.; Zhang, L.; Zheng, Z. Understanding the formation of CuS concave superstructures with peroxidase-like activity. *Nanoscale* **2012**, *4*, 3501–3506.
80. Song, Y.; Qu, K.; Zhao, C.; Ren, J.; Qu, X. Graphene oxide: intrinsic peroxidase catalytic activity and its application to glucose detection. *Adv. Mater.* **2010**, *22*, 2206–2210.
81. Song, Y.; Wang, X.; Zhao, C.; Qu, K.; Ren, J.; Qu, X. Label-free colorimetric detection of single nucleotide polymorphism by using single-walled carbon nanotube intrinsic peroxidase-like activity. *Chem.−Eur. J.* **2010**, *16*, 3617–3621.
82. Jv, J.; Li, B.; Cao, R. Positively-charged gold nanoparticles as peroxidase mimic and their application in hydrogen peroxide and glucose detection. *Chem. Commun.* **2010**, *46*, 8017–9.
83. Wang, X; Wu, Q; Shan, Z; Huang, Q. BSA-stabilized Au clusters as peroxidase mimetics for use in xanthine detection. *Biosens Bioelectron.* **2011**, *26*, 3614–9.
84. Wang, S.; Chen, W.; Liu, A.; Hong, L.; Deng, H.; Lin, X. Comparison of the peroxidaselike activity of unmodified, amino-modified, and citrate-capped gold nanoparticles. *Chem. Phys. Chem.* **2012**, *13*, 1199–204.
85. He, W.; Zhou, Y.-T.; Wamer, W. G.; Hu, X.; Wu, X.; Zheng, Z.; Boudreau, M. D.; Yin, J.-J. Intrinsic catalytic activity of Au nanoparticles with respect to hydrogen peroxide decomposition and superoxide scavenging. *Biomaterials* **2013**, *34*, 765–773.
86. Endo, T.; Yanagida, Y.; Hatsuzawa, T. Quantitative determination of hydrogen peroxide using polymer coated Ag nanoparticles. *Measurement* **2008**, *41*, 1045–1053.
87. Filippo, E.; Serra, A.; Manno, D. Poly(vinyl alcohol) capped silver nanoparticles as localized surface plasmon resonance-based hydrogen peroxide sensor. *Sens. Actuators, B* **2009**, *138*, 625–630.
88. Vasileva, P.; Donkova, B.; Karadjova, I.; Dushkin, C. Synthesis of starch-stabilized silver nanoparticles and their application as a surface

plasmon resonance-based sensor of hydrogen peroxide. *Colloids Surf., A* **2011**, *382*, 203–210.

89. Bhatia, P.; Yadav, P.; Gupta, B. D. Surface plasmon resonance based fiber optic hydrogen peroxide sensor using polymer embedded nanoparticles. *Sens. Actuators, B* **2013**, *182*, 330–335.
90. Zhang, Y.; Zhang, Y.-J.; Xia, X.-D.; Hou, X.-Q.; Feng, C.-T.; Wang, J.-X.; Deng, L. A quantitative colorimetric assay of H_2O_2 and glucose using silver nanoparticles induced by H_2O_2 and UV. *Chin. Chem. Lett.* **2013**, *24*, 1053–1058.
91. Salimi, A.; Hallaj, R.; Soltanian, S.; Mamkhezri, H. Nanomolar detection of hydrogen peroxide on glassy carbon electrode modified with electrodeposited cobalt oxide nanoparticles. *Anal. Chim. Acta* **2007**, *594*, 24–31.
92. Lei, C. X.; Hu, S. Q.; Shen, G. L.; Yu, R. Q. Immobilization of horseradish peroxidase to a nano-Au monolayer modified chitosan-entrapped carbon paste electrode for the detection of hydrogen peroxide. *Talanta* **2003**, *59*, 981–988.
93. Zhao, Y. D.; Bi, Y. H.; Zhang, W. D.; Luo, Q. M. The interface behavior of hemoglobin at carbon nanotube and the detection for H_2O_2. *Talanta* **2005**, *65*, 489–494.
94. Deng, Z.; Gong, Y.; Luo, Y.; Tian, Y. WO_3 nanostructures facilitate electron transfer of enzyme: application to detection of H_2O_2 with high selectivity. *Biosens. Bioelectron.* **2009**, *24*, 2465–2469.
95. Kicela, A.; Daniele, S. Platinum black coated microdisk electrodes for the determination of high concentrations of hydrogen peroxide in phosphate buffer solutions. *Talanta* **2006**, *68*, 1632–1639.
96. Tang, Y. H.; Cao, Y.; Wang, S. P.; Shen, G. L.; Yu, R. Q. Surface attached-poly(acrylic acid) network as nanoreactor to in situ synthesize palladium nanoparticles for H_2O_2 sensing. *Sens. Actuators, B* **2006**, *137*, 736–740.
97. Yao, S. J.; Xu, J. H.; Wang, Y.; Chen, X. X.; Xu, Y. X.; Hu, S. S. A highly sensitive hydrogen peroxide amperometric sensor based on MnO_2 nanoparticles and dihexadecyl hydrogen phosphate composite film. *Anal. Chim. Acta* **2006**, *557*, 78–84.
98. Batchelor-McAuleya, C.; Du, Y.; Wildgoosea, G. G.; Compton, R. G. The use of copper(II) oxide nanorod bundles for the non-enzymatic voltammetric sensing of carbohydrates and hydrogen peroxide. *Sens. Actuators, B* **2008**, *135*, 230–235.
99. Miao, X. M.; Yuan, R.; Chai, Y. Q.; Shi, Y. T.; Yuan, Y. Y. Direct electrocatalytic reduction of hydrogen peroxide based on Nafion and copper oxide nanoparticles modified Pt electrode. *J. Electroanal. Chem.* **2008**, *612*, 157–163.
100. Jia, W. Z.; Guo, M.; Zheng, Z.; Yu, T.; Wang, Y.; Rodriguez, E. G.; Lei, Y. Vertically Aligned CuO Nanowires Based Electrode for Amperometric Detection of Hydrogen Peroxide. *Electroanalysis* **2008**, *20*, 2153–2157.
101. Ping, J.; Ru, S.; Fan, K.; Wu, J.; Ying, Y. Copper oxide nanoparticles and ionic liquid modified carbon electrode for the non-enzymatic electrochemical sensing of hydrogen peroxide. *Microchim. Acta* **2010**, *171*, 117–123.

102. Zhao, W.; Wang, H.; Qin, X.; Wang, X.; Zhao, Z.; Miao, Z.; Chen, L.; Shan, M.; Fang, Y.; Chen, Q. A novel nonenzymatic hydrogen peroxide sensor based on multi-wall carbon nanotube/silver nanoparticle nanohybrids modified gold electrode. *Talanta* **2009**, *80*, 1029–1033.
103. Wang, Q.; Yun, Y. B.; Zheng, J. B. Nonenzymatic hydrogen-peroxide sensor based on a polyaniline-single walled carbon nanotubes composite in a room temperature ionic liquid. *Microchim. Acta* **2009**, *167*, 153–157.
104. Maleki, H.; Safavi, A.; Tajabadi, F. High-Performance Carbon Composite Electrode Based on an Ionic Liquid as a Binder. *Anal. Chem.* **2006**, *78*, 3820–3826.
105. Rosi, N. L.; Mirkin, C. A. Nanostructures in Biodiagnosis. *Chem. Rev.* **2005**, *105*, 1547–1562.
106. He, P.; Shen, L.; Liu, R.; Luo, Z.; Li, Z. Direct Detection of β-Agonists by Use of Gold Nanoparticle-Based Colorimetric Assays. *Anal. Chem.* **2011**, *83*, 6988–6995.
107. Ghadiali, J. E.; Stevens, M. M. Enzyme-responsive nanoparticle systems. *Adv. Mater.* **2008**, *20*, 4359–4363.
108. Xu, L.; Han, G.; Hu, J.; He, Y.; Pan, J.; Li, Y.; Xiang, J. Hydrophobic coating- and surface active solvent-mediated self-assembly of charged gold and silver nanoparticles at water-air and water-oil interfaces. *Phys. Chem. Chem Phys.* **2009**, *11*, 6490–6497.
109. Halliwell, B.; Gutteridge, J. M. C. Oxygen toxicity, oxygen radicals, transition metals, and disease. *Biochem. J.* **1984**, *219*, 1–14.

Chapter 17

Proton-Transfer-Reaction Time-of-Flight Mass Spectrometry (PTR-TOFMS) for Aroma Compound Detection in Real-Time: Technology, Developments, and Applications

J. Beauchamp*,[1] and J. Herbig[2]

[1]Fraunhofer Institute of Process Engineering and Packaging IVV, Department of Sensory Analytics, Giggenhauser Str. 35, 85354 Freising, Germany
[2]IONICON Analytik GmbH, Eduard-Bodem-Gasse 3, 6020 Innsbruck, Austria
***E-mail: jonathan.beauchamp@ivv.fraunhofer.de.**

Proton transfer reaction-mass spectrometry, PTR-MS, is an established tool in flavour research for the real-time detection of volatile aroma compounds in both *in vitro* and *in vivo* applications. The latter development of a PTR-MS system coupled to a time-of-flight mass spectrometer, PTR-TOFMS, provides unprecedented time and mass resolution in the real-time analysis of odorants. This chapter reviews the technology and reports on latest developments of PTR-TOFMS, including a fast gas-chromatographic pre-separation stage, autosampling capabilities and sensitivity improvements. Selected food-flavour applications are also reviewed and discussed.

Technology

Proton transfer reaction-mass spectrometry, PTR-MS, is a technique based on soft chemical ionization via proton transfer for the detection of volatile organic compounds, VOCs. It was developed approximately two decades ago and has seen application in many disciplines, predominantly atmospheric chemistry, medicine, and flavour research (*1*). PTR-MS conventionally utilizes a quadrupole mass filter for selectively transferring reagent and analyte ions to the secondary electron multiplier, SEM, detection system. Although this configuration – often referred to as PTR-QMS – is sufficient and even beneficial in certain applications, the rapid measurement time potentially achievable is restricted when measuring multiple compounds by the inherent need for these to be individually selected in succession by the quadrupole mass filter, which necessarily imposes a required period for measurement and thereby prolongs a single analytical cycle. The coupling of the PTR ion source and flow-drift tube, FDT, reaction chamber to a time-of-flight, TOF, mass spectrometer, which was first achieved approximately ten years ago (*2*), has resulted in a PTR-TOFMS system that can overcome these time restrictions whilst additionally offering a greater mass resolution and analytical range.

The operating principle of PTR-TOFMS in terms of the ionization of neutral analytes is identical to that of PTR-QMS, and its construction is depicted in Figure 1. The system comprises a front-end ion source, typically a hollow cathode discharge source (*1*) or a radioactive α-particle emitting ^{241}Am strip (*2*), to produce hydronium ions, H_3O^+, i.e. protonated water, from a continuous flow of water vapour that is fed into the ion source. Recent developments have broadened the range of reagent ion use to include NO^+ or O_2^+ (*3*), or Kr^+ (*4*), but these have limited applications in flavour research and will not be discussed in detail here.

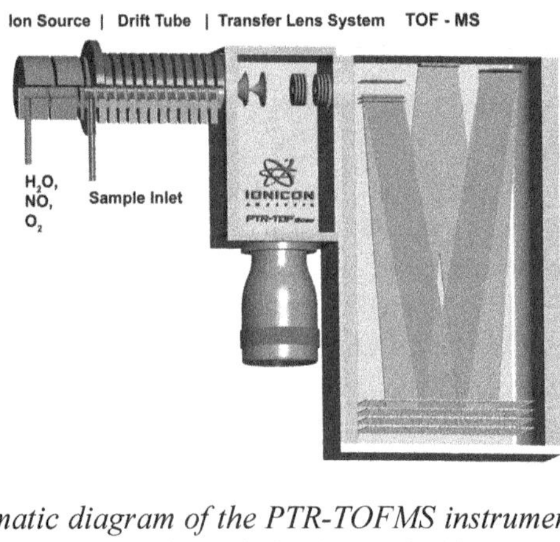

Figure 1. Schematic diagram of the PTR-TOFMS instrument consisting of a hollow cathode ion source, a (flow) drift tube, and a detection system comprising a transfer lens system and a time-of-flight, TOF, chamber. © IONICON Analytik GmbH

The reagent hydronium ions are transferred to a FDT reaction chamber in which sample gas containing the volatile target molecules is directly introduced. These neutral molecules become ionized via proton transfer upon 'collision' with the hydronium ions, provided the conditions are energetically favourable for such reactions. A particular advantage of this protonation reaction is that it proceeds at the collisional rate and at relatively low energy, thus the fragmentation of the neutral target is limited and invariably the quasi-molecular parent ion, i.e. MH^+, is generated.

At the downstream end of the FDT the protonated analytes and precursor ions are transferred into the detection system, which differ between PTR-QMS and PTR-TOFMS. The PTR-QMS detection system comprises a quadrupole mass filter followed by an axially-orthogonal SEM for quantitation of incoming ions. By comparison, the PTR-TOFMS system incorporates a series of transfer lenses – or alternatively a multipole ion guide, as discussed later – to focus the ion beam into the time-of-flight chamber. The transfer system also serves as a multi-stage differential pumping system to reduce the defined pressure from within the FDT to the high vacuum conditions required in the detection system.

Upon entering the TOF region the ion swarm is given a transverse electronic impulse – or more accurately, rapid, successive pulses at ~30 μs intervals – to inject or 'shoot' a cluster of ions into the flight chamber. The cluster moves in this field-free region with its initial axial momentum and the additional transverse momentum received by the electronic impulse, and is only reflected and refocused at the far end of the chamber by a reflector, thus describing a v-shaped path. After reflection the swarm returns to the axial position where the individual ions are detected by a multichannel plate, MCP (*5*).

All of the ions within the pulsed swarm are imparted the same initial impulse into the flight chamber. Their flight time through the chamber is directly proportional to their kinetic energy and, by definition, to the square root of their mass. Thus, lighter ions travel faster and reach the detector sooner than heavier ions; exact knowledge of this timing and the length of the flight chamber allow precise calculation of the mass-to-charge ratio, *m/z*, of the incoming ions, thus providing the high mass resolution afforded by the PTR-TOFMS system.

The rapid mass spectral analysis and high mass resolving power offered by PTR-TOFMS have been a driver for its increasing use and popularity in recent years, especially in flavour analysis. The ensuing sections of this chapter introduce selected applications of PTR-TOFMS for the analysis of aroma compounds in food science and discuss recent notable technical developments.

Applications

PTR-MS, like atmospheric pressure chemical ionization MS, APCI-MS (*6*), has become the technology of choice for investigating fast processes in flavour research, such as the *in vivo* release of aroma compounds during food mastication or the development of volatiles in the headspace of food over time. The advantages offered by PTR-TOFMS over PTR-QMS greatly broaden the possibilities achievable in on-line flavour analysis. First and foremost, the rapid detection ability of a complete mass spectrum enables the entire complex mixture of aroma compounds to be monitored almost simultaneously, which is not achievable with quadrupole mass filter based instruments such as PTR-QMS and APCI-MS. It should be noted here that APCI-MS employing ion-trap systems also exist: these similarly allow full mass spectral scans to be performed within a fraction of a second and additionally offer the advantage of MS^n capabilities for compound structure elucidation, but such systems are not widespread and the latter feature is mainly reserved for specific scientific study rather than routine use for flavour release applications (*7*). Secondly, the high mass resolving power afforded by the PTR-TOFMS, which is in the region of *m*/*m* 5000-6000, offers the possibility for seperating isobaric compounds and improving compound identification certainties in the analysis of complex gas matrices, as often encountered with food.

PTR-TOFMS has been used for varied studies in flavour research. Due to its complex flavour composition, coffee in particular has been the subject of many PTR-TOFMS early and current investigations that have included tentatively identifying compounds in the headspace of speciality coffee varieties that were distinguishable by chemometric analysis of PTR-QMS data (*8*), characterizing flavour formation in roasted coffee of different geographic origin (*9*, *10*), and analysing *in vivo* aroma release of coffee during consumption (*11*). Other diverse food-related studies utilizing PTR-TOFMS have included attempts to correlate pig rearing systems on volatile profiles of the resulting dry-cured ham (*12*) and following post-harvest ripening of apples based on their VOC profiles (*13*). Clearly, PTR-TOFMS has wide-reaching applicability in flavour science in view of the diversity of the aforementioned topics.

Two specific areas of application of PTR-TOFMS will be highlighted here, namely its use in nosespace analysis for flavour release studies and its implementation as a method for ascertaining limits of detection of aroma compounds in relation to the headspace analysis of liquid samples.

Nosespace Analysis

Nosespace analysis is the technique by which aroma compounds that are released in the mouth during food mastication or beverage consumption are detected in the nose-exhaled breath. Notably, the analytical detection of these compounds is generally considered to directly relate to the perceived aroma of food being consumed, albeit with many influencing factors (*6*, *14*, *15*).

Sensory studies on foods via on-line nosespace analysis can be performed in diverse fashions. In relation to consumption behaviour, broadly speaking these can be executed either via a free sampling procedure or by imposing a consumption protocol. There is currently no general consensus amongst flavour scientists as to which mode yields better results; although the free-form approach more closely reflects the natural eating behaviour and thereby more realistically emulates perception, the inherent and extensive inter- and intra-individual variability in consumption behaviour and breathing patterns can create substantial challenges in aligning the PTR-MS nosespace data for comparison within and between subjects. To tackle this issue, Frank and co-workers proposed using a breathing protocol for aiding PTR-MS nosespace analysis of aroma release by presenting test participants with a visual animation to enforce and ensure a uniform breathing rate across all panellists (*16*). The resulting PTR-MS data could be more easily and directly compared due to the uniform time intervals of exhalation and facilitated an accurate examination of the pre-swallow and post-swallow phases of volatile release with respect to concurrent time intensity perceptual measurements.

This approach of using a visual breathing protocol was recently adopted for PTR-TOFMS measurements of the release of selected aroma compounds from an aromatized confectionary foam via nosespace analysis (*17*). Like in the aforementioned study, participants conformed to a prescribed consumption and breathing protocol. Figure 2 demonstrates the uniformity of the data achievable between a single subject and the group mean for the release of the compound 2-methylbutan-1-ol (detected at *m/z* 89.09) from a confectionary foam.

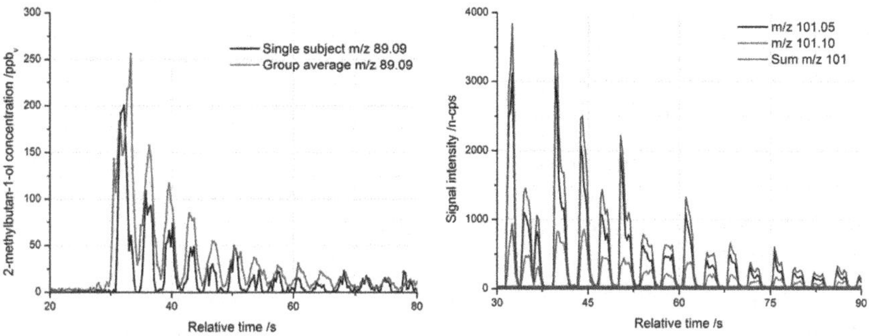

Figure 2. Real-time PTR-TOFMS nosespace analysis of selected aroma compounds released from aromatized confectionary foams during consumption following a strict sampling and breathing protocol. Left: The retronasal release pattern of 2-methylbutan-1-ol (detected at m/z 89.09) from an individual test subject in comparison with the group average. Right: The release of 2,3-pentanedione (m/z 101.05) and cis-3-hexen-1-ol (m/z 101.10) in comparison to the total trace for m/z 101 (representative of PTR-QMS analysis) for a single test subject. (Note that the right-hand graph is plotted in signal intensity units of normalized counts per second, n-cps.). © Fraunhofer IVV

In addition, these analyses further highlight the strength of PTR-TOFMS in its ability to separate isobaric compounds. The right-hand graph of Figure 2 plots the PTR-TOFMS *in vivo* detection of the concordant release of two isobaric flavour compounds, namely 2,3-pentanedione ($C_5H_8O_2$ at *m/z* 101.05) and *cis*-3-hexen-1-ol ($C_6H_{12}O$ at *m/z* 101.10), from the same confectionary foam. Clearly, these two compounds have differing release profiles, with the former dominating in terms of absolute abundance. Nevertheless, cumulative intensity, plotted as the sum of the two signals (at nominal *m/z* 101), closely reflects the signal of the dominant compound, as expected, and indicates the potential error in PTR-QMS analysis due to the inability for isobaric compound separation. In this case, if PTR-QMS analysis were to be made under the assumption that only *cis*-3-hexen-1-ol is present and responsible for the signal at *m/z* 101, then the release profile would be falsely characterized.

Clearly, PTR-TOFMS analysis offers great advantages in this respect. The ability to distinguish between many isobaric compounds additionally increases the amount of information generated from analysis; in PTR-QMS the analysis of any nominal *m/z* by definition only returns one signal, but analysis by PTR-TOFMS of the same *m/z* might deliver several individual signals, each attributable to a different compound.

A further advantage of using PTR-TOFMS for nosespace analysis is that complete mass spectra are recorded in quick succession, thus there is no need to restrict the analysis to a small handful of compounds *a priori*, as is required in PTR-QMS nosespace analysis. This has been achieved to good effect in the headspace and nosespace analysis of coffee volatiles from different coffee types and for different roasting degrees (*11*), which is an excellent example of the strength of this technique for such purposes.

Limits of Detection and Quantitation for Headspace Analysis

Any analytical procedure for the measurement of individual compounds has an associated limit of detection, LOD, which represents the minimum concentration at which a compound can be detected with the specified confidence. It is common and appropriate practice when reporting measurements performed using analytical instrumentation to determine and disclose the LOD of the system and set-up in order to provide a measure of the quality and suitability of the equipment and analytical procedure. The LOD is typically taken as the instrumental signal associated with a particular compound that is a factor three higher than the standard deviation of the background noise of that signal, i.e. the signal intensity in the absence of the substance. Further, a limit of quantitation, LOQ, which relates to the minimum concentration at which a compound can be quantified, is typically given as a factor 3.3 higher than the LOD.

In PTR-MS applications, including PTR-TOFMS, the LOD and LOQ values are reported as gas-phase volume mixing ratios, typically in the range of parts-per-billion or parts-per-trillion by volume (ppb_v and ppt_v, respectively). Technological advances in recent years have led to further reductions in the PTR-MS LOD, demonstrated to sub-pptv levels (*13*, *18*). It should be noted, however, that the reported values are instrument LODs that were measured under ideal conditions and that the analysers themselves are rarely the limiting factor. By contrast, method-related LODs are more relevant, since they take into account the analytical procedure, such as the sampling and/or transfer of headspace gas into the analyser. In addition, in flavour research often the concentration of aroma compounds in the food matrix is of greater interest than the headspace concentrations. As such, it is desirable to establish a measure for the minimum concentration at which a compound in a food matrix can be detected or quantified in its headspace.

Recent studies investigated this concept by using PTR-TOFMS to detect different flavour compounds in the headspace of model aqueous solutions containing these compounds (*19*). Solutions containing different dilutions of isobaric pairs of flavour compounds were individually analysed via PTR-TOFMS static headspace analysis and the detected concentrations were correlated with the aqueous concentrations in solution. Figure 3 (left) shows the detected gas-phase headspace concentration of 2,3-pentanedione in relation to its concentration in the aqueous solution.

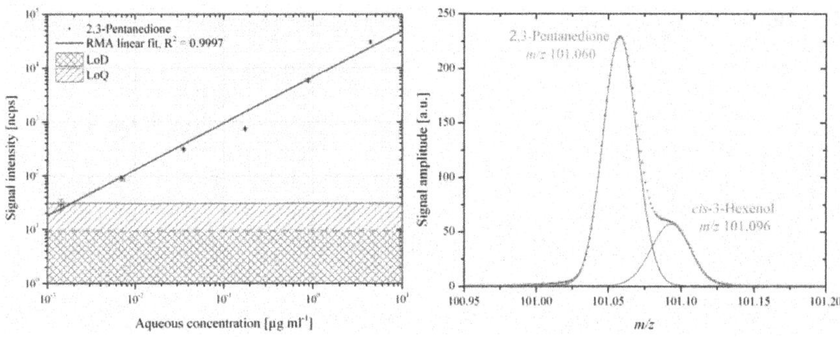

Figure 3. PTR-TOFMS static headspace analysis of selected aroma compounds from aqueous solutions. Left: Determination of the limits of detection, LOD, and quantitation, LOQ, in solution for 2,3-pentanedione via headspace analysis of aqueous solutions containing this compound at 0.0014, 0.007, 0.035, 0.18, 0.89, and 4.43 µg mL^{-1} concentration. A reduced major axis, RMA, fit was applied to the data to estimate the limits. Right: PTR-TOFMS spectrum of 2,3-pentanedione (m/z 101.06) and cis-3-hexen-1-ol (m/z 101.10) from an aqueous solution containing these two compounds. The data were fitted with Gaussian peak functions. Reproduced with permission from (19).

Based on the definitions given above for LOD and LOQ, the investigations yielded values for the limits of detection and quantitation in solution, indicating the aqueous concentration at which this compound can still be detected and quantified with certainty. Similar analyses for other flavour compounds are reported in the related study (*19*).

In addition to determining LOD and LOQ, the discrimination levels of isobaric (flavour) compounds by PTR-TOFMS was also investigated in this study. The right-hand plot of Figure 3 depicts the mass spectrum of two isobaric compounds, namely (similarly for the nosespace analyses depicted in Figure 2) 2,3-penatanedione and *cis*-3-hexen-1-ol, as measured via headspace analysis of an aqueous solution containing the two compounds. In this example the two discrete peaks of the individual compounds are clearly discernible, but below a certain threshold aqueous concentration they are no longer distinguishable, even if this concentration is above the gas-phase LOQ (*19*). Characterizing these parameters of LOD and LOQ in solution, as well as the isobaric discrimination level, should be made for the specific food matrix and flavour compounds under investigation to ensure that their detection in subsequent applications is accurately reported.

Developments

Since its inception, PTR-TOFMS has undergone several evolutions in an effort to lower its limits of detection and improve its sensitivity. Several further key developments have been made in broadening the analytical range that are of particular value in flavour analysis. These major developments are reviewed here.

Improvement in Sensitivity

The sensitivity of any analytical instrument is broadly defined as the signal response to a given stimulus quantity. In PTR-MS – including PTR-TOFMS – the signal intensity relates to the number of (product) ions reaching the detector, which is generally quantified as counts-per-second, cps. This signal is directly proportional to the abundance, more precisely, the volume mixing ratio, of neutral target molecules in the sample gas, which are quantified in parts-per unit volume, typically ppb_V. The sensitivity of PTR-MS is therefore usually stated as cps ppb_V^{-1}. In PTR-MS it is common practice to normalize the signal intensity of the product ions to that of the primary hydronium ions, typically with a normalization factor of 10^6, to give normalized counts-per-second, ncps, thus sensitivity data are often alternatively given as ncps ppb_V^{-1} (*20, 21*). However, since the true sensitivity is reflected only in the raw cps signal, comparison between instruments of sensitivities reported as normalized signals is inaccurate, thus papers reporting only ncps-based units will be not be included here.

PTR-TOFMS instruments have undergone significant improvement in terms of their sensitivities. The prototype instrument was reported in 2004 to have a sensitivity of 0.017 cps ppb_V^{-1} for hydrogen sulphide (*2*). An alternative system was built in 2005 by Ennis and co-workers, but the publication reported sensitivities in ncps ppb_V^{-1} units and provided no information relating to the primary ion signal intensity, thus making a conversion of these data to un-normalized cps ppb_V^{-1} impossible (*22*). The first commercial PTR-TOFMS system available in 2009 reported sensitivities ranging from 13 cps ppb_V^{-1} for benzene to 26 cps ppb_V^{-1} for tricholorobenzene (*5*).

Recently the PTR-TOFMS system was modified to replace the series of transfer lenses within the detection system with a quadrupole ion guide, which greatly enhanced ion transfer and thus detection sensitivity. The prototype of this commercial PTR-QiTOF system was compared to the existing commercial PTR-TOFMS (PTR-TOF 8000 model from IONICON Analytik) and achieved the following sensitivities: 2900 cps ppb_V^{-1} for benzene (compared to 100 cps ppb_V^{-1} for the PTR-TOF 8000) and 4660 cps ppb_V^{-1} for dichlorobenzene (compared to 210 cps ppb_V^{-1} for the PTR-TOF 8000) (*18*). It should be noted that the PTR-QiTOF was equipped with a different ion detection system and operated at higher FDT conditions than the PTR-TOF 8000, namely 3.8 mbar pressure (compared to 2.3 mbar) and 1 kV drift voltage (compared to 600 V), which contributed to the sensitivity improvements achieved. Further, the two aromatic compounds selected are particularly suited to PTR-MS analysis: they are detected exclusively at their parent ions, i.e. they do not fragment upon ionization and, moreover, the ion transmission efficiency of PTR-TOFMS increases proportionally with *m/z*, hence the higher sensitivity of dichlorobenzene. Other compounds at lower *m/z* or those that undergo greater fragementation are not expected to be detected with such high sensitivity. Nevertheless, this new development of a PTR-QiTOFMS offers unsurpassable sensitivities for on-line VOC detection systems currently on the market. Notably, the favourable conditions for injecting ions into the TOF chamber with this new system additionally result in a higher mass resolution of *m*/*m* 6900 (and up to *m*/*m* 10400 for dichlorobenzene at *m*/*z* 149, albeit after tuning for maximum resolution at a cost of sensitivity) (*18*).

An alternative method for improving the sensitivity of PTR-MS instruments involves utilizing a radio frequency, RF, ion funnel within the FDT, which acts to channel reagent and analyte ions into the detection system with greater efficiency (*23*). Sensitivity improvements for a PTR-TOFMS system equipped with this RF funnel were shown to be one to two orders of magnitude higher than without use of the funnel. The highest sensitivity reported was for acetone at 1162 cps ppb_V^{-1} (cf. 5.8 cps ppb_V^{-1} in standard operating mode; i.e. a factor 200 increase), but the detection efficiencies were strongly compound-dependent. Nevertheless, this configuration offers a promising alternative – or perhaps in future, an addition – to the PTR-QiTOFMS approach.

Rapid Sample Throughput

In flavour research, gas chromatography coupled with mass spectrometry, GC-MS, is the gold standard for the analysis of volatile aroma compounds. GC-MS is a comprehensive technique that typically offers unequivocal identification and accurate quantitation of flavour compounds. Despite these strengths, sample workup for GC-MS analysis is often laborious and time consuming, requiring diverse preparative steps such as solvent extraction, enrichment of volatiles in headspace analysis, or dehydration of gas-phase samples. The on-line analytical capability of PTR-TOFMS overcomes these issues by offering direct and non-destructive headspace analysis for rapid throughput of individual samples.

Until recently, the fast analysis time offered by PTR-TOFMS was not exploited to its full extent due to the necessity for manual intervention in the analyses in terms of exchanging samples and/or reconnection of the sampling inlet. In GC-MS, routine analysis of multiple samples utilizes an automated sampling system, thereby reducing manual procedures and allowing samples to be run continuously, for example overnight. This approach was recently successfully incorporated in PTR-TOFMS analysis by coupling an autosampler to the inlet system of the PTR-TOFMS (*9*). The coupled system was used to analyse volatile flavour compounds in the headspace of 108 coffee samples, which was achieved at a rate of 30 s per sample and an acquisition rate of one spectrum per second. This method of rapid, non-destructive and automated sampling opens up new possibilities in real-time analysis in flavour research. The same system has also been successfully implemented in the PTR-QMS analysis of the volatile compounds produced by bakery yeast starters (*24*) and in the detection of ethylene emissions from biological samples using PTR-TOFMS with selectable reagion ion, SRI, feature, i.e. PTR-SRI-TOFMS (*25*).

This coupled system has particular strength when used in combination with multivariate analysis. Data mining approaches including principle component analysis, PCA, partial least squares, PLS, or analysis of variance, ANOVA, are already routinely used to good success in food research applications of PTR-MS to discriminate between different samples based on their volatile headspace profiles, e.g. in relation to geographic origin (*26*) or cultivar (*13*), or nosespace profiles, e.g. for discriminating coffee types and roasting degrees (*11*). Use of an autosampler with PTR-MS will offer more efficient execution of such studies in the future.

Separation of Isomers

An inherent problem in PTR-MS, as indicated earlier, is its inability to discriminate between most isobaric and isomeric compounds due to identical parent ions and similarities in fragmentation patterns, i.e. the compounds that elicit a signal at identical *m/z*. The development of PTR-TOFMS has made it possible for most isobaric compounds to be separated, provided their centre of mass separation, or more correctly, the separation between the full-width at half maximum, FWHM, of their centre of mass, is sufficiently large (*5*, *19*, *27*). By contrast, isomeric compounds cannot be separated by their mass because such species have identical chemical composition and thus, by definition, identical mass. This issue has been partly addressed with the introduction of a selectable reagent ion, SRI, add-on to utilize charge transfer reactions from either NO^+ or O_2^+ primary ions, which often produce discriminable product ions and/or fragments (*3*). However, this solution can be impractical due to the higher complexity of the resulting mass spectra and is thus far from ideal.

Another avenue to tackle this shortcoming is to make use of a gas-phase chromatographic separation step, as used in GC-MS. A pre-separation of volatiles in gas samples using this method was achieved with good success in previous studies using PTR-QMS (*21*, *28*, *29*). However, although this newly-developed hyphenated instrument offered discriminative analysis of isomeric species, the sample analysis time was compromised and the system effectively reduced the PTR-MS to a GC system with subsequent PTR ionization and MS separation and detection.

To address this issue, recent investigations have concentrated on the implementation of a more rapid GC pre-separation stage in combination with PTR-TOFMS. This development aims to offer rapid GC separation in near real-time and in a compact peripheral unit that is small enough for direct installation inside a PTR-TOFMS instrument. With this system the same PTR-TOFMS sample inlet can be operated in its normal, real-time detection mode but can alternatively be switched to activate the GC mode, whereby the sample gas is intermittently flushed through the GC column. A set of valves are configured to load the sample gas into a sampling loop and subsequently inject a defined volume into the GC column for sample pre-separation. A regulated carrier gas flow is used to push the gas through the column. Compounds eluting from the GC column, which are separated according to their retention time, are then indivudally detected by the PTR-TOFMS instrument.

The first approach that explored this concept made use of a micro-capillary column, MCC (*30*). The MCC is typically comprised of about 1000 single capillaries that are tightly packed together and allow sufficient gas flow for subsequent connection to PTR-MS, whereby the established gas flow can be set to match the flow required by the FDT. The column is mounted outside the heated compartment that houses the FDT of the PTR-MS and can thereby be maintained at a different, defined temperature. A typical GC run with the MCC is on the order of 5-10 min. A prototype MCC-PTR-TOFMS system demonstrated a good performance for selected isomeric compounds – including food-relevant flavour compounds – and is a promising development in realizing the unequivocal detection of such compounds by PTR-TOFMS (*30*).

The second approach utilizes a standard – albeit short – GC column (6 m) with direct resistive heating (*31*). The carrier gas flow through this column is less than 10 mL min^{-1} and therefore additional make-up gas is required to reach the gas flow of the PTR-TOFMS inlet, typically 30 mL min^{-1}, in order to maintain the operating pressure in the FDT. Although this leads to a dilution of the sample gas, a major advantage of this set-up is the resistive heating concept by sending an electric current directly through the column. While variations of the temperature of the pencil-sized MCC are very limited, the short column can be heated at a rate of more than 30 °C s^{-1}, which greatly exceeds the heating rates of conventional GC systems. Moreover, the cooling rate is in the same range. Adapted temperature ramps are the key technique to optimize compound separation in GC. The temperature ramp speeds of this GC system allow optimized compound separation and spectral runs in less than one minute, which is significantly shorter than the MCC version and is therefore referred to as fastGC.

The prototype fastGC-PTR-TOFMS has been successfully tested in the analysis of aroma compounds in the headspace of wine samples to discriminate between isobaric flavour compounds (*31*), leading to an enhanced sample separation in the data analysis. This experiment also demonstrated another advantage of the fastGC step, namely the pre-separation of compounds present in the headspace at high concentrations; in this case ethanol, which is known to be problematic in PTR-MS analysis of wine (*32*, *33*).

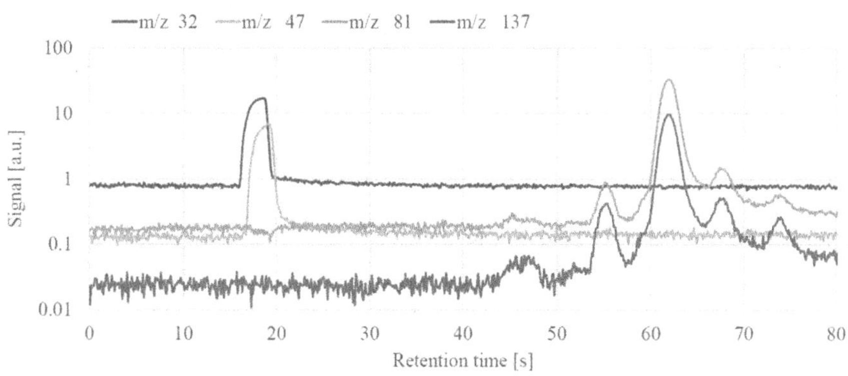

Figure 4. FastGC-PTR-TOFMS headspace analysis of a shandy beverage (a 'Radler', containing beer and lemonade in an approximate 50:50 v/v mixture). Sample injection was made at 12 s. Note the multiple signal peaks at m/z 81 and m/z 137, relating to isomeric monoterpene compounds. © IONICON Analytik GmbH

The capabilities of the fastGC pre-separation in combination with PTR-TOFMS are shown here for the headspace analysis of a shandy beverage – the German 'Radler' drink comprising an approximate 50:50 v/v mixture of beer and lemonade. The selected traces shown in the plot of Figure 4 demonstrate the advantages of analysis via fastGC-PTR-TOFMS.

As indicated above, the high concentration of ethanol in the headspace of alcoholic beverages are usually problematic for PTR-MS analysis due to a resulting depletion in the primary reagent ion; in this example using the fastGC pre-separation, ethanol elutes from the column immediately after sample injection and rapidly decays to background levels, making subsequent detection of other constituent aroma compounds of the beverage unproblematic. The presence of isomeric compounds in the present example is indicated by several distinct signal peaks in short succession, relating to the time at which each isomer elutes from the column. The trace at *m/z* 137 corresponds to a multitude of monoterpenes that would not be distinguishable without GC pre-separation. In standard PTR-MS analysis the signal at *m/z* 137 for this beverage might quickly be associated with limonene, since this is an aroma compound known to be present in lemonade at high concentrations; in this example, however, at least five different compounds (cf. five consecutive peaks) are visible for this trace, with limonene being only one.

An additional advantage of the fastGC system is that it allows closer inspection of fragmentation ratios, which can be carried out using the real sample without the necessity to revert to analysing pure compounds. In a fastGC spectrum the fragment ions exactly coincide at identical retention times; in Figure 4 this is clearly seen at *m/z* 81, which is a known PTR-MS fragment of monoterpenes detected at the parent ion at *m/z* 137. Thus fragment ions can be easily identified and their ratios to the parent ion can aid in compound identification. Oxygen, which elutes from the column after a minimal transit time, can be monitored via the parasitic pre-cursor ion O_2^+ at *m/z* 32 (cf. Figure 4). As a non-interacting compound its retention time can serve as a reference and the peak shape can be used to check for correct injection of the sample gas into the column.

Clearly the fastGC approach adds an additional dimension to PTR-TOFMS data, not only separating isomeric compounds but also providing a much deeper insight into reaction products.

Calibration of Aroma Compounds

PTR-MS is a semi-quantitative technique, meaning that the concentration of a detected target compound can be calculated from its signal intensity based on a theoretical consideration of reaction kinetics (*21*, *34*). This approach relies on exact knowledge of several parameters, including the reaction rate, which can be either calculated from known parameters of the compound or taken from literature reports. Due to uncertainties in the reaction rate, amongst other parameters, the accuracy of concentrations derived from theoretical calculations is not always optimal. The quantitative accuracy of reported concentrations can typically be improved by calibrating the instrument to the compounds of interest.

Traditionally this can be made using certified gas standards containing those compounds, whereby these are diluted at known concentration steps and analysed by PTR-MS to generate instrument-specific sensitivities for each compound (*20*). In flavour research, however, the aroma compounds of interest are typically not readily available as certified gas standards.

A recent endeavour that has aimed to overcome this problem and make calibration of PTR-MS possible with a far broader range of volatiles is the development of a calibration system in which liquid mixtures of compounds can be used. This liquid calibration unit, LCU, offers the possibility to generate accurate gas-phase concentrations of custom suites of volatile compounds (*35*).

The operating premise of the LCU, which is shown schematically in Figure 5, is an efficient evaporation of the volatile constituents of the aqueous calibration solution via nebulization at 5-50 µL min^{-1} into a 1 L min^{-1} flow of volatile-free gas, or zero-air. Microdroplets, which have a large surface area for evaporation, are sprayed into a chamber that is heated to 100 °C, and evaporate into the carrier gas stream that is then sampled by the PTR-MS instrument. Exact knowledge of specific physicochemical properties of the target (flavour) compounds, as well as the nebulization and flow parameters, allows for accurate calculation of the compound concentrations in the carrier gas, thereby enabling PTR-MS calibration of these compounds. The system has been demonstrated with great success, even for problematic compounds such as acids, semi-volatiles with high vapour pressure, and non-polar compounds, and offers a robust and reliable system for calibrating PTR-MS instruments for user-defined compounds that were previously inaccessible, including most aroma compounds.

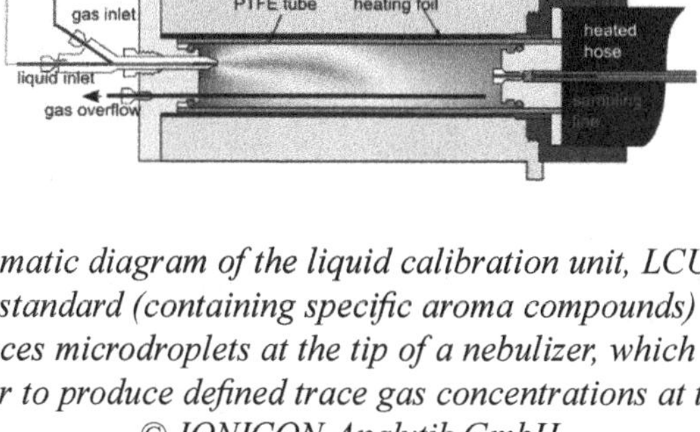

Figure 5. Schematic diagram of the liquid calibration unit, LCU. A liquid flow of an aqueous standard (containing specific aroma compounds) into a zero-air gas flow produces microdroplets at the tip of a nebulizer, which evaporate in a heated chamber to produce defined trace gas concentrations at the LCU outlet. © *IONICON Analytik GmbH*

Conclusions

Proton-transfer-reaction time-of-flight mass spectrometry, PTR-TOFMS, has found footing in food research as a non-destructive, direct analytical tool for monitoring fast processes relating to flavour compounds. The ability of the instrument to carry out rapid mass spectral analysis, in combination with its high mass resolving power for the discrimination of many isobaric compounds, make PTR-TOFMS the ideal tool for following flavour release of complex gas matrices containing multiple volatiles and it is fast becoming the technology of choice for on-line applications of aroma detection; unique amongst the real-time detection system. The key applications of PTR-TOFMS in aroma analysis reviewed in this chapter, namely retronasal aroma release via nosespace analysis, and quantitation of aqueous-phase concentrations based on careful calibration, clearly demonstrate the utility of this technology in this field of research. Improvements to the system in terms of increasing its sensitivity, and developments of peripheral additions for rapid sample throughput by use of an autosampler, calibration of user-defined compounds via a liquid calibration system, or distinguishing between isomers by use of a fast gas-chromatographic pre-separation stage, greatly extend the sampling speed and analytical range of PTR-TOFMS. The fast spectral runs allow repeated measurements to also study dynamic systems. The implementation of these new features open up extensive possibilities in the field of on-line aroma analysis, as future publications will no doubt demonstrate.

Acknowledgments

The authors would like to thank the development team at IONICON Analytik GmbH for supplying data in relation to the fastGC-PTR-TOFMS and the liquid calibration unit, LCU, in particular Lukas Fischer and Philipp Sulzer. Oxana Typakova is acknowledged for her work relating to PTR-TOFMS in vivo flavour release analysis from confectionary foams.

Conflict of Interest

JB declares no conflict of interest. JH is chief technology officer, CTO, of IONICON Analytik GmbH, manufacturer of PTR-QMS and PTR-TOFMS instrumentation and peripheral devices including the liquid calibration unit, LCU, and fastGC interface. It is noted that other manufacturers of PTR-MS instruments exist.

References

1. Hansel, A.; Jordan, A.; Holzinger, R.; Prazeller, P.; Vogel, W.; Lindinger, W. Proton-Transfer Reaction Mass-Spectrometry - Online Trace Gas-Analysis at the ppb Level. *Int. J. Mass Spectrom. Ion Processes* **1995**, *149/150*, 609–619.
2. Blake, R. S.; Whyte, C.; Hughes, C. O.; Ellis, A. M.; Monks, P. S. Demonstration of Proton-Transfer Reaction Time-of-Flight Mass

Spectrometry for Real-Time Analysis of Trace Volatile Organic Compounds. *Anal. Chem.* **2004**, *76*, 3841–3845.

3. Jordan, A.; Haidacher, S.; Hanel, G.; Hartungen, E.; Herbig, J.; Märk, L.; Schottkowsky, R.; Seehauser, H.; Sulzer, P.; Märk, T. D. An online ultra-high sensitivity proton-transfer-reaction mass-spectrometer combined with switchable reagent ion capability (PTR+SRI-MS). *Int. J. Mass Spectrom.* **2009**, *286*, 32–38.
4. Sulzer, P.; Edtbauer, A.; Hartungen, E.; Jürschik, S.; Jordan, A.; Hanel, G.; Feil, S.; Jaksch, S.; Märk, L.; Märk, T. D. From conventional proton-transfer-reaction mass spectrometry (PTR-MS) to universal trace gas analysis. *Int. J. Mass Spectrom.* **2012**, *321*, 66–70.
5. Jordan, A.; Haidacher, S.; Hanel, G.; Hartungen, E.; Märk, L.; Seehauser, H.; Schottkowsky, R.; Sulzer, P.; Märk, T. D. A high resolution and high sensitivity proton-transfer-reaction time-of-flight mass spectrometer (PTR-TOF-MS). *Int. J. Mass Spectrom.* **2009**, *286*, 122–128.
6. Taylor, A. J.; Linforth, R. S. T.; Harvey, B. A.; Blake, A. Atmospheric pressure chemical ionisation mass spectrometry for in vivo analysis of volatile flavour release. *Food Chem.* **2000**, *71*, 327–338.
7. Le Quéré, J.-L.; Gierczynski, I.; Sémon, E. An atmospheric pressure chemical ionization–ion-trap mass spectrometer for the on-line analysis of volatile compounds in foods: a tool for linking aroma release to aroma perception. *J. Mass Spectrom.* **2014**, *49*, 918–928.
8. Özdestan, Ö.; van Ruth, S. M.; Alewijn, M.; Koot, A.; Romano, A.; Cappellin, L.; Biasioli, F. Differentiation of specialty coffees by proton transfer reaction-mass spectrometry. *Food Res. Int.* **2013**, *53*, 433–439.
9. Yener, S.; Romano, A.; Cappellin, L.; Märk, T. D.; Sánchez del Pulgar, J.; Gasperi, F.; Navarini, L.; Biasioli, F. PTR-ToF-MS characterisation of roasted coffees (C. arabica) from different geographic origins. *J. Mass Spectrom.* **2014**, *49*, 929–935.
10. Gloess, A. N.; Vietri, A.; Wieland, F.; Smrke, S.; Schönbächler, B.; López, J. A. S.; Petrozzi, S.; Bongers, S.; Koziorowski, T.; Yeretzian, C. Evidence of different flavour formation dynamics by roasting coffee from different origins: On-line analysis with PTR-ToF-MS. *Int. J. Mass Spectrom.* **2014**, *365–366*, 324–337.
11. Romano, A.; Cappellin, L.; Ting, V.; Aprea, E.; Navarini, L.; Gasperi, F.; Biasioli, F. Nosespace analysis by PTR-ToF-MS for the characterization of food and tasters: The case study of coffee. *Int. J. Mass Spectrom.* **2014**, *365–366*, 20–27.
12. del Pulgar, J. S.; Soukoulis, C.; Biasioli, F.; Cappellin, L.; García, C.; Gasperi, F.; Granitto, P.; Märk, T. D.; Piasentier, E.; Schuhfried, E. Rapid characterization of dry cured ham produced following different PDOs by proton transfer reaction time of flight mass spectrometry (PTR-ToF-MS). *Talanta* **2011**, *85*, 386–393.
13. Soukoulis, C.; Cappellin, L.; Aprea, E.; Costa, F.; Viola, R.; Märk, T.; Gasperi, F.; Biasioli, F. PTR-ToF-MS, A Novel, Rapid, High Sensitivity and Non-Invasive Tool to Monitor Volatile Compound Release During Fruit

Post-Harvest Storage: The Case Study of Apple Ripening. *Food Bioprocess Technol.* **2013**, *6*, 2831–2843.

14. Taylor, A. J. Release and Transport of Flavors In Vivo: Physicochemical, Physiological, and Perceptual Considerations. *Compr. Rev. Food Sci. Food Saf.* **2002**, *1*, 45–57.
15. Buettner, A.; Beauchamp, J. Chemical input - Sensory output: Diverse modes of physiology-flavour interaction. *Food Qual. Prefer.* **2010**, *21*, 915–924.
16. Frank, D.; Appelqvist, I.; Piyasiri, U.; Wooster, T. J.; Delahunty, C. Proton Transfer Reaction Mass Spectrometry and Time Intensity Perceptual Measurement of Flavor Release from Lipid Emulsions Using Trained Human Subjects. *J. Agric. Food Chem.* **2011**, *59*, 4891–4903.
17. Tyapkova, O.; Siefarth, C.; Schweiggert-Weisz, U.; Beauchamp, J.; Buettner, A.; Bader-Mittermaier, S. Unpublished data.
18. Sulzer, P.; Hartungen, E.; Hanel, G.; Feil, S.; Winkler, K.; Mutschlechner, P.; Haidacher, S.; Schottkowsky, R.; Gunsch, D.; Seehauser, H.; Striednig, M.; Jürschik, S.; Breiev, K.; Lanza, M.; Herbig, J.; Märk, L.; Märk, T. D.; Jordan, A. A Proton Transfer Reaction-Quadrupole interface Time-Of-Flight Mass Spectrometer (PTR-QiTOF): High speed due to extreme sensitivity. *Int. J. Mass Spectrom.* **2014**, *368*, 1–5.
19. Zardin, E.; Tyapkova, O.; Buettner, A.; Beauchamp, J. Performance assessment of proton-transfer-reaction time-of-flight mass spectrometry (PTR-TOF-MS) for analysis of isobaric compounds in food-flavour applications. *LWT−Food Sci. Technol.* **2014**, *56*, 153–160.
20. Beauchamp, J.; Herbig, J.; Dunkl, J.; Singer, W.; Hansel, A. On the performance of proton-transfer-reaction mass spectrometry for breath-relevant gas matrices. *Meas. Sci. Technol.* **2013**, *24*, 125003.
21. de Gouw, J.; Warneke, C.; Karl, T.; Eerdekens, G.; van der Veen, C.; Fall, R. Sensitivity and specificity of atmospheric trace gas detection by proton-transfer-reaction mass spectrometry. *Int. J. Mass Spectrom.* **2003**, *223-224*, 365–382.
22. Ennis, C. J.; Reynolds, J. C.; Keely, B. J.; Carpenter, L. J. A hollow cathode proton transfer reaction time of flight mass spectrometer. *Int. J. Mass Spectrom.* **2005**, *247*, 72–80.
23. Barber, S.; Blake, R. S.; White, I. R.; Monks, P. S.; Reich, F.; Mullock, S.; Ellis, A. M. Increased Sensitivity in Proton Transfer Reaction Mass Spectrometry by Incorporation of a Radio Frequency Ion Funnel. *Anal. Chem.* **2012**, *84*, 5387–5391.
24. Makhoul, S.; Romano, A.; Cappellin, L.; Spano, G.; Capozzi, V.; Benozzi, E.; Märk, T. D.; Aprea, E.; Gasperi, F.; El-Nakat, H.; Guzzo, J.; Biasioli, F. Proton-transfer-reaction mass spectrometry for the study of the production of volatile compounds by bakery yeast starters. *J. Mass Spectrom.* **2014**, *49*, 850–859.
25. Cappellin, L.; Makhoul, S.; Schuhfried, E.; Romano, A.; Sanchez del Pulgar, J.; Aprea, E.; Farneti, B.; Costa, F.; Gasperi, F.; Biasioli, F. Ethylene: Absolute real-time high-sensitivity detection with PTR/SRI-MS. The example of fruits, leaves and bacteria. *Int. J. Mass Spectrom.* **2014**, *365–366*, 33–41.

26. Araghipour, N.; Colineau, J.; Koot, A.; Akkermans, W.; Rojas, J. M. M.; Beauchamp, J.; Wisthaler, A.; Märk, T. D.; Downey, G.; Guillou, C.; Mannina, L.; Ruth, S. v. Geographical origin classification of olive oils by PTR-MS. *Food Chem.* **2008**, *108*, 374–383.
27. Herbig, J.; Muller, M.; Schallhart, S.; Titzmann, T.; Graus, M.; Hansel, A. On-line breath analysis with PTR-TOF. *J. Breath Res.* **2009**, *3*, 027004.
28. Warneke, C.; de Gouw, J. A.; Kuster, W. C.; Goldan, P. D.; Fall, R. Validation of Atmospheric VOC Measurements by Proton-Transfer-Reaction Mass Spectrometry Using a Gas-Chromatographic Preseparation Method. *Environ. Sci. Technol.* **2003**, *37*, 2494–2501.
29. Lindinger, C.; Pollien, P.; Ali, S.; Yeretzian, C.; Blank, I.; Märk, T. Unambiguous Identification of Volatile Organic Compounds by Proton-Transfer Reaction Mass Spectrometry Coupled with GC/MS. *Anal. Chem.* **2005**, *77*, 4117–4124.
30. Ruzsanyi, V.; Fischer, L.; Herbig, J.; Ager, C.; Amann, A. Multi-capillary-column proton-transfer-reaction time-of-flight mass spectrometry. *J. Chromatogr. A* **2013**, *1316*, 112–118.
31. Romano, A.; Fischer, L.; Herbig, J.; Campbell-Sills, H.; Coulon, J.; Lucas, P.; Cappellin, L.; Biasioli, F. Wine analysis by FastGC proton-transfer reaction-time-of-flight-mass spectrometry. *Int. J. Mass Spectrom.* **2014**, *369*, 81–86.
32. Boscaini, E.; Mikoviny, T.; Wisthaler, A.; Hartungen, E. v.; Märk, T. D. Characterization of wine with PTR-MS. *Int. J. Mass Spectrom.* **2004**, *239*, 215–219.
33. Spitaler, R.; Araghipour, N.; Mikoviny, T.; Wisthaler, A.; Via, J. D.; Märk, T. D. PTR-MS in enology: Advances in analytics and data analysis. *Int. J. Mass Spectrom.* **2007**, *266*, 1–7.
34. Cappellin, L.; Probst, M.; Limtrakul, J.; Biasioli, F.; Schuhfried, E.; Soukoulis, C.; Märk, T. D.; Gasperi, F. Proton transfer reaction rate coefficients between H_3O^+ and some sulphur compounds. *Int. J. Mass Spectrom.* **2010**, *295*, 43–48.
35. Fischer, L.; Winkler, K.; Gutmann, R.; Singer, W.; Herbig, J.; Hansel, A. 2013 Liquid Calibration Unit – a versatile tool for trace gas calibration. Breath Analysis Summit 2013, Saarbruecken/Wallerfangen.

Chapter 18

Potential Use of Food Synthetic Colors as Intrinsic Luminescent Probes of the Physical State of Foods

Ariella Kashi, Sarah M. Waxman, Jennifer S. Komaiko,[1] Andrew Draganski, Maria G. Corradini, and Richard D. Ludescher*

Department of Food Science, School of Environmental and Biological Sciences, Rutgers, The State University of New Jersey, 65 Dudley Road, New Brunswick, New Jersey, 08901, U.S.A.
[1]Current address: Department of Food Science, University of Massachusetts - Amherst, 100 Holdswoth Way, Amherst, Massachusetts, 01002, U.S.A.
***E-mail: ludescher@aesop.rutgers.edu. Ph: 848-932-3516.**

Several synthetic food colors, such as azorubine, exhibit presumptive molecular rotor behavior. They consist of two parts that can easily rotate relative to each other and their photoexcitation generates an excited singlet state that can deactivate through a) a non-radiative decay process that involves internal conversion to the ground state through rotation or b) a radiative decay process that results in emission of a photon. Any environmental restriction to twisting in the excited state, e.g., high viscosity, can have a dramatic effect on the emission intensity of these fluorophores. The sensitivity of azorubine's photophysical properties to the concentration and temperature-dependent viscosity of model systems (glycerol, glycerol-water, sucrose-water, hydrocolloid-water solutions) was investigated. The molecular mass of the thickening agent affected the photophysical response of the dye, with lesser probe sensitivity found with the larger hydrocolloids; fluorescence studies of pyranine hydration suggest these differences were due to local effects of solute crowding. Advantages and limitations of using food colors as intrinsic luminescent sensors of physical properties related to food quality are discussed.

Introduction

Luminescence spectroscopy uses optical chromophores as molecular probes of the structure, chemical nature, and physical properties of the local environment (*1*). Due to the specific structure of the chromophore (aromatic ring, presence of hetero-atoms or specific groups such as alcohol, aldehyde, amino, acid, etc.) and the nature of the electronic excitation (change in bond order in the excited state, change in permanent dipole moment orientation and/or magnitude in the excited state), each molecule exhibits unique photophysical properties (*2*). Upon proper interpretation, these properties can report on the local physical environment around the chromophores; that is, can act as sensitive, site-specific molecular probes of the material.

Optical chromophores that are routinely added to foods, such as synthetic organic colors, can be used as intrinsic and safe luminescent probes of important physical properties in foods, for example, viscosity or structural organization. The physical properties of many foods play an important direct role in quality assurance and consumer acceptance. They are also important for the effective delivery of specific nutrients, which are often contained in organized molecular structures (emulsions, microemulsions, liposomes, organogels, etc.) to enhance stability and bioaccesibility. Therefore, in principle, synthetic food dyes can act as intrinsic optical probes and be used to monitor food quality and stability at various stages of the production and distribution chain using highly sensitive, inexpensive and minimally invasive luminescence techniques.

Synthetic azo dyes are widespread in food products due to their strong absorbance and high stability. Worldwide, Allura red AC, Azorubine, Sunset Yellow and Tartrazine are the most commonly found monoazo dyes in edible goods. Due to limited functionality and/or regulatory requirements, the use of additional azo dyes such as Amaranth, Citrus Red 2, Lithol Rubine BK, Orange B, Ponceau 4R and Red 2G have been limited to specific applications. For example, Ponceau 4R and Citrus Red 2 are applied to provide color to hot dog casings and to enhance the external appearance of fresh oranges, respectively. A recent study has explored the potential link between childhood hyperactivity and consumption of food synthetic colors (*3*). Based on this study and additional toxicological data on these dyes, a re-evaluation of their safety was conducted by the European Food Safety Authority (*4*) and the Food Drug Administration (*5*). Both organizations concluded that no alteration of the acceptable daily intake (ADI) for the most commonly used azo dyes (Allura Red, Azorubine, Sunset Yellow and Tartrazine) was required based on the current available information.

In terms of their molecular structure, all monoazo dyes contain one functional azo group with structure R-N=N-R′; in the case of the food dyes, the pendant R groups are bulky aromatic rings functionalized with large charged groups (Figure 1). The presence of the azo group potentially confers molecular rotor properties and, consequently, environmental sensitivity to this kind of dye.

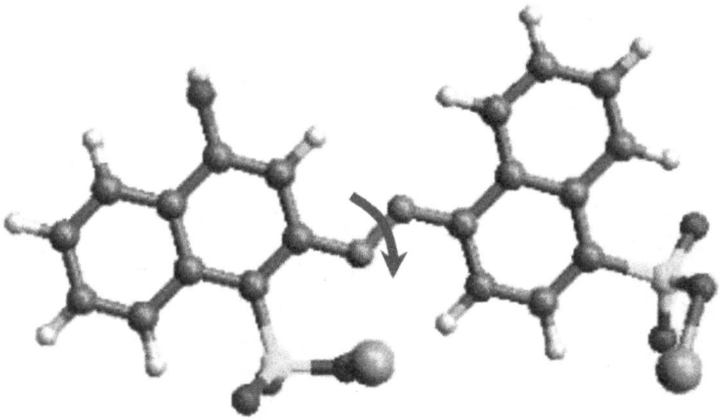

Figure 1. Structure and possible intramolecular rotation in Azorubine (disodium 4-hydroxy-2-[(E)-(4-sulfonato-1-naphthyl)diazenyl]naphthalene-1-sulfonate).

The term molecular rotor was first used by Ichimura, Hayashi, Akiyama, Ikeda, and Ishizuki (*6*) to define molecules that consist of two or more segments that can easily rotate relative to each other (*7*, *8*). Photoexcitation of fluorescent molecular rotors can result in the formation of twisted intramolecular charge transfer (TICT) states. Deactivation from the TICT state is predominantly through a nonradiative pathway. Alternatively, return to ground state from the local excited state can occur through a radiative decay process that results in emission of a photon. Since TICT state formation rate is lower in more viscous or restricted environments, these two competing decay pathways determine the sensitivity of the probe to the microviscosity of the surrounding environment (*9–11*). Consequently, any change in the medium rigidity or mobility, such as increase in high local and bulk viscosity (*2*), cross linking, aggregation (*12*), polymerization and phase transitions (*9*), will hinder the formation of TICT states and will result in an increase in fluorescence emission.

Although several extrinsic synthetic molecular rotors, such as benzopyrones, stilbenes and benzylidene malononitriles, have been successfully used to measure physical properties in biological systems such as cell membranes and biofluids containing proteins (*13–15*), the use of this class of molecules in foods has not been explored. Toxicity, price, selective solubility, or inedible nature of the available probes may be in part responsible for the limited use of molecular rotors for that purpose.

The photophysical properties of edible azo dyes have rarely been reported. However, research on the luminescence of non edible azo-compounds supports the molecular rotor behavior of the synthetic food dyes (*16*) and, consequently, their potential applicability as noninvasive probes for physical properties of foods. Additionally, a recent study by Du, Corradini, Ludescher and Rogers (*17*) has reported the efficacy of a liposoluble edible azo dye (Citrus Red 2) as a sensor for the micro viscosity of oils confined in colloidal fat crystal networks. These authors observed that the luminescent response of the azo dye could not be unequivocally related to solids content but to a decrease in the void volume where the liquid oil was entrapped, expressed in terms of the fractal dimension of the fat network. This observation emphasizes that the basis for environmental sensitivity of molecular rotors is hindrance of intramolecular rotation. Therefore, understanding the molecular scale interaction of the molecular rotor and its environment (*9*) is crucial to determine its applicability as an effective sensor of the physical state of foods and to establish limiting operating conditions for these luminescent probes.

This chapter will, thus, discuss the sensitivity of the photophysical properties, specifically fluorescence intensity, of hydrosoluble monoazo dyes towards micro (and bulk) viscosity of their surrounding medium. Also, the advantages, limitations and applicability of this methodology will be covered.

Experimental Procedures

Azorubine (AZ) has been chosen to illustrate the photophysical behavior and sensitivity of hydrosoluble edible azo dyes to solution viscosity in this chapter.

Disodium 4-hydroxy-2-[(E)-(4-sulfonato-1-naphthyl) diazenyl] naphthalene-1-sulfonate (azorubine, AZ) was purchased from Alfa Aesar (Ward Hill, MA) and used without further purification. Stock solutions (1mM) of this food dye were prepared in water and glycerol (Alfa Aesar, spectrophotometric quality – 99.5% purity) and used to obtain excitation and emission spectra.

Stock solutions of azorubine were diluted to 30 μM to obtain the fluorescence excitation and emission spectra of the dye in both solvent solutions, i.e., pure water and pure glycerol. The absorption spectra of AZ at 30 μM in both solvents were also recorded over the range of 200-800 nm using a Cary 60 UV-VIS spectrophotometer (Agilent Technologies, Santa Clara, CA).

Steady state fluorescence excitation and emission spectra were obtained using a Fluoromax-3 spectrofluorometer (Horiba Scientific Inc., Edison, NJ) equipped with a thermoelectric temperature controller (Wavelength Electronics, Inc. Bozeman, MT). 30 μM AZ solutions were tested in 1 cm lightpath quartz cuvettes (NSG Precision Cells, Farmingdale, NY). The initial studies of AZ in pure solvents were conducted at a fixed temperature of 20°C. The samples were equilibrated at the target temperature for 5 min to avoid temperature gradients within the cuvette. Preliminary runs were conducted to identify the optimal excitation wavelength and emission and excitation slits for each sample. Peak emission, excitation wavelength and fluorescence emission intensity in counts per second were determined. The effect of dye concentration on fluorescence intensity was further tested in order to minimize inner filter effects in subsequent studies.

Fluorescence Emission of Azorubine (AZ) in Solutions of Different Viscosities

In order to assess AZ sensitivity to viscosity, the fluorescence emission intensity of 30 μM solutions of azorubine in mixtures of water and glycerol was recorded using a Fluoromax-3. As reported in the literature (*18*, *19*), variations in viscosities within an ~0.5-12,000 mPa s range can be obtained by modifying the water (low viscosity solvent)-glycerol (high viscosity solvent) ratio and varying the temperature between 10-100°C. Studies were thus done using either pure glycerol over this temperature range (modulating viscosity between 15-4,000 mPa s) and glycerol-water (0-100% glycerol) solutions at 10°C (modulating viscosity between 1-4000 mPa s).

To evaluate the effect of molecular crowding on the photophysical properties of the selected azo dye, three additional compounds, namely sucrose, methyl cellulose (MC) and carboxymethyl cellulose (CMC) were used to produce solutions of AZ (30 μM) of different viscosities at 20°C. 60-70 °Brix solutions of sucrose (99% purity, obtained from Sigma Aldrich, St. Louis, MO) were prepared and the concentration of each sucrose solution was corroborated using a hand held refractometer (N4 E, Atago USA Inc., Bellevue, WA). Concentrated solutions (3% w/v) of both hydrocolloids (methyl cellulose and carboxymethyl cellulose – Sigma Aldrich) were prepared overnight and further diluted with distilled water within the range of 0-3% w/v. The concentrations were selected to obtain MC and CMC solutions with viscosities from 10 to 1800 mPa s at 20°C. A Discovery hybrid rheometer equipped with a 60 mm cone plate geometry (TA instruments, New Castle, DE) was used to verify the viscosity of all samples. The rheometer was employed under a flow sweep mode and the recorded viscosity at a shear rate of 20 s^{-1} was used for data analysis.

The fluorescence spectra of samples with and without the addition of the probe were collected over the range from 530 to 800 nm, with excitation at 510 nm. Excitation and emission monochromators were set at 5 nm bandpass. Although the background signal was very low for all samples, the emission spectrum of the control was subtracted from each corresponding AZ spectrum and then normalized towards the higher concentration. All samples were run in triplicate.

The relationship between the fluorescent quantum yield, Φ_F , of a molecular rotor to the viscosity (η) of the surrounding solution can be described by the Förster and Hoffman equation (*20*) and expressed as follows:

$$log\ \Phi_F = C + x\ log\ \eta \tag{1}$$

where C and x are solvent and dye dependent constants.

Since fluorescence emission intensity, I_F, and quantum yield are proportional, the relationship between the measured fluorescence intensity and viscosity can be reworked from Eq. 1 and is often expressed by the nearly identical power law model (*9*) (Eq. 2):

$$I_F = \alpha\eta^x \tag{2}$$

where α (not equal to exp(C) due to the unknown and instrumentally determined relationship between Φ and I_F) can be considered a measure of the probe's brightness and x a measure of its sensitivity to local viscosity (*14*). The relationship between maximum fluorescence intensity and viscosity of all AZ solutions was fitted using Eq. 2 to reduce systematic fitting bias and plotted using the lograrithmic expression to provide a direct visual verification of the model. The sensitivity of AZ to viscosity, represented by the parameter x, was compared to reported values of commonly used molecular rotors analyzed using the same model. A comparison of the performance of the selected molecular rotor, AZ, in all the studied solutions was also done based on values of the *x* parameter.

Local versus Bulk Viscosity - Fluorescence Emission of Azorubine (AZ) and Pyranine in Solutions of Equal Viscosity

Although the agents used to modulate bulk viscosity in this study (i.e., glycerol, sucrose, methyl cellulose and carboxymethyl cellulose) are able to produce solutions of similar viscosity, each compound affects the local environment around the azo dye in a different way. In order to evaluate how crowding around the molecular rotor determines the sensitivity of the azo dye to local and bulk viscosity, an additional fluorescent probe, pyranine (8-hydroxypyrene-1,3,6-trisulfonic acid), was used to indirectly assess the physical properties of the micro-environment. The sensitivity of pyranine fluorescence to the composition (and mobility) of its surroundings is due to the dissociation/recombination of a single ionizable group in this fluorophore (*21*). Upon photoexcitation in a free-water rich environment, the hydroxyl group in the pyranine molecule readily dissociates and is transferred to the surrounding water molecules; the fluorescence emission of the unprotonated pyranine yields a single band at 513 nm. When the pyranine molecule is surrounded primarily by weak proton acceptors such as glycerol or sugars, the proton transfer is hindered and emission is shifted to 435 nm characteristic of protonated pyranine (*22*).

Pyranine was obtained from Sigma Aldrich and used as received. Two sets of glycerol-water, sucrose and hydrocolloids solutions of equal viscosity at 20°C (90 mPa s) were prepared with the addition of AZ (30 uM) and pyranine (0.5 μM), respectively. All the tested solutions exhibited a viscosity of 90 ± 1 mPa s at 20 s^{-1} as determined using the Discovery hybrid rheometer under a flow sweep mode. The fluorescence emission spectra of the molecular rotor (AZ) containing solutions were recorded as described in the previous section. In the case of the pyranine solutions, the fluorescence spectra were obtained using an excitation wavelength of 350 nm and 5 nm emission and excitation slits. All samples were run in triplicate.

The ratios of the 435 nm to 510 nm emission peaks of the pyranine spectra were calculated and analyzed alongside the maximum fluorescence intensity of the azo dye in the different solutions.

Results and Discussion

In addition to azorubine, five hydrosoluble azo dyes (Allura Red, Amaranth, Ponceau 4R, Sunset Yellow, and Tartrazine) have been tested using the described methodology. Although only azorubine data are discussed in detail as an exemplar, the datasets obtained from all edible azo dyes followed the same trends.

Azorubine, also known as carmoisine, Food Red 3 or E 122, has the ability to provide a bright red color to a large variety of edible goods. It is widely used in several countries of the European Union, where its ADI has been set to 0-4 mg/kg bw/day. The approval of this colorant for use in the USA has not been pursued due to the widespread use of two alternative red synthetic colors (i.e., Allura Red and erythrosine) in food products.

As seen in Figure 1, azorubine contains one functional azo group, as well as two sulfate moieties and a hydroxyl group that facilitate its dissolution in aqueous solution or polar matrices. The presence of the azo group potentially confers molecular rotor properties to this dye such that it exhibits low fluorescence emission intensity in low viscosity fluid solutions and a significant enhancement of fluorescence intensity in highly viscous solvents.

Excitation and emission spectra of AZ in water and glycerol solutions are shown in Figure 2. The data were normalized towards that of AZ in glycerol to illustrate the effect of the surrounding viscosity on the fluorescence intensity of AZ. As can be observed in Figure 2, AZ is practically non-emissive in water, which can explain the limited information available on its photophysical properties. Excited state tautomerization and/or internal twisting, already reported for synthetic non-edible dyes in low viscosity solutions (16), can also constitute the predominant non-radiative relaxation pathway of this food color in a fluid environment such as water. A large Stokes shift (λ_{em}-λ_{exc}) of about 80 nm is also evident from the excitation and emission spectra of AZ in both pure water and glycerol solutions.

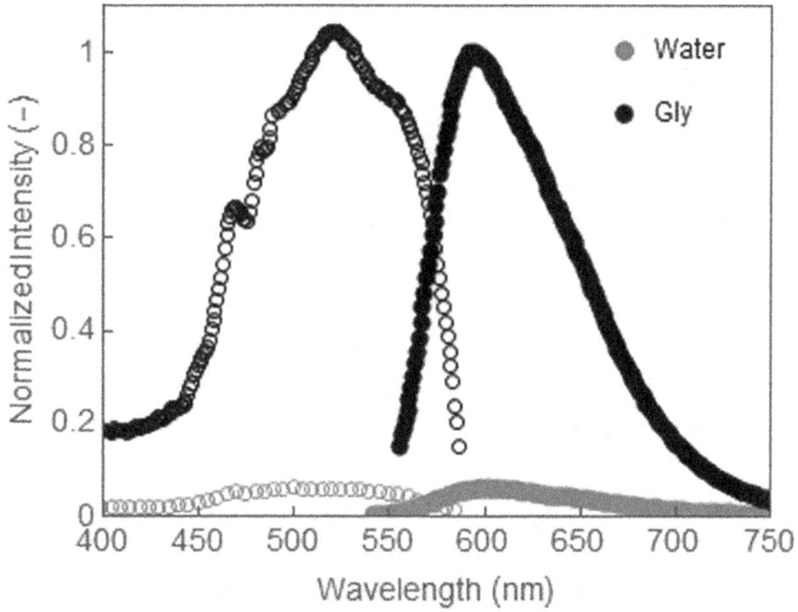

Figure 2. Excitation (lower wavelength, empty circles) and emission (higher wavelength, solid circles) spectra of AZ in solvents of similar polarity and different viscosity (water and glycerol at 20°C).

The medium's rigidity surrounding the azo dye was altered by changing temperature and composition in selected solvents. The emission spectra of AZ in all solutions were obtained by subtracting the background spectra of each respective dye-free control and correcting for scattering. To facilitate comparison within each experiment, the data were normalized so that the emission from the highest viscosity corresponds to unity. In the case of AZ in pure glycerol, glycerol-water, and sucrose solutions, as the viscosity of the surrounding medium increased, the dyes' fluorescent emission intensity increased in a similar manner (see Figure 3-left). In contrast, when hydrocolloids, either methyl cellulose (MC) or especially carboxymethyl cellulose (CMC), were used to modulate the physical properties of the surrounding medium (See Figure 4-left) less differentiation was detected between systems of different bulk viscosity.

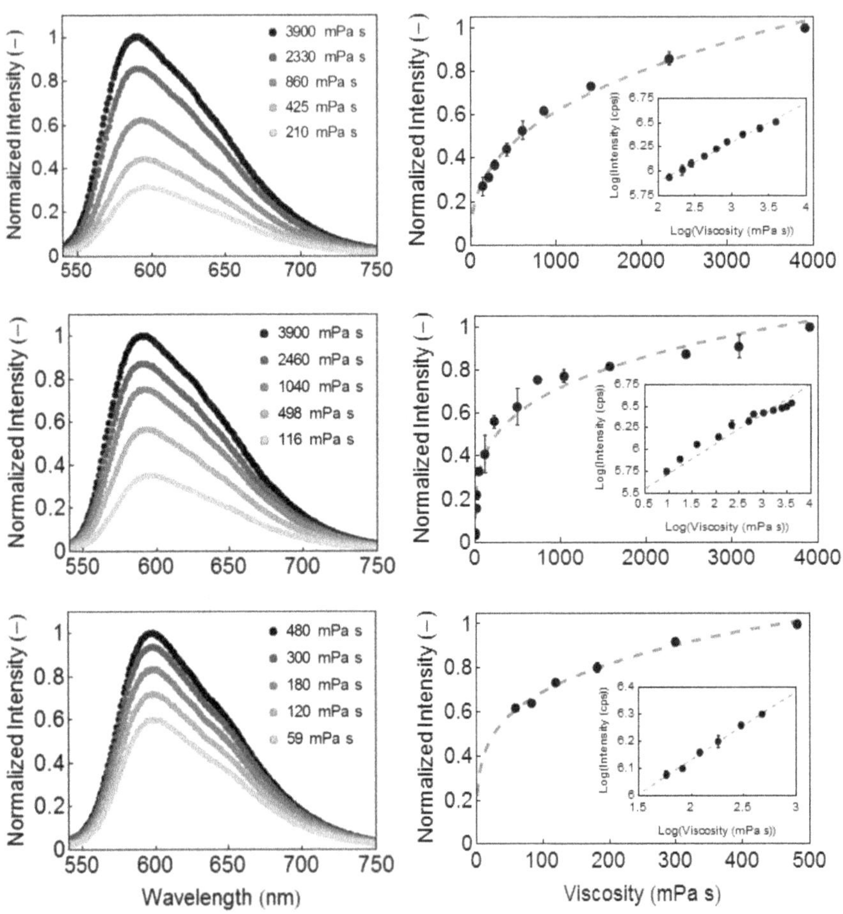

Figure 3. Left - Normalized emission spectra of AZ in glycerol at different temperatures (top), in glycerol-water (middle) at 10°C and sucrose-water (bottom) mixtures of different composition at 20°C. Right - Dependence of AZ normalized fluorescence intensity on viscosity fitted with Eq. 2. Inset: Relationship between measured fluorescence intensity (counts per second, cps) and solution viscosity presented as a log-log plot to illustrate linearity.

The response of the maximum fluorescence intensity of azorubine vs. viscosity of the media was characterized by a power law model (Eq. 2) for each type of solution (Figures 3-right and 4-right). Although the data have been plotted here in double logarithmic coordinates to visually verify the applicability of Eq. 1 (see inserts to these figures in which the log of the experimentally measured fluorescence intensity in counts per second was plotted versus log viscosity), the parameters of the power law relationship were obtained by analyzing the linear data with Eq. 2 using a nonlinear regression procedure to minimize bias in the estimation (*23*). The goodness of fit (mean square error, MSE) of Eq. 2 and the sensitivity of azorubine's fluorescence intensity to changes in viscosity for each solution, expressed in terms of the parameter x, are summarized in Table 1.

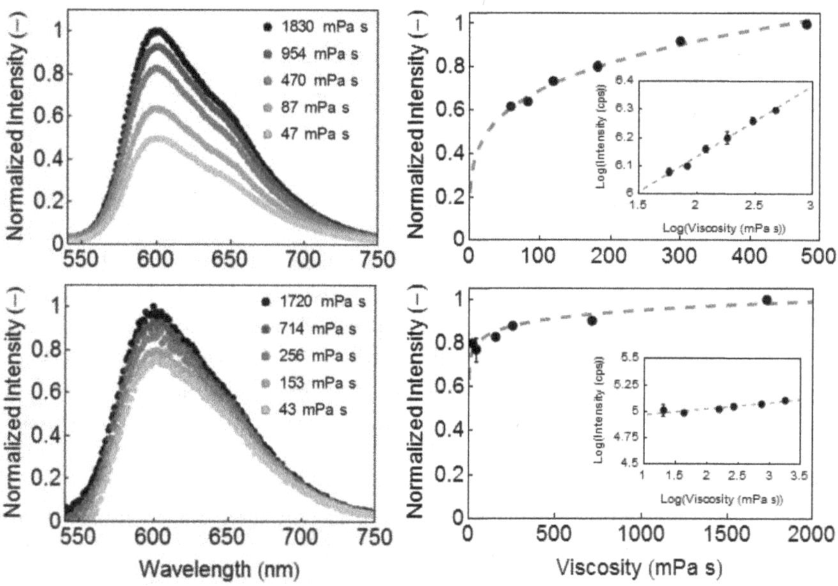

Figure 4. Left - Normalized emission spectra of AZ in hydrocolloid solutions of different concentration at 20°C (MC—top and CMC—bottom). Right - Dependence of AZ normalized fluorescence intensity on viscosity fitted with Eq. 2. Inset: Relationship between measured fluorescence intensity (cps) versus viscosity presented as a log-log plot to illustrate linearity.

Sensitivity values (x) of novel and commonly used molecular rotors in glycerol-based solutions have been reported in the range of 0.25 to 0.6 [i.e., 0.26-0.4 for modified nucleosides (*24*), 0.53 for 9-(2,2-dicyanovinyl)-julolidine (DCVJ), and 0.52 for 9-(2-carboxy-2-cyano)vinyl julolidine (CCVJ) (*25*) and 2-cyano-3-(4-dimethylaminophenyl) acrylic acid methyl ester (CMAM) (*26*)]. In the case of AZ in similar solutions, the viscosity sensitivity was established to be 0.38, which is within the range reported for these other molecular rotors. Sutharsan et al. (*14*) have stated that two basic requirements of an adequate molecular rotor are a large Stokes shift and high sensitivity to viscosity (or molecular rigidity) change. The results for AZ in glycerol-based solutions, which meet the above mentioned requirements, can establish AZ's molecular rotor character and supports its potential use as a probe of physical properties in foods.

Table 1. Sensitivity of Azorubine'S Fluorescence Intensity to Viscosity of Different Solutions Determined by the Parameter x Calculated Using Eq. 2, with Mean Square Error (MSE)

Solution	x	MSE
Glycerol/Temperature	0.38	0.0007
Glycerol/Water Mixtures	0.38	0.0020
Sucrose/Water Mixtures	0.25	0.0003
Methyl cellulose (MC)	0.20	0.0230
Carboxymethyl cellulose (CMC)	0.05	0.0100

Although the fluorescence intensity of this monoazo dye is sensitive to solution viscosities in the range from about 10 to nearly 10^4 mPa s in pure glycerol, glycerol/water, sucrose/water, and to some extent in hydrocolloid/water solutions, a clear reduction in sensitivity is observed when macromolecules were used to modulate bulk viscosity. Hydrocolloids due to their large effective hydrodynamic volumes have the ability to increase bulk viscosity even at very low concentrations when added to a solution. Less molecular crowding and extensive hydration around the azo dye in hydrocolloid solutions might explain its lower sensitivity to changes in bulk viscosity. To test for this hypothesis, pyranine was used to probe the characteristics of the water microenvironment. Pyranine was selected as a sensor for molecular crowding due to its ability to assess the extent to which an added thickener can invade and thus replace water in its hydration shell (*22*). Additionally pyranine's molecular size and footprint is similar to those of AZ, which reduces bias in data interpretation due to additional artifacts; we can thus expect that the solvent behavior around pyranine provides direct insight into the solvent behavior around azorubine.

Pyranine and AZ were dissolved in solutions of equal viscosity at 20°C (90 mPa s) and different composition, i.e., glycerol/water, sucrose/water, MC/water and CMC/water. The fluorescence spectra of both probes in all solutions are presented in Figure 5. The magnitude of the 435 nm pyranine peak in glycerol and sucrose solutions indicates the extent of penetration of glycerol and sucrose molecules into the hydration shell of the pyranine (*22*). This extensive crowding of solute around the probe explains the sensitivity of AZ to changes in bulk viscosity in solutions containing low molecular mass solutes as the local environment of the probe includes an appreciable number of solute molecules. Conversely, significantly higher 510 nm pyranine peaks in the hydrocolloid solutions suggest more free water and consequently less molecular crowding by polymeric solute around the probe in the hydrocolloid solutions. This, in turn, explains the lower fluorescence intensity of AZ in both hydrocolloids in comparison to its luminescent response in glycerol and sucrose solutions of identical bulk viscosity as the mechanism of azo dye quenching ensures that the intensity of fluorescence emission is primarily sensitive to the resistance to intramolecular rotation offered by the local molecular environment surrounding the probe. In other words, the probe always senses local (molecular) viscosity; for small molecules, the local viscosity correlates directly with bulk (macroscopic) viscosity whereas for larger polymeric hydrocolloids, the correlation between local molecular and bulk macroscopic viscosity is less direct.

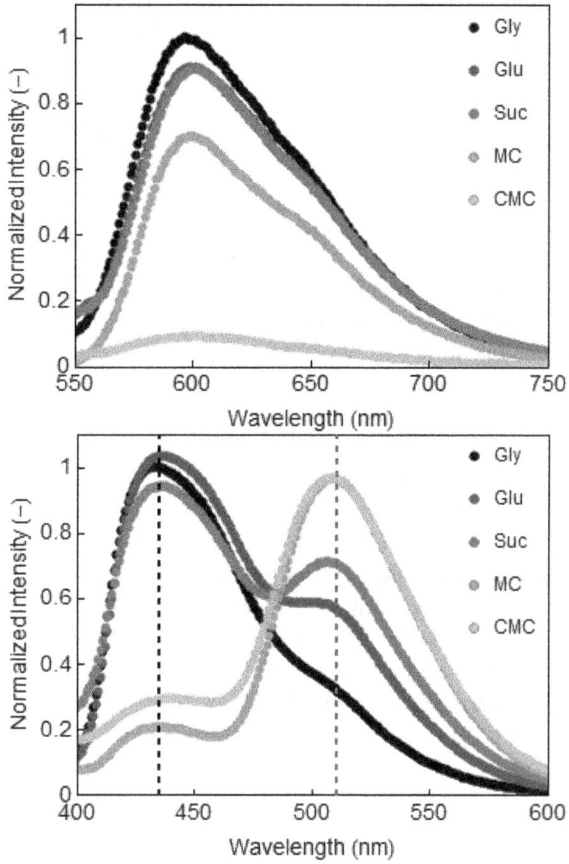

Figure 5. Top - Normalized fluorescence intensity of AZ in solutions of different composition (glycerol, sucrose, MC and CMC) and equal viscosity at 20°C (90 mPa s). Bottom - Changes in normalized fluorescence intensity of pyranine in solutions of different composition and equal viscosity (90 mPa s).

The maximum normalized fluorescence intensity of AZ and the ratio of the 435 nm to 510 nm emission peaks of the pyranine spectra are reported side by side in Table 2 along with the sensitivity parameter x from Eq. 2.

Table 2. Correlation between Solute Crowding (Expressed as the Ratio of Pyranine Fluorescence Intensity (FI) at 435 and 513 nm) and Sensitivity of Azorubine'S Fluorescence Emission to Viscosity Expressed as Relative Maximum Intensity or Sensitivity Value (Parameter x from Eq. 2)

Solution	Pyranine Fluorescence	Azorubine Fluorescence	
	$FI_{435\ nm}/FI_{513\ nm}$	Normalized Maximum Fluorescence Intensity	Sensitivity Value (x)
Glycerol – Water	2.95	1.00	0.38
Sucrose	1.32	0.91	0.25
Methyl cellulose (MC)	0.22	0.70	0.20
Carboxymethyl cellulose (CMC)	0.30	0.09	0.05

Differences in the performance of AZ as a sensor for bulk viscosity in both hydrocolloids might be associated to the molecular size of the macromolecules used. The molecular weight of MC was one order of magnitude smaller than that of CMC (i.e., 41,000 vs. 250,000 g mol^{-1}), which contributes to differential crowding around the probe. Haidekker and Theodorakis (*27*) have reported changes in the emission intensity of CCVJ in aqueous colloidal solutions of dextran of similar molecular weight (40,000 g mol^{-1} and lower) as the MC used in this study. Additional studies should thus be conducted to identify the sensitivity of azorubine fluorescence to the addition of a specific hydrocolloid thickener and the specific molecular mass ranges that allow effective correlations between local and bulk properties. Although molecular mass, structure (extent of branching) and intrinsic rigidity (all factors that modulate the effective hydrodynamic volume) of the thickener might play a role in probe sensitivity, other aspects such as interactions between the probe and the viscosity modulating molecule cannot be disregarded. All of these issues are under investigation.

Conclusions

The use of molecular rotors offers a non-invasive, non-disruptive and highly sensitive alternative to conventional analytical methodologies to evaluate the physical properties in foods. Synthetic azo-dye food colors exhibit molecular rotor behavior and, in the case of azorubine, their sensitivity to changes in local viscosity are comparable to commonly reported molecular rotors, such as DCVJ, in similar environments. In food systems where the viscosity modulating agent affects the bulk and local environment to the same extent these luminescent compounds can constitute feasible probes for bulk viscosity; determination of exactly what foods this includes remains to be done. Due to the sensitivity of these probes towards the structure and properties of the microenvironment, not only can they provide bulk viscosity measurements but they might also facilitate studies elucidating the relationship between a bulk property, like consistency, and the material's microstructural characteristics.

Additional studies aiming to understand the molecular scale interaction of the molecular rotor and its environment (*9*) can enhance the applicability of edible azo dyes as effective sensors of the physical state of foods and elucidate their limiting operating conditions.

Due to their sensitivity to molecular crowding, they can become effective sensors for additional physical properties of foods, such as microstructural organization.

Determining physical properties such as viscosity using embedded (intrinsic) probes offers operational flexibility and rapid means for screening food quality. The availability of this methodology can facilitate real-time condition-based monitoring of food products, which in turn might increase safety and quality along the food distribution chain.

Acknowledgments

This project was supported by the Agriculture and Food Research Initiative Grant no. 2014-67017-21649 from the USDA National Institute of Food and Agriculture, Improving Food Quality –A1361.

References

1. Slavik, J. *Fluorescent Probes in Cellular and Molecular Biology*; CRC Press: Boca Raton, FL, 1994.
2. Turro, N. J.; Ramamurthy, V.; Scaiano, J. C. *Modern Molecular Photochemistry of Organic Molecules*; University Science Books: Herndon, VA, 2010; pp 912–915.
3. McCann, D.; Barrett, A.; Cooper, A.; Crumpler, D.; Dalen, L.; Grimshaw, K.; Kitchin, E.; Lok, K.; Porteous, L.; Prince, E.; Sonuga-Barke, E.; Warner, J. O.; Stevenson, J. Food additives and hyperactive behaviour in 3-year-old and 8/9-year-old children in the community: a randomised, double-blinded, placebo-controlled trial. *Lancet* **2007**, *370*, 1560–1567.
4. European Food Safety Authority (EFSA) Panel on Food Additives and Nutrient Sources added to Food. Scientific opinion on the re-evaluation of azorubine/carmoisine (E 122) as a food additive. *EFSA J.* **2009**, *7*, 1332.
5. Food and Drug Administration- FDA/CFSAN Food Advisory Committee. *Certified color additives in food and possible association with attention deficit hyperactivity disorder in children*; 2011; URL http://www.fda.gov/downloads/AdvisoryCommittees/CommitteesMeetingMaterials/FoodAdvisoryCommittee/UCM248549.pdf (20-Nov-2014).
6. Ichimura, K.; Hayashi, Y.; Akiyama, H.; Ikeda, T.; Ishizuki, N. Photo-optical liquid crystal cells driven by molecular rotors. *Appl. Phys. Lett.* **1993**, *63*, 449–451.

7. Kottas, G. S.; Clarke, L. I.; Horinek, D.; Michl, J. Artificial Molecular Rotors. *Chem. Rev.* **2005**, *105*, 1281–1376.
8. Simkovitch, R.; Huppert, D. Optical Spectroscopy of Molecular-Rotor Molecules Adsorbed on Cellulose. *J. Phys. Chem. A.* **2014**, *118*, 8737–8744.
9. Haidekker, M. A.; Theodorakis, E. A. Environment-sensitive behavior of fluorescent molecular rotors. *J. Biol. Eng.* **2010**, *4*, 1–14.
10. Loutfy, R. O.; Arnold, B. A. Effect of Viscosity and Temperature on Torsional Relaxation of Molecular Rotors. *J. Phys. Chem.* **1982**, *86*, 4205–4211.
11. Law, K. Y. Fluorescence probe for microenvironments: anomalous viscosity dependence of the fluorescence quantum yield of p-N,N-dialkyl-amino-benzylidene-malononitrile in 1-alkanols. *Chem. Phys. Lett.* **1980**, *75*, 545–549.
12. Uzhinov, B. M.; Ivanov, V. L.; Ya Melnikov, M. Molecular rotors as luminescence sensors of local viscosity and viscous flow in solutions and organized systems. *Russ. Chem. Rev.* **2011**, *80*, 1179–1190.
13. Akers, W.; Haidekker, M. A. A molecular rotor as viscosity sensor in aqueous colloid solutions. *J. Biomech. Eng.* **2004**, *126*, 340–345.
14. Sutharsan, J.; Lichlyter, D.; Wright, N. E.; Dakanali, M.; Haidekker, M. A.; Theodorakis, E. A. Molecular rotors: synthesis and evaluation as viscosity sensors. *Tetrahedron* **2010**, *66*, 2582–2588.
15. Lopez-Duarte, I.; Thanh, T. V.; Izquierdo, M. A.; Bulla, J. A.; Kuimova, M. K. A molecular rotor for measuring viscosity in plasma membranes of live cells. *Chem. Commun.* **2014**, *50*, 5282–5284.
16. Abbott, L. C.; Batchelor, S. N.; Jansen, L.; Oakes, J.; Smith, J. R.; Moore, J. N. Picosecond deactivation of azo dye excited states in solution and in cellulose. *J. Photochem. Photobiol., A* **2011**, *218*, 11–16.
17. Du, H.; Kim, C.; Corradini, M. G.; Ludescher, R. D.; Rogers, M. A. Micro-viscosity of liquid oil confined in colloidal fat crystal networks. *Soft Matter* **2014**, *10*, 8652–8658.
18. Chen, Y. M.; Pearlstein, A. J. Viscosity temperature correlation for glycerol water solutions. *Ind. Eng. Chem. Res.* **1987**, *26*, 1670–1672.
19. Segur, J. B.; Oberstar, H. E. Viscosity of glycerol and its aqueous solutions. *Ind. Eng. Chem.* **1951**, *43*, 2117–2120.
20. Förster, T.; Hoffmann, G. Die viskositätsabhängigkeit der fluoreszenzquantenausbeuten einiger farbstoffsysteme. *Z. Phys. Chem. Neue Folge* **1971**, *75*, 63–76.
21. Flora, K. K.; Dabrowski, M. A.; Musson, S. P.; Brennan, J. D. The effect of preparation and aging conditions on the internal environment of sol-gel derived materials as probed by 7-azaindole and pyranine fluorescence. *Can. J. Chem.* **1999**, *77*, 1617–1625.
22. Roche, C. J.; Guo, F.; Friedman, J. M. Molecular level probing of preferential hydration and its modulation by osmolytes through the use of pyranine (HPT) complexed to hemoglobin. *J. Biol. Chem.* **2006**, *281*, 38757–38768.
23. Van Boekel, M. A. J. S. Kinetic modeling of food quality: A critical review. *Compr. Rev. Food Sci. Food. Saf.* **2008**, *7*, 144–158.
24. Sinkeldam, R. W.; Wheat, A. J.; Boyaci, H.; Tor, Y. Emissive nucleosides as molecular rotors. *Chem. Phys. Chem.* **2011**, *12*, 567–570.

25. Haidekker, M. A.; Brady, T. P.; Lichlyter, D.; Theodorakis, E. A. Effects of solvent polarity and solvent viscosity on the fluorescent properties of molecular rotors and related probes. *Bioorg. Chem.* **2005**, *33*, 415–425.
26. Yasuhara, K.; Sasaki, Y.; Kikuchi, J. Fluorescent sensor responsive to local viscosity and its application to the imaging of liquid-ordered domain in lipid membranes. *Colloids Surf., B* **2008**, *67*, 145–149.
27. Haidekker, M. A.; Theodorakis, E. A. Molecular rotors: fluorescent biosensors for viscosity and flow. *Org. Biomol. Chem.* **2007**, *5*, 1669–1678.

Chapter 19

Application of Untargeted LC/MS Techniques (Flavoromics) To Identify Changes Related to Freshness of Food

I. G. Ronningen and D. G. Peterson*

**Flavor Research and Education Center,
Department of Food Science and Nutrition,
University of Minnesota, 1334 Eckles Avenue,
Saint Paul, Minnesota
*E-mail: dgp@umn.edu.**

Freshness in food is a highly desirable attribute that has proven difficult to chemically characterize using traditional targeted methods. This work focuses on applying untargeted chemometric techniques to investigate differences in the chemical composition of orange extracts as they age as a strategy to identify compounds that contribute to the "fresh" flavor character. Ethanol extracts of oranges products were aged and sampled every 48 hours. RP-UPLC-MS (ESI-NEG) was used for data collection and two modeling techniques including the projection to latent variables (PLS) and Random Forest analysis were utilized for data analysis. Random forest and PLS provide different modeling criteria and identified common as well as unique features in the data set. Future work will focus on the compound identification and further sensory characterization of the selected markers. In summary, a method was developed to chemically profile the changes in a food product during aging to provide a unique basis to investigate changes in flavor profiles, identifying chemical attributes that may relate to freshness perception in food.

Introduction

From the point of harvest to the processing and distribution of foodstuffs, these materials are continually undergoing complex chemical changes that impact the product quality. Often these reactions lead to a reduction in flavor quality with different foods having a unique time point of optimal flavor quality, and a set of chemical reactions that lead to flavor degradation. Strategies to successfully mitigate undesirable flavor changes are more likely to be realized by understanding the chemistry involved with the loss of quality. Historically, targeted approaches have been implemented to chemically characterize flavor attributes of food products (*1*). Although these analytical approaches have provided valuable insight into the identities of compounds that contribute to flavor they are not without limitation. Often they ignore food matrix complexity by evaluating compounds singularly or out of the context of a food system. Commonly utilized methods include Aroma Extract Dilution Analysis (AEDA), CHARM, "sensomics" and LC-taste (*1*). The identification of flavor compounds are generally made with small panels identifying aroma through gas chromatography techniques (*2*) or taste through liquid chromatography (*3*).

With the successful development of untargeted chemical fingerprinting analytical methods coupled with multivariate analysis for research and discovery in biological sciences, a new opportunity has emerged for flavor analysis to identify and understand chemical drivers of flavor, termed Flavoromics (*4*). Drivers of 'fresh' flavor have been previously associated with the volatile aroma compounds which are rapidly released from food systems (*5*). The non-volatile composition of food flavor is commonly overlooked and it is the focus of the current study. Through a better understanding of non-volatile changes novel tastants may be discovered or precursors of volatile flavor compounds elucidated. Shedding light on compounds related to 'freshness' can lead to enhanced understanding of formulation tools to better meet consumer demands. This work will differ from other applications of untargeted research that emphasize prediction of sensory results with analytical instrumentation (*4*, *6*, *7*).

The underlying goal of this work is to develop untargeted methods to identify chemical differences in a food isolate (orange) during aging and to develop data management practices that maximize the likelihood for flavor identification through leveraging varietal composition for screening of common chemistries. In a subsequent study the impact of the identified markers on the flavor quality was defined and identified (*8*).

Experimental Methods

Citrus Model Systems

Oranges was purchased at local markets, the varietys purchased included Navel, Valencia, and Mineola Tangelo. The fruit was washed, lightly scrubbed, and dried. The entire citrus fruit was sliced to a thickness less than 5 mm. For each 500 g of fruit 200 g of ethanol was added. After a 24 hour extraction on a shaker table the fruit was removed and aliquots were taken. Aliquots were aged for either 0, 48, 96, or 144 hours before freezing at -80°C for further preparation.

Sample Clean up

Orange juice samples were passed through a 3,000 MW cutoff ultrafiltration membrane (Millipore, MA). The filtrate was further cleaned up with solid phase extraction using a 3 mL tube, 500 mg packing C18 phase (Supelco, St. Louis MO). Isolates were eluted with 600 µL UPLC grade acetonitrile (JT Baker) and to this 400 µL Nanopure water (Barnsted, Waltham, MA) was added.

UPLC-MS Conditions

A Waters I-class sample manager and binary solvent manager were coupled to a Waters Xevo G2 Q-TOF. A Waters BEH C18 (2.1 x 50 mm) was kept at 45°C in a Waters Column Manager. A flow rate of 0.55 mL/min was used with initial gradient conditions of 3% acetonitrile (ACN) and 97% Water (0.1% Formic Acid), which was held for 0.5 min. A linear gradient and raised ACN content to 15% at 1.5 min, 45% ACN at 8 min followed by a 1 min column wash (100% ACN) and re-equilibration. Electrospray Spray ionization was run in negative mode with source temperature of 120°C, desolvation temperature of 350°C, capillary set to 1.75 kV, sample cone of 25 V, TOF scan range was 100-1200 m/z, with lock mass corrected automatically. The reference compound was reserpine and 6 traces were used for correction per injection.

Each sample was injected 5 times in a randomized block design. Injection volume was 1 µL, with randomized blank and standard injections added to each replicate block.

Pre-Processing Optimization

To the citrus extract, methyl, ethyl, propyl and butyl parabens were added at 5 parts per billion and 10 parts per billion. Pre-processing techniques were optimized piecewise to achieve differentiation though PLS modeling. Once fully accurate classification was achieved, the variable of importance was used to ensure that the 4 doped compounds were identified as the most statistically significant compounds present in the modeling. The pre-processing conditions that gave the highest quality model were used to for the full aging study. The doping experiment was repeated, with the addition of a control sample that had been aged at room temperature for 24 hours. This ensured that modeling would still be sensitive to small differences as well as being robust for larger chemical variation. 100 ppm parabens were run in acetonitrile as a standard to generate expected mass and retention time values. LC-MS conditions were improved to reduce runtime and increase throughput for the aging experiment.

Pre-processing

UPLC-MS peak processing was done using Markerlynx software (Waters, Milford, CT). Peak detection was performed between 0.2-8 min, to exclude column wash and dead volume, m/z range was 100-1200 m/z using a m/z step of 0.01, noise elimination was set at 15, and spectra smoothing was applied. Peaks were matched if the retention time was within 0.1 min, established by the variation observed in elution of standard runs across analysis, the mass within 0.03 m/z and the peak detected in four of five injections. Peak lists were exported in CSV format for further analysis. Unit variance scaling was applied. Variables with a coefficient of variation of zero (no significant change) were eliminated, to reduce model over fitting and reduce time required to generate models.

Model Generation

Projection to latent structures (PLS) models were generated using SIMCA-P+ 12 (Umetrics, Umeå, Sweden), random forest (RF) models were generated using the R (R v.3.0.1 "Good Sport", University of Aukland) package "randomForest". Data was divided into training and test sets (70% training, and 30% test), and was sampled randomly. Sample groups were identified by age of extract and binning varietys across an age point. Before analysis PCA was used to screen for outliers in data sets to prevent over leveraging. PCA and PLS models went through permutation testing and Leave One Out Cross Validation (LOOCV) to ensure model quality. For random forest generation, the forest depth was optimized based on minimization of the classification error. Model optimization produced a model with 2.38% out of bag error using 110 trees and 110 variables tried at each split. Forest depth and variable tried at each split was piecewise optimized for model quality and speed of analysis. An importance plot provides insight into the variables that provided the most powerful leverage into classification of the samples.

Results and Discussion

A key component of untargeted chemical fingerprinting methods involve statistical modeling of data for discovery. To enhance the goal of compound discovery, model sensitivity was evaluated and optimized to ensure analytical and statistical robustness. Initial doping studies helped establish baselines for pre-processing conditions and established analytical thresholds for limits of detection. These steps helped ensure that pre-processing methods were reliable and provided detection of compounds at low concentrations typical of taste compounds, while limiting noise as much as possible. As any pre-processing approach includes noise to some level, ensuring that the statistical framework is still able to detect small differences given a specific method and food matrix helps ensure that the extracted data is as noise free as possible. Comparison of the doped samples at 5 or 10 ppb to the control system produced a model that allowed for complete classification (data not shown), indicating strong sensitivity and selectivity of the method. To further test the model, a sample that had been aged for 24 hours was added for analysis.Once again full classification was achieved, this is critical as the model is sensitive to small changes in chemical composition (differentiation of 5 ppb and 10 ppb samples), while still being able to handle and differentiate more chemical complexity (aged sample) (Figure 1). The model presented in Figure 1 was comprised of 3 factors, with a a cumulative 59.9% of variation of X variables and 99.3% of Y variables explained in 3 factors.

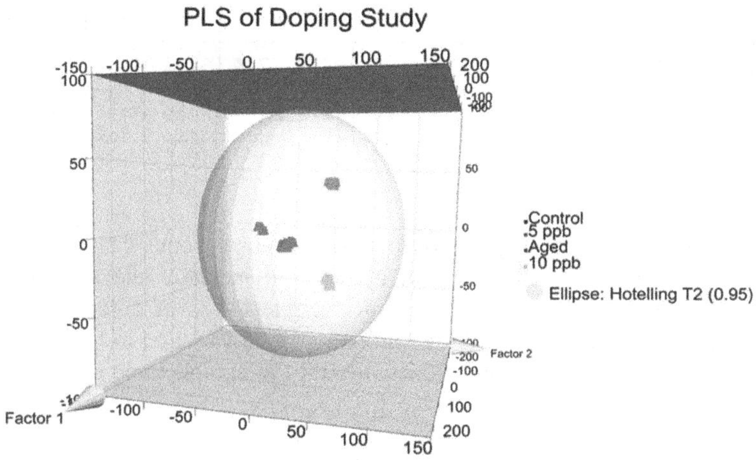

Figure 1. PLS Model of doped samples and aged control show good model sensitivity and selectivity.

The VIP were extracted from the model that indicated the compounds with the largest leverage on the model were the parabens doped compounds (see Table 1).

For the orange samples aged at different time points, the model generation was first conducted on collected data using varietal and time as the sample identifier. Initial model generation focused on the citrus variety (Figure 2), which would be expected given the large differences in the fruit varieties selected.

The model was subsequently adjusted to emphasize the sample age for unique grouping by time. Driving classification of samples using time as the sample identifier, rather than time and varietal, allowed for reduced significance of varietal differences and provided a different model for chemical driver classification (*9*). The model generated from this approach showed a high quality multiple correlation coefficient (R^2= 0.95) and a Q^2 of 0.981. Additionally this approach effectively filtered varietal specific compounds, ensuring that the model emphasized chemical changes that are associated with the aging of oranges. Thus the model effectively differentiated aging chemistry that is ubiquitous to this food platform. Figure 3 shows tight grouping of the samples based off time, rather than variety.

Table 1. Top Four Compounds from PLS Model Generated for Doping Experiment. Retention and m/z Match What Was Present for the Standards Run.

Expected Retention Time	Expected m/z	Marker ID	Retention Time	Mass	VIP
10.37	179.0009	10.37_178.9949	10.37	178.9949	1.59713
8.66	165.0028	8.66_164.9838	8.66	164.9838	1.59426
11.87	193.0014	11.87_193.0104	11.87	193.0104	1.5915
6.91	151.0156	6.91_150.9717	6.91	150.9717	1.58538

Although this modeling approach may eliminate some compounds that contribute to the unique freshness of a variety it increases the likelihood of discovering compounds that relate to the freshness loss of a food platform. Furthermore this experimental design utilizes sample variation to help filter the data and provides great variable reduction utility in complex and often varying systems like food.

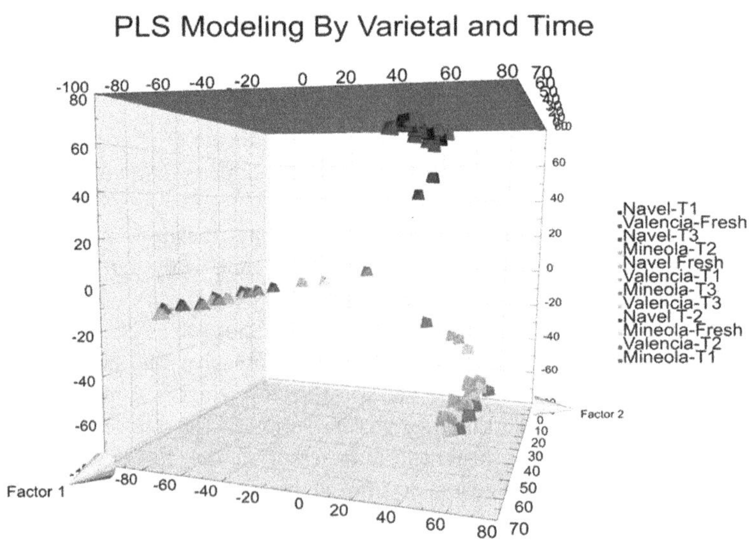

Figure 2. PLS model generated with varietal chemistry dominationg latent classification. Model generation does not focus on desired research outcome, in order to address aging model generation has to classify by time.

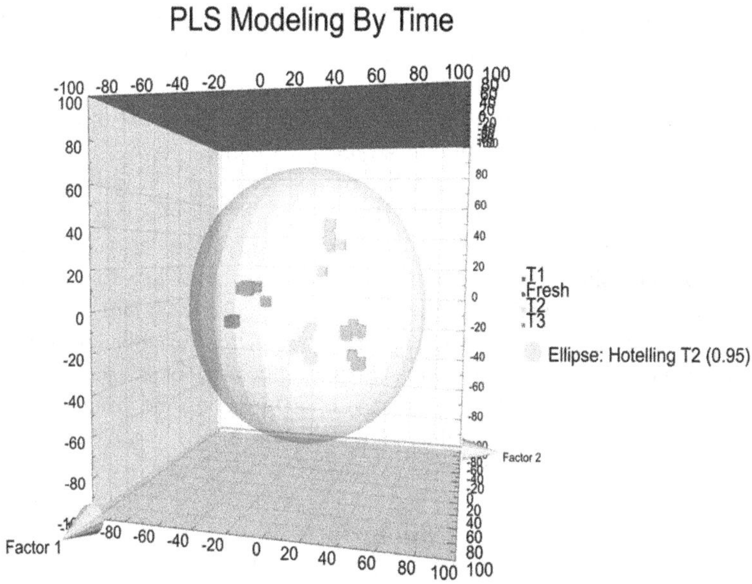

Figure 3. PLS model generated to focus on aging rather than varietal chemistry. From this model features that relate to aging are more likely identified.

A PLS-Differential Analysis (PLS-DA) was also conducted between each aging point. The four comparisons: Time 0 versus 48 hours (R^2=0.982, Q^2=0.942), 48 hours versus 96 hours (R^2=0.987, Q^2=0.941), 96 hours versus 144 hours (R^2=0.985, Q^2=0.963), produced 3 data sets through selecting the top 1200 markers based on Variable of Importance (VIP). These four data sets were compared and of the collective 3600 markers, 97 were identified to be present in all comparisons. A second-order analysis such as this helps focus on active compounds across a number of sample classes (*9*). Albeit the application of this metafilter is not without limitation, it does provide a focused view of common orange chemistry changes induced by aging that are associated with loss of freshness. The VIP from the original model was used to give a systematic order for which compounds were isolated first. For each of the 97 compounds the VIP was of significance (over 1.0), indicating importance to both the overall model and the individual comparison models.

Since Random Forest algorithms model data in a different manner than PLS, meta filtering was not required. PLS typically favor data trends whereas Random Forest does not necessarily emphasize trends, and can identify markers that may have maxima or minima inflections at an intermediate time points. Numerous collinear variables can increase the challenge of interpretation and variable selection in PLS, combined with noisey data sets can lead to challenging variable selection. Additional benefits to random forest include explicit noise elimination, not present in PLS (*10, 11*). As the data is iterated through poorly classifying compounds are eliminated from decision nodes and only strong classifiers are kept. Noise elimintation and identifying trend independent compounds are beneficial for this experimental design. Consequently Random Forest analysis can nicely compliment chemical markers identified through PLS modeling. This is also the reason classification Random Forest modeling was used rather than regression. With an out of bag error rate at 2.38%, the model is of high quality showing very low misclassification of sample identifiers.

Both statistical modeling approaches proved successful in identifying small chemical differences down to part per billion levels for sample classification. The benefit of using both approaches comes in the ability to identify supplemental information that using a singular approach would miss. Random forest modeling is robust against noise and colinearity and non-parametric variables, while PLS allows for effective implemtnation of meta-analysis for variable selection. Big data approaches are frequently outcome sparse, especially considering the amount of data collected. Reliance on a single approach to modeling and variable selection simply obscures other possible valuable information.

Conclusions

In this work, the application of chemometrics was applied to identify chemical changes that happen as food systems age. Based on this variable reduction technique, multiple modeling approaches were able to successfully identify specific changes in large scale chemical data sets that were ubiquitous to changes in orange juice isolated over time. Both PLS and Random Forest modeling effectively modeled aging of the food system. Subsequent fractionation and recombination models of identified compounds led to sensory active compounds in a majority of the statistically significant variables. Future work will address chemical characterization and identify structure of compounds of interest.

References

1. Bruemmer, J. H., Aroma substances of citrus fruits and their biogenesis. In *Symposium on Fragrance and Flavour Substances. Geruch und Geschmackstoffe Internationales Symposium*; 1975.
2. Deibler, K., Delwiche, J. *Handbook of Flavor Characterization: Sensory Analysis, Chemistry, and Physiology*; CRC Press: New York, 2003.
3. Stark, T.; Bareuther, S.; Hofmann, T. Sensory-Guided Decomposition of Roasted Cocoa Nibs (Theobroma cacao) and Structure Determination of Taste-Active Polyphenols. *J. Agric. Food Chem.* **2005**, *53* (13), 5407–5418.
4. Charve, J. I. M. Prediction of mandarin juice flavor: a flavoromic approach. Dissertation, University of Minnesota, 2011.
5. Bruemmer, J. H. Aroma substances of citrus fruits and their biogenesis. In *Symposium on Fragrance and Flavour Substances. Geruch und Geschmackstoffe Internationales Symposium*; 1975.
6. Ebeler, S. Sensory Analysis and Analytical Flavor Chemistry: Missing Links. In *Handbook of Flavor Characterization: Sensory Analysis, Chemistry, and Physiology*; Deibler, K., Delwiche, J., Eds.; CRC Press: New York, 2003.

7. Patti, G. J.; Tautenhahn , R.; Siuzdak, G. Meta-analysis of untargeted metabolomic data from multiple profiling experiments. *Nat. Protoc.* **2012**, *7* (3), 508–516.
8. Ronningen, I., Peterson, D. *Untargeted Flavoromics for Discovery of Flavor Active Compounds Related To age*; Unpublished work.
9. Tautenhahn, R.; Patti, G. J.; Kalisiak, E.; Miyamoto, T.; Schmidt, M.; Lo, F. Y; McBee, J.; Baliga, N.; Siuzdak, G. metaXCMS: second-order analysis of untargeted metabolomics data. *Anal. Chem.* **2010**, *83* (3), 696–700.
10. Menze, B. H.; Kelm, B. M.; Masuch, R.; Himmelreich, U.; Bachert, P.; Petrich, W.; Hamprecht, F. A. A comparison of random forest and its Gini importance with standard chemometric methods for the feature selection and classification of spectral data. *BMC Bioinf.* **2009**, *10* (1), 213.
11. Breiman, L. Random forests. *Mach. Learn* **2001**, *45* (1), 5–32.

Chapter 20

Exploratory Chemometric Analysis on Source of Origin and Cultivar of Chinese Wolfberries Using Liquid Chromatograph–Mass Spectrometric and Flow Injection Mass Spectrometric (FIMS) Fingerprints

Weiying Lu,[1] Haiming Shi,[1] and Liangli (Lucy) Yu[*,1,2]

[1]Institute of Food and Nutraceutical Science, School of Agriculture and Biology, Shanghai Jiao Tong University, Shanghai 200240, China
[2]Department of Nutrition and Food Science, University of Maryland, College Park, Maryland 20742, United States
[*]E-mail: liangli.yu@sjtu.edu.cn. Tel: (86)-21-3420-5828. Fax: (86)-21-3420-4107.

Lycium barbarum L. fruits (Chinese wolfberries) from different cultivation locations and with different cultivar types may vary in nutritional composition and market value. The relationships between different wolfberry cultivation locations and cultivars were studied by ultra-performance liquid chromatography coupled with mass spectrometry (UPLC−MS) and flow injection mass spectrometric (FIMS) fingerprinting techniques combined with chemometrics techniques. The principal component analysis (PCA) was able to obtain graphical overviews of the spectrometric data sets. The similarity analysis and the hierarchical clustering analysis (HCA) were also performed to study the similarities or differences between the phytochemical fingerprints of wolfberries produced from different locations, as well as different wolfberries cultivars. All results by the exploratory chemometric analyses indicated that the fingerprints of wolfberries were closely related to their producing locations and cultivars. The results might be utilized for Chinese wolfberry authentication.

Introduction

The practical use of *Lycium barbarum* L. fruits (Chinese wolfberries) as a traditional Chinese functional food ingredient has been lasted for thousands of years. The estimated annual production of the Chinese wolfberry was 1.7 million tons in 2010, according to the Chinese National Bureau of Statistics. The high production and consumption of wolfberries indicate the popularity of wolfberry and its related products nowadays (*1*). Chinese wolfberries have many validated health properties, for example, antioxidant, anti-age, and immune-enhancement, according to recent studies (*2–5*). Among all wolfberry products, the wolfberries produced in Ningxia province is considered to offer the best nutraceutical quality and thus with the highest market value (*6*, *7*). Because it is difficult distinguish wolfberries produced in different locations or with different cultivar types by their visual appearance or smell without expert knowledge, the economically motivated counterfeiting may occur by using wolfberries grown at other locations as authentic Ningxia wolfberries. On the other hand, a lot of new wolfberry cultivars have been developed in Ningxia with different nutritional, nutraceutical and cultivation properties. Therefore, the comparisons of Ningxia-cultivated wolfberries from those grown at other locations, as well as that of different Ningxia wolfberry cultivars are important to promote the commercial production of wolfberries.

Fingerprinting techniques combined with chemometrics analysis have been increasingly popular in the phytochemical researches of botanical and food materials. There are fingerprinting studies related to the wolfberries using various analytical techniques. Two-dimensional infrared spectroscopic fingerprints were utilized to distinguish eight *Lycium* species (*8*). The clustering analysis on the high performance liquid chromatographic (HPLC) fingerprints was reported to identify different cultivation locations of eight wolfberry samples (*9*). The ultra-performance liquid chromatography coupled with mass spectrometry (UPLC−MS) followed by the partial least squares-discriminant analysis (PLS-DA) differentiated wolfberries from Tibet, northern China, and Mongolia (*10*). In our previous works, different genotypes, growing environment, or cultivation modes (organic versus conventional farming) were demonstrated using HPLC and flow injection mass spectrometric (FIMS) fingerprints in the analyses of peppermints, basil and *Gynostemma pentaphyllum* (*11–14*). The differentiation of Chinese wolfberries produced in Ningxia from those grown in other locations, and the different wolfberry cultivars from a selected growing region is achieved by both the UPLC−MS and the FIMS fingerprints with PLS-DA as a supervised learning method (*15*). All these researches indicated that with the rapid development of novel analytical techniques and the many successful chemometrics applications, fingerprinting techniques have great potential in food research to obtain not only detailed compositional profiles, but also mathematical models and plots that is easy to interpret the complex relationships between different samples.

This research focused on the inter-group relationships of different Chinese wolfberries, including wolfberries produced in Ningxia and those from those grown in five other locations, as well as four Ningxia wolfberry cultivars using exploratory analysis methods. Three chemometric analysis techniques including principal component analysis (PCA), similarity analysis, and hierarchical clustering analysis (HCA) were applied to the UPLC−MS and FIMS fingerprints. These methods were able to provide information about whether there is closer similarity of wolfberries produced in Ningxia as compared with that produced in other locations, as well as which wolfberry producing locations or cultivars yielded greater component similarity. The results may provide a scientific foundation for the characterization and authentication of Chinese wolfberries.

Materials and Methods

Materials and Reagents

Twenty nine Chinese wolfberry samples were collected. The sample set included 19 and 10 wolfberry samples from Ningxia and five other provinces in China, respectively. Specifically, three Inner Mongolia, three Qinghai, two Gansu, one Hebei, and one Xinjiang samples were included as samples produced outside Ningxia. Except three Ningxia samples had no cultivar information, the cultivars of 16 Ningxia samples, including five cultivar No. 1, five cultivar No. 4, two cultivar No. 5, and four cultivar No. 7 samples were recorded. The cultivar information of wolfberries produced outside Ningxia was not recorded. All samples were *L. barbarum* fruits, except one Qinghai sample was *Lycium ruthenicum* Murr. fruits.

Methanol and acetonitrile were purchased from Merck (LC−MS grade, Darmstadt, Germany). Formic acid was purchased from Sigma-Aldrich (LC−MS grade, St. Louis, MO, U.S.A.). Deionized water was obtained using a Milli-Q purification system (Millipore Laboratory, Bedford, MA, U.S.A.). Other chemicals were of the highest commercial grade and used without further purification.

Analysis Methods

The detailed experimental procedure was described in a previous work (*15*). Briefly, 0.5 g of each dried fruit was extracted by 80% (v/v) ethanol. UPLC−MS and FIMS were performed by an Acquity UPLC coupled with a Xevo G2 quadrupole time-of-flight (QTOF) mass spectrometer (Waters, Milford, MA, U.S.A.). The mobile phases were 0.1% (v/v) formic acid in water (solvent A) and 0.1% (v/v) formic acid in acetonitrile (solvent B). For UPLC−MS, the elution program was: 0−8 min, 10−100% B with a concave gradient (curve parameter was 8); 8−13 min, 100% B; 13−15 min, 100−10% B; and 15−16.5 min, 10% B with linear gradients (curve parameter was 6). The column temperature was 40 °C and the flow rate was 0.4 mL/min. The column was Acquity UPLC BEH C18 (2.1 mm i.d. × 100 mm, 1.7 μm, Waters) with a VanGuard BEH C18 preolumn (2.1 mm i.d. × 5 mm, 1.7 μm, Waters). The FIMS applied no column separation. An isocratic elution of 50% B with 0.5 mL/min flow rate was applied. The capillary voltage was 2.5 kV and the cone voltage was 40 V. The ionization temperature was 120 °C and the desolvation temperature was 250 °C. Spectra from 0.5 to 8.5 min and 0-0.5 min for UPLC−MS and FIMS were recorded with 0.3 s/scan, respectively. Two MS channels, including one primary MS channel (*m/z*: 100 to 1500 Da) without collision energy, and the other mass fragmentation channel (*m/z*: 100 to 1000 Da) with a 35 eV collision energy. Each sample was injected in triplicate.

Data Analysis

The peak areas were generated by MarkerLynx 4.1 (Waters). For UPLC–MS, twenty characteristic peaks appeared in all Ningxia samples were selected. The peaks were extracted by a 0.01 Da chromatogram mass window and a 0.2 min retention time window. For FIMS, the peaks were extracted by a mass window of unit Da. Each triplicate measurement was averaged. The PCA and other data processing methods were performed using in-house programs written in MATLAB R2013b (The MathWorks, Natick, MA, U.S.A.). The dataset were preprocessed by the autoscaling transformation in PCA, and were directly averaged to mean peak area or spectra by same type of samples without normalization or autoscaling when calculating the similarity and the HCA.

Results and Discussion

UPLC–MS Chromatograms and FIMS Fingerprints

Figure 1 is the UPLC–MS base peak chromatograms (BPCs) of nine representative Chinese wolfberries. BPC presents the intensity of the most abundant ion in each MS scan. In Figure 1, samples from Ningxia and other locations were plotted separately. As shown in Figures 1A and 1B, the BPCs of Ningxia wolfberries is unique compared with that from other provinces. The peak profiles between 2-5 min of a sample from Qinghai significantly differed from samples grown in other locations, probably was because this sample was *L. ruthenicum* fruit with a unique phytochemical composition (*16*). The profiles of Hebei were also different from the Ningxia samples. This result was consistent with a previous UPLC–UV fingerprinting study, which concluded that samples cultivated from Hebei furnished the largest dissimilarity than that from eight other locations (*9*). Comparing all Ningxia wolfberries, most of the peaks were identical. However, the relative concentrations of components were slightly different. The QTOF mass spectrometer offered *m/z* measurement at high resolution (<10 ppm). Consequently, a part of the compound can be tentatively identified. Twenty characteristic compounds found in Ningxia wolfberries were listed in Table 1. In a preliminary PCA study, peak selection using automatic thresholds yielded 10,220 peaks, although similar discrimination power was obtained compared to differentiation by only the major peaks included in Table 1. The phenolic acids listed in Table 1, including quercetin-rhamno-di-hexoside, quercetin-di-(rhamno)-hexoside, quercetin-3-*O*-rutinoside, kaempferol-3-*O*-rutinoside, and isorhamnetin-3-*O*-rutinoside were studied previously to be major functional components of wolfberries for antioxidant activities (*3*). Therefore, the peak areas of compounds in Table 1 were used for further analyses.

Table 1. Tentatively Identified Compounds by UPLC−MS*

	RT	*[M-H]⁻*			
	(min)	*exptl*	*theor*	*formula*	*compound*
1	0.71	191.0187	191.0192	$C_6H_8O_7$	citric acid
2	0.92	794.3354	794.3361	$C_{39}H_{55}O_{17}$	citric acid glycoside
3	1.03	796.3470	796.3459	$C_{46}H_{53}O_{12}$	unknown
4	1.15	163.0393	163.0395	$C_9H_8O_3$	coumaric acid glycoside
5	1.32	771.1987	771.1984	$C_{33}H_{40}O_{21}$	quercetin-rhamno-dihexoside
6	1.95	325.0928	325.0923	$C_{15}H_{17}O_8$	coumaric acid glycoside
7	2.51	193.0500	193.0501	$C_{10}H_{10}O_4$	ferulic acid glycoside
8	2.64	191.0340	191.0344	$C_{10}H_8O_4$	scopoletin
9	3.62	119.0495	119.0497	C_8H_8O	1,3-dihydro-2-benzofuran
10	4.21	609.1456	609.1456	$C_{27}H_{30}O_{16}$	quercetin-3-*O*-rutinoside
11	4.48	245.0926	245.0966	$C_8H_{14}O$	unknown
12	4.73	593.1519	593.1506	$C_{27}H_{30}O_{15}$	kaempferol-3-*O*-rutinoside
13	4.81	623.1610	623.1612	$C_{28}H_{32}O_{16}$	isorhamnetin-3-*O*-rutinoside
14	6.66	329.2320	329.2328	$C_{18}H_{34}O_5$	trihydroxy octadecenoic acid
15	7.47	311.2224	311.2222	$C_{18}H_{32}O_4$	dihydroxy octadecadienoic acid
16	7.53	309.2058	309.2066	$C_{18}H_{30}O_4$	hydroperoxy octadecatrienoic acid
17	7.72	311.2220	311.2220	$C_{18}H_{32}O_4$	dihydroxy octadecadienoic acid
18	7.95	295.2271	295.2273	$C_{18}H_{32}O_3$	hydroxy linoleic acid
19	8.03	814.5580	814.5595	$C_{48}H_{78}O_{10}$	unknown
20	8.27	595.2890	595.2907	$C_{34}H_{44}O_9$	unknown

* RT, retention time; exptl [M-H]⁻ and theor [M-H]⁻ were experimental and theoretical *m/z* of molecular ions, respectively.

FIMS fingerprinting is a novel technique without chromatographic separation. It is suitable for a rapid analysis of food materials (*11*, *17*). Figure 2 reported the FIMS fingerprints of the representative nine wolfberry samples. Similar to Figure 1, the unique spectra of wild grown Qinghai and Hebei samples were shown in Figure 2. Compared with the conventional UPLC−MS base peak fingerprints under the same instrumentation configurations, the FIMS fingerprints can effectively detect a part of characteristic mass spectrometric features using wolfberries from different cultivation locations or different cultivars.

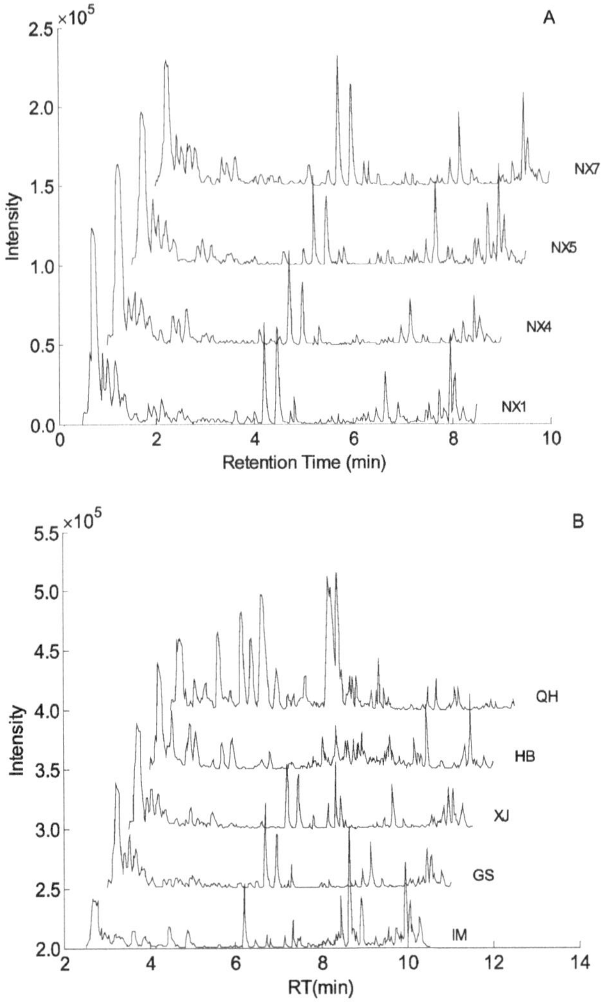

Figure 1. UPLC−MS base peak chromatograms of Chinese wolfberries produced in Ningxia (A) and other provinces (B). NX1, NX4, NX5, and NX7 represent Ningxia wolfberry cultivars No. 1, No. 4, No. 5, and No. 7, respectively. QH, HB, XJ, GS, and IM represent samples from Qinghai, Hebei, Xinjiang, Gansu, and Inner Mongolia, respectively.

Principal Component Analysis (PCA)

Figure 3 is the principal component scores plots of UPLC−MS peak areas and FIMS fingerprints. Wolfberries produced in Ningxia and other locations were plotted in different markers. The principal component scores achieved certain degree of separation between the two classes. Comparing Figure 3A with Figure 3B, FIMS achieved slightly worse separation. Since many complicated and confounding factors such as cultivation modes, growing environment, or genotype may alter the phytochemical profile of a botanical (*9*, *18*), it is challenging to collect the samples needed for differentiation with every experimental factor under control. Consequently, the relatively poor separations when differentiating wolfberries produced inside or outside Ningxia using PCA were indicated in Figure 3.

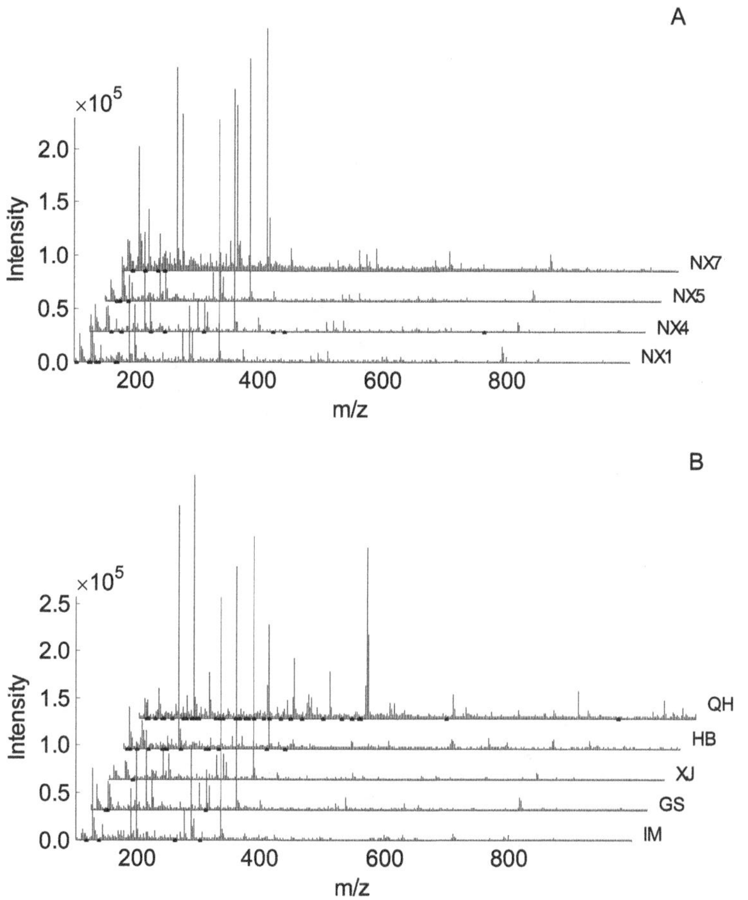

Figure 2. FIMS fingerprints of Chinese wolfberries produced in Ningxia (A) and other provinces (B). NX1, NX4, NX5, and NX7 represent Ningxia wolfberry cultivars No. 1, No. 4, No. 5, and No. 7, respectively. QH, HB, XJ, GS, and IM represent samples from Qinghai, Hebei, Xinjiang, Gansu, and Inner Mongolia, respectively.

The principal component scores plots of Ningxia cultivars using UPLC−MS peak areas and FIMS fingerprints were reported Figure 4A and 4B, respectively. In the plots of both methods, different wolfberry cultivars were slightly separated from each other and positioned following a similar tendency. Data points representing Ningxia cultivar No. 1 were located in the center of both plots. The location of cultivar No. 1 suggested that there are less characteristic components detected in cultivar No. 1 compared with others. Because cultivar No. 1 is the earliest commercial breed of Ningxia wolfberry, cultivar Nos. 4, 5, and 7 were derived from this cultivar. It can be assumed that the characteristic components appeared during the cultivation process of wolfberries.

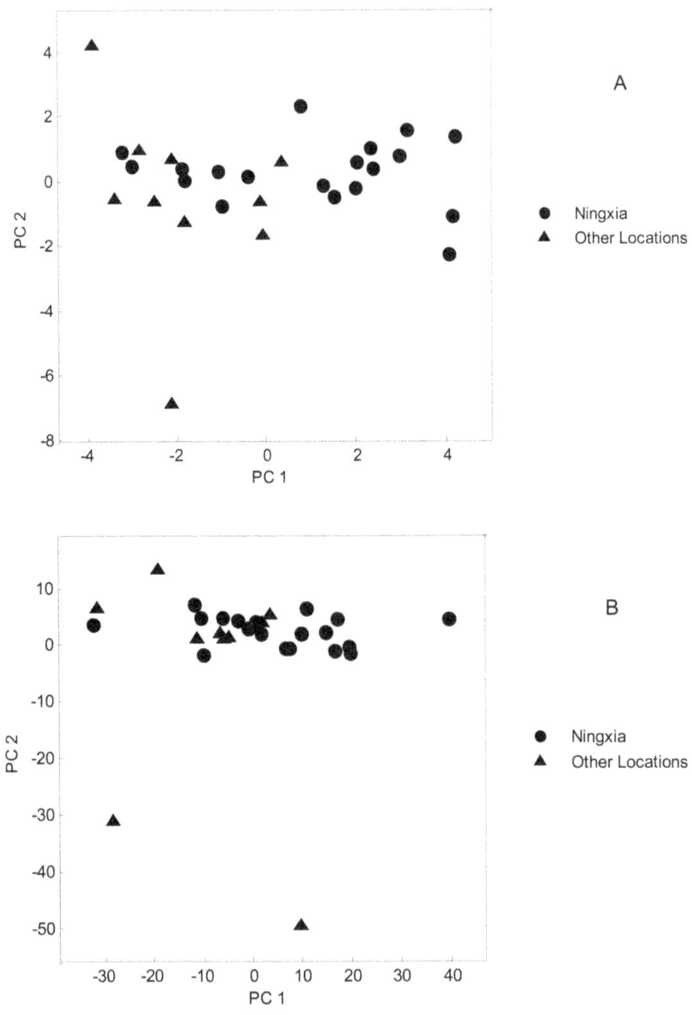

Figure 3. Principal component scores plots of wolfberries produced in Ningxia and other locations using UPLC−MS peak areas (A) and FIMS fingerprints (B).

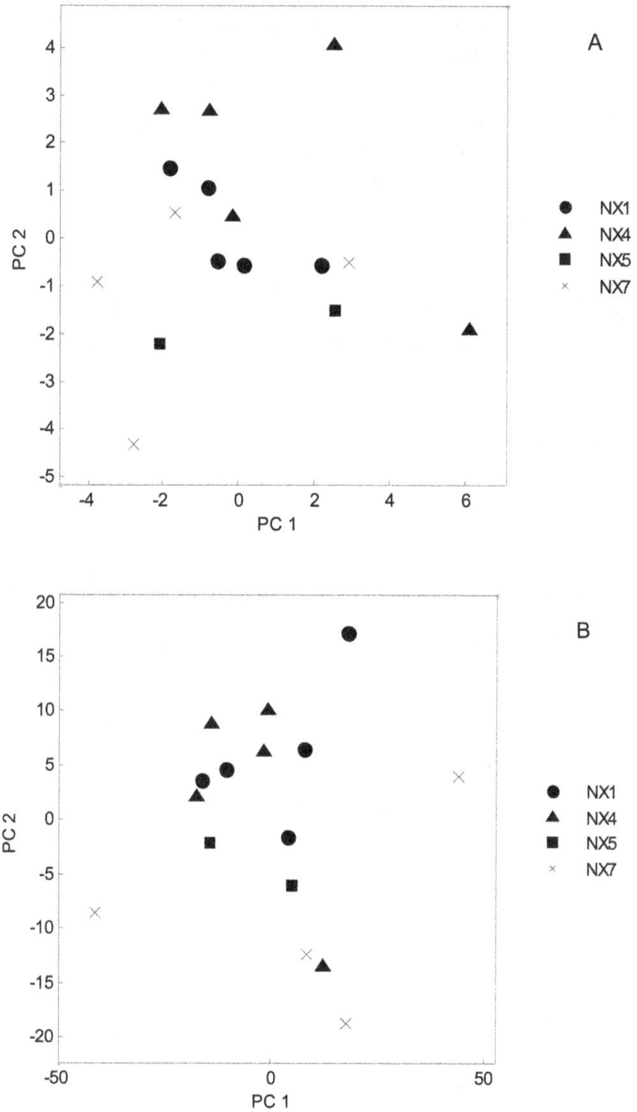

Figure 4. Principal component scores plots of Ningxia cultivars using UPLC−MS peak areas (A) and FIMS fingerprints (B). NX1, NX4, NX5, and NX7 represent Ningxia wolfberry cultivars No. 1, No. 4, No. 5, and No. 7, respectively.

Similarity Analysis

Similarity analysis calculates correlation coefficient to determine the relationship between a pair of samples (*19*). The correlation coefficient is a simple and quantitative metric that ranges between 0 and 1. A correlation coefficient close to 1 indicates more similarity between the pair of samples, and vice versa.

When dealing with multiple classes of samples, a similarity matrix can be constructed. Table 2 is the similarities of wolfberries with respect to different producing locations and different cultivars by UPLC–MS. All similarities were high (>0.89), indicating that all wolfberries furnished very similar phytochemical profiles in general. The coefficients between Ningxia samples, especially those with designated cultivars, were larger than Ningxia samples versus samples in other provinces. It is consistent with the fact that the Ningxia samples have a close similarity than samples outside Ningxia. Samples from Qinghai and Hebei were less similar compared to any other samples. On the contrary, part of samples from Xinjiang, Gansu, and Inner Mongolia yielded comparable correlations to samples from Ningxia. Among different Ningxia cultivars, cultivar Nos. 1 and 4 were the most similar pair of cultivars (the similarity was 0.997) compared to cultivar Nos. 5 and 7 (the similarities were ranged from 0.980 to 0.991 with respect to other Ningxia cultivars). The similarity analyses were consistent with the results of PCA.

Table 2. Similarities of Wolfberries by UPLC–MS*

	NX1	*NX4*	*NX5*	*NX7*	*NX*	*GS*	*XJ*	*IM*	*HB*	*QH*
NX1	1.000	0.997	0.991	0.986	0.983	0.986	0.990	0.970	0.935	0.971
NX4		1.000	0.985	0.980	0.985	0.987	0.992	0.974	0.939	0.981
NX5			1.000	0.980	0.972	0.967	0.976	0.954	0.944	0.950
NX7				1.000	0.987	0.986	0.974	0.979	0.909	0.958
NX					1.000	0.995	0.988	0.991	0.933	0.971
GS						1.000	0.989	0.989	0.924	0.977
XJ							1.000	0.974	0.939	0.967
IM								1.000	0.908	0.971
HB									1.000	0.892
QH										1.000

* NX1, NX4, NX5, and NX7 represent Ningxia wolfberry cultivars No. 1, No. 4, No. 5 and No. 7, respectively. QH, HB, XJ, GS, and IM represent samples from Qinghai, Hebei, Xinjiang, Gansu, and Inner Mongolia, respectively.

Table 3 presents the similarities of wolfberries by FIMS. Sample from Hebei was particularly different from other samples. The similarities were ranged from 0.556 to 0.799, significantly lower than any other coefficient in the matrix. Except that, the results were in good agreement with those calculated by UPLC–MS data, thus similar conclusions about the sample relationships could be obtained.

Table 3. Similarities of Wolfberries by FIMS*

	NX1	*NX4*	*NX5*	*NX7*	*NX*	*GS*	*XJ*	*IM*	*HB*	*QH*
NX1	1.000	0.995	0.992	0.988	0.988	0.991	0.969	0.981	0.697	0.972
NX4		1.000	0.976	0.973	0.990	0.996	0.977	0.978	0.705	0.975
NX5			1.000	0.997	0.975	0.973	0.952	0.978	0.670	0.953
NX7				1.000	0.974	0.972	0.952	0.979	0.663	0.949
NX					1.000	0.998	0.995	0.995	0.614	0.949
GS						1.000	0.990	0.988	0.658	0.962
XJ							1.000	0.991	0.556	0.924
IM								1.000	0.570	0.930
HB									1.000	0.799
QH										1.000

* NX1, NX4, NX5, and NX7 represent Ningxia wolfberry cultivars No. 1, No. 4, No. 5, and No. 7, respectively. NX represent Ningxia wolfberry with no cultivar information available. QH, HB, XJ, GS, and IM represent samples from Qinghai, Hebei, Xinjiang, Gansu, and Inner Mongolia, respectively.

Hierarchical Clustering Analysis (HCA)

HCA is a fundamental chemometrics clustering technique that advantages in organizing and presenting data in a hierarchical binary tree structure. In the HCA tree, the distances of samples are calculated, and then samples more correlated to each other are organized in nearby branches in the HCA tree. In this study, the Euclidian distance metric were used for similarity measure. The HCA trees using UPLC−MS peak areas and FIMS fingerprints were presented in Figure 5. It can be observed that samples from Qinghai and Hebei were the most dissimilar samples compared to the rest of the samples. Samples from Inner Mongolia and Xinjiang were also grouped outside samples from Ningxia. Most wolfberries from Ningxia, especially those with designated cultivar types, were grouped together in the HCA tree. Samples from Gansu were closest to the samples from Ningxia in both UPLC−MS peak areas and FIMS fingerprints, probably because both provinces have similar climatic conditions. Cultivar Nos. 1 and 4 were close to each other, compared to cultivar Nos. 5 and 7. These grouping tendencies were consistent with the results from PCA (Figure 4) and similarity analysis (Tables 2 and 3).

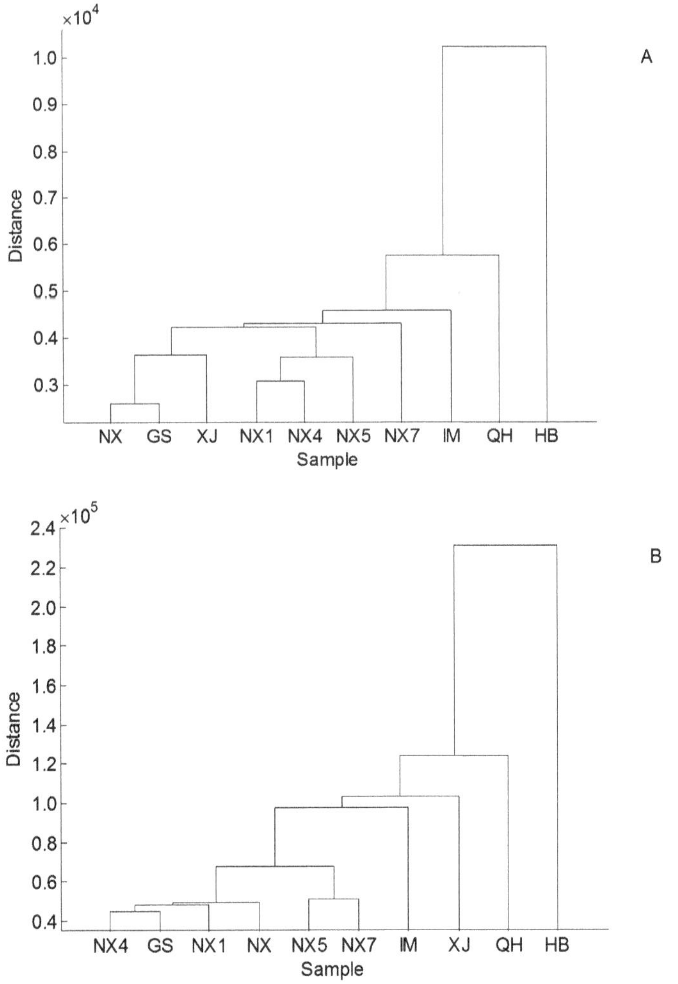

Figure 5. Hierarchical clustering trees using UPLC−MS peak areas (A) and FIMS fingerprints (B). NX1, NX4, NX5, and NX7 represent Ningxia wolfberry cultivars No. 1, No. 4, No. 5, and No. 7, respectively. NX represent Ningxia wolfberry with no cultivar information available. QH, HB, XJ, GS, and IM represent samples from Qinghai, Hebei, Xinjiang, Gansu, and Inner Mongolia, respectively.

In summary, the chemometric exploratory techniques applied to UPLC−MS chromatogram and the FIMS fingerprints analyzed the relationships among wolfberries produced in six different locations, as well as four different Ningxia wolfberry cultivars. Generally, wolfberries produced in Qinghai and Hebei differed from those produced in Ningxia, Inner Mongolia, and Xinjiang. Regarding wolfberries produced in Ningxia, cultivar Nos. 1 and 4 differed from cultivar Nos. 5 and 7. The UPLC−MS and FIMS achieved comparable analysis results. PCA, similarity analysis, and HCA processed the original data by three respective metrics of the principal component scores, the correlation coefficients, and the Euclidean distances. The principal component scores plot advantages in achieving visual overviews of the samples; the similarity analysis advantages in offering detailed pairwise and quantitative comparisons of samples; while the HCA advantages in organizing data to a hierarchical structure that indicates the tendencies of sample grouping. Combining the exploratory techniques together, they could be helpful in the quality assurance, quality control and authentication of wolfberry-containing foods.

Acknowledgments

This research was funded by National High Technology Research and Development Program of China (Grant Nos. 2013AA102202; 2013AA102207); SJTU startup fund for young talent (Grant No. 13X100040047); a special fund for Agro-scientific Research in the Public Interest (Grant No. 201203069); SJTU 985-III disciplines platform and talent fund (Grants TS0414115001; TS0320215001). Prof. Jinming Gao (Shaanxi Engineering Center of Bioresource Chemistry & Sustainable Utilization, College of Science, Northwest A&F University) is thanked for kindly providing us the wolfberry samples.

Part of this work was adapted with permission from Lu, W.; Jiang, Q.; Shi, H.; Niu, Y.; Gao, B.; Yu, L., Partial least-squares-discriminant analysis differentiating Chinese wolfberries by UPLC–MS and flow injection mass spectrometric (FIMS) fingerprints. *J. Agric. Food. Chem.* **2014,** *62*, 9073-9080.

References

1. Wang, P. Wolfberry industrial competitiveness of Zhongning Ningxia. Ph.D. Thesis, Northwest A&F University, Yangling, Shaanxi, China, 2013.
2. Zhang, X.; Li, Y.; Cheng, J.; Liu, G.; Qi, C.; Zhou, W.; Zhang, Y. Immune activities comparison of polysaccharide and polysaccharide-protein complex from *Lycium barbarum* L. *Int. J. Biol. Macromol.* **2014**, *65*, 441–445.
3. Wang, C.; Chang, S.; Stephen Inbaraj, B.; Chen, B. Isolation of carotenoids, flavonoids and polysaccharides from *Lycium barbarum* L. and evaluation of antioxidant activity. *Food Chem.* **2010**, *120* (1), 184–192.
4. Stephen Inbaraj, B.; Lu, H.; Hung, C.; Wu, W.; Lin, C.; Chen, B. Determination of carotenoids and their esters in fruits of *Lycium barbarum* Linnaeus by HPLC–DAD–APCI–MS. *J. Pharm. Biomed. Anal.* **2008**, *47* (4–5), 812–818.
5. Chang, R.; So, K. Use of anti-aging herbal medicine, *Lycium barbarum*, against aging-associated diseases. what do we know so far? *Cell. Mol. Neurobiol.* **2008,** *28* (5), 643–652.
6. Zheng, G.; Zheng, Z.; Xu, X.; Hu, Z. Variation in fruit sugar composition of *Lycium barbarum* L. and *Lycium chinense* Mill. of different regions and varieties. *Biochem. Syst. Ecol.* **2010**, *38* (3), 275–284.

7. Zhao, L.; Qiu, Z.; Narasimhamoorthy, B.; Greaves, J. Development of a rapid, high-throughput method for quantification of zeaxanthin in Chinese wolfberry using HPLC–DAD. *Ind. Crop. Prod.* **2013**, *47* (0), 51–57.
8. Yao, X.; Peng, Y.; Zhou, Q.; Xiao, P.; Sun, S. Distinction of eight *Lycium* species by Fourier-transform infrared spectroscopy and two-dimensional correlation IR spectroscopy. *J. Mol. Struct.* **2010**, *974* (1–3), 161–164.
9. Li, X.; Li, R.; Xiang, H.; Zhao, Z.; Liu, X. Study on quality evaluation of *Lycium chinense* from different areas by HPLC fingerprint and cluster analysis. *Mod. Food Sci. Technol.* **2012**, *28* (9), 1251–1253.
10. Bondia-Pons, I.; Savolainen, O.; Törrönen, R.; Martinez, J. A.; Poutanen, K.; Hanhineva, K. Metabolic profiling of Goji berry extracts for discrimination of geographical origin by non-targeted liquid chromatography coupled to quadrupole time-of-flight mass spectrometry. *Food Res. Int.* **2014**, *63* (Part B) (0), 132–138.
11. Gao, B.; Lu, Y.; Qin, F.; Chen, P.; Shi, H.; Charles, D.; Yu, L. Differentiating Organic from Conventional Peppermints Using Chromatographic and Flow Injection Mass Spectrometric (FIMS) Fingerprints. *J. Agric. Food. Chem.* **2012**, *60* (48), 11987–11994.
12. Lu, Y.; Gao, B.; Chen, P.; Charles, D.; Yu, L. Characterisation of organic and conventional sweet basil leaves using chromatographic and flow-injection mass spectrometric (FIMS) fingerprints combined with principal component analysis. *Food Chem.* **2014**, *154* (0), 262–268.
13. Wang, Z.; Chen, P.; Yu, L.; Harrington, P. Authentication of Organically and Conventionally Grown Basils by Gas Chromatography/Mass Spectrometry Chemical Profiles. *Anal. Chem.* **2013**, *85* (5), 2945–2953.
14. Xie, Z.; Shi, H.; Yu, L. HPLC/MS fingerprinting techniques for quality control of *Gynostemma pentaphyllum* (Thunb.) Makino samples. In *Physical Methods in Food Analysis*; Tunick, M. H., Onwulata, C.I., Eds.; ACS Symposium Series 1138; American Chemical Society: Washington, DC, 2013; pp 31–47.
15. Lu, W.; Jiang, Q.; Shi, H.; Niu, Y.; Gao, B.; Yu, L. Partial Least-Squares-Discriminant Analysis Differentiating Chinese Wolfberries by UPLC–MS and Flow Injection Mass Spectrometric (FIMS) Fingerprints. *J. Agric. Food. Chem.* **2014**, *62* (37), 9073–9080.
16. Zheng, J.; Ding, C.; Wang, L.; Li, G.; Shi, J.; Li, H.; Wang, H.; Suo, Y. Anthocyanins composition and antioxidant activity of wild *Lycium ruthenicum* Murr. from Qinghai-Tibet Plateau. *Food Chem.* **2011**, *126* (3), 859–865.
17. Chen, P.; Harnly, J. M.; Lester, G. E. Flow Injection Mass Spectral Fingerprints Demonstrate Chemical Differences in Rio Red Grapefruit with Respect to Year, Harvest Time, and Conventional versus Organic Farming. *J. Agric. Food. Chem.* **2010**, *58* (8), 4545–4553.
18. Whent, M.; Hao, J.; Slavin, M.; Zhou, M.; Song, J.; Kenworthy, W.; Yu, L. Effect of Genotype, Environment, and Their Interaction on Chemical Composition and Antioxidant Properties of Low-Linolenic Soybeans Grown in Maryland. *J. Agric. Food. Chem.* **2009**, *57* (21), 10163–10174.
19. Xie, Z.; Zhao, Y.; Chen, P.; Jing, P.; Yue, J.; Yu, L. Chromatographic Fingerprint Analysis and Rutin and Quercetin Compositions in the Leaf and Whole-Plant Samples of Di- and Tetraploid Gynostemma pentaphyllum. *J. Agric. Food. Chem.* **2011**, *59* (7), 3042–3049.

Chapter 21

Flavoromics for Determining Markers of *Cooked* and *Fermented* Flavor in Strawberry Juices

I. Andujar-Ortiz,*[,1] T. L. Peppard,[2,3] and G. Reineccius[1]

[1]Flavor Research and Education Center, Department of Food Science and Nutrition, University of Minnesota, 1334 Eckles Avenue, Saint Paul, Minnesota 55108, U.S.A.
[2]Robertet Flavors Inc., 10 Colonial Drive, Piscataway, New Jersey 08854, U.S.A.
[3]Current affiliation: Apposite Science Consulting, Warren, New Jersey 07059
***E-mail: iandujar@umn.edu.**

Flavoromics aims to elucidate the molecules contributing to flavor perception by collecting as much chemical information as possible by adapting concepts and tools taken from the field of metabolomics. In this study, we have applied flavoromics to find markers of *cooked* and *fermented* flavor in strawberry juices submitted to different treatments (heat, storage, and freeze-drying). Chemical information from the samples was obtained by analysis of both volatile and non-volatile constituents, followed by processing of chromatograms and reduction in the number of variables using specialized software. By using Partial Least Squares Regression analysis, chemical data were correlated with the *cooked* and *fermented* flavor notes present in the strawberry juices (determined by sensory analysis). Some variables were selected as diagnostic markers, based on having a high impact in the statistical models developed. Regarding *cooked* flavor, most of the markers were chemical compounds known to be formed during heating of fruit juices; others were important volatile compounds of strawberries in their own right. On the other hand, the markers of *fermented* flavor were mainly esters, together with some alcohol oxides and compounds known generally for causing unpleasant flavor. The sensory contributions of these marker compounds will be further elucidated by targeted analysis and compound addition studies.

Introduction

Flavor perception is a complex phenomenon, which is difficult to study since it involves a variety of stimuli resulting from the stimulation of our senses while eating or drinking (*1*). Chemical stimuli from the food contribute the most to flavor perception and then, understanding which chemicals are responsible for flavor perception is of primary interest for many flavor chemists. Traditionally, a targeted approach has been used involving isolation of the chemical fraction of interest and evaluation of its contribution to the sensory attribute under investigation. However, this approach has not resulted in the chemical characterization of the flavor of any food (a knowledge of all chemical compounds needed, and their upper and lower concentrations, to perfectly match the flavor of that food using stringent sensory evaluation). While much has been learned using targeted approaches, additional knowledge is needed that this technique cannot provide.

In order to consider a wide range of chemical compounds present in food, instrumental-sensory correlations have been used since the late 1970s (*2*). The main idea was to establish associations between the chemical stimuli (measured instrumentally) and the flavor perceived (measured by sensory analysis) by mathematical models. For example, flavor compounds have been related to sensory attributes in olive oil (*3*), wine (*4–6*), and tomato (*7*). However, the studies carried out so far have been subject to several limitations. For example, most of them identify and quantify all chemical compounds, *before* their correlation with sensory attributes, which is time-consuming. In addition, most of these studies only included volatiles in the mathematical models. Moreover, when including the non-volatiles, these are often limited to "traditional" physicochemical measurements (pH, titrable acidity, etc.) as previously reported in the case of tomato (*7*); or to a specific group of compounds, such as amino acids and peptides in Iberian hams (*8*), or polyphenols in wine (*9*).

In order to overcome these limitations, a new non-targeted approach called flavoromics has emerged (*10*, *11*). It aims to elucidate the chemical compounds contributing to flavor perception by adapting concepts and tools typically used in metabolomics-based investigations. This approach considers all low molecular weight compounds (volatiles and non-volatiles) in food systems as candidate chemical stimuli in human flavor perception, instead of focusing just on compounds already known to influence flavor quality. Thus it provides an unbiased view of the food system. Characteristically it uses mass spectrometry for sample analysis, also extensively employed in metabolomics (*12*). Due to the large amount of information generated, specialized software is used to preprocess the chromatograms (noise reduction, peak detection, retention time alignment, etc.) and to reduce the number of variables (to those that are statistically significant). Finally, a range of chemometric techniques are utilized to visualize and interpret data.

One of the applications of this approach is the prediction of flavor. More specifically, this consists of obtaining predictive models based on chemical and sensory data, and then using such models to predict the flavor attributes of additional samples based on chemical data alone, i.e., without using sensory evaluation. As flavoromics includes inputs from a larger number of chemical compounds, it opens the possibility to achieve better prediction of flavor. The models obtained can be especially useful to monitor food flavor, or to design foods with a specific flavor profile. Using this approach, Charve (*11*) constructed models to predict the flavor of mandarin juices.

Another application of flavoromics is to find characteristic markers of particular flavor attributes. As this approach considers both volatiles and non-volatiles, and all of the information is potentially valuable *a priori* (not restricted to earlier thinking), this presents a better opportunity to find new compounds important for flavor perception. Another advantage is that it is not necessary to identify all of the chemical compounds detected, only those that are related to the sensory attribute of interest (markers). An important caveat is that the markers are merely *correlated* with the sensory attributes, which does not necessarily mean that they are actually responsible for the sensory perception. To confirm the role of the marker in the sensory perception, i.e., to establish cause and effect, subsequent experiments using a targeted approach are needed, and ideally also compound addition-sensory studies.

Strawberries (*Fragaria* x *ananassa* Duch) have a unique, highly desirable odor and taste and are one of the most popular summer fruits. Consumers mainly purchase strawberries and its derived products for their sensory characteristics and nutritional value. However, different processes such as storage and heat treatment applied to strawberries or juices can impact the concentrations of flavor compounds, and affect their sensory characteristics (*13*, *14*). In this study, strawberry juices ranging in sensory characters were created utilizing various conditions of heat, storage, and freeze-drying and flavoromics has been used to determine flavor markers of selected sensory attributes (*cooked* and *fermented* flavor) in these strawberry juices.

Materials and Methods

1. Preparation of Strawberry Juices

For the preparation of the samples, a total of 150 kg of strawberries from 2 varieties (Wendy, Jewel) were manually picked at a local farm (Pine Tree Orchard, White Bear Lake, MN) over the course of 5 days during June 2012. Each day after picking, strawberries were stored at 8 °C until their treatment, either on the same day or the next. Juices were obtained using a fruit juicer (Omega, model 1000, Harrisburg, PA). In addition, commercial organic strawberry puree was used (Nature's Flavor, Orange, CA). Table 1 shows the different treatments applied to the strawberries and juices and the sample identification used in this study.

Table 1. Treatments Applied to Strawberries and Strawberry Juices

Origin	*Treatment*	*Conditions*	*Sample ID*
Commercial strawberry puree	As received	-	C
	Heated	60 °C or 90 °C, 1 h	C60, C90
	Freeze-dried	-	Cfd
Strawberries Jewel	Fresh	-	J
	Storage, as fruit	2 °C; 3 or 6 d	Jf23, jf26
	Storage, as fruit	8 °C, 4 d	Jf84
	Storage, as juice	2 °C; 1, 2 or 3 d	Jj21, Jj22, Jj23
	Heated	60 °C or 90 °C, 1 h	J60, J90
	Freeze-dried	-	Jfd
	Mixture	Fresh and heated 90 °C	Jm90
	Mixture	Fresh and fruit stored 2 °C, 3 d	Jm23
Strawberries Wendy	Fresh	-	W
	Storage, as fruit	2 °C, 3 d	Wf23
	Storage, as fruit	8 °C, 4 d	Wf84
	Storage, as juice	2 °C; 1 ,2 or 3 d	Wj21, Wj22, Wj23
	Heated	60 °C or 90 °C, 1 h	W60, W90
	Freeze dried	-	Wfd
	Mixture	Fresh and heated 90 °C	Wm90
	Mixture	Fresh and fruit stored 2 °C, 3 d	Wm23

As indicated, the treatments mainly consisted of heating, storage, freeze-drying and mixing of juices obtained by different methods. More specifically, juices submitted to heat treatment were placed in quart glass bottles and heated in a water bath for 1 hour, at 60 or 90 °C. To obtain freeze-dried samples, juices were frozen and then placed in a freeze dryer until dry. The samples were then reconstituted with the same volume of potable water as was removed. For the storage treatment, both strawberries and juices were used. The former were placed in 3 kg cardboard boxes whereas the juices were stored in quart glass bottles. After storage, the strawberries were made into juice. As can be seen, some of the treatments were also applied to the commercial strawberry organic puree, namely, heat (60 and 90 °C) and freeze drying. A total of 27 strawberry juices were prepared for our study.

Juices were subsequently transferred to quart freezer bags for freezing. Samples for the analysis of volatile and non-volatile compounds, and for sensory analysis (in different sessions) were stored in different bags (a separate bag for each session). The samples were stored at -30 °C to minimize changes as much as possible. All equipment, labs and glassware used for the preparation of samples were food grade.

2. Collection of Chemical Data

Compositional data from the volatile and non-volatile constituents of the samples were obtained. An additional "quality control" sample (QC) was analyzed; QC was prepared by mixing the 27 samples in equal proportions (*15*). Analyses were performed continuously, in random order and based on 5 replicates per sample.

2.1. Volatiles Analysis by Stir Bar Sorptive Extraction (SBSE) and Gas Chromatography with Time-of-Flight Mass Spectrometry (GC-TOFMS)

For the isolation of volatile compounds, Twister™ stir bars (1 cm long, 0.5 mm film thickness) with a polydimethylsiloxane phase (Gerstel, Müllheim an der Ruhr, Germany) were introduced into 6 mL of strawberry juices in 20 mL-vials. The conditions of the extraction were stirring at 1,000 rpm during 1 hour at room temperature. Afterwards, the Twisters were rinsed three times in water to remove non-adsorbed compounds, such as sugars and then dried carefully with a wipe.

For the analyses, an Agilent 6890 gas chromatograph (Agilent Technologies, Inc., Wilmington, DE) equipped with a time-of-flight mass detector (Leco Co., St. Joseph, MI) and a Gerstel MPS-2 multipurpose TDU autosampler with a CIS-4 cooled injection system (Gerstel, Baltimore, MD) was used. Twisters were thermally desorbed in the TDU in solvent vent mode, ramping from 40 to 300 °C at 60 °C/min. The CIS was cooled to -100 °C with liquid nitrogen during sample injection, then heated at 12 °C/s to 300 °C, and held for 1 min. Splitless mode was used during injection. Separation was achieved using a DB-5 column (30 m x 0.25 mm x 0.25 μm i.d.). The oven temperature was held at 40 °C for 3 min, then ramped to 170 °C at a rate of 5 °C/min, then increased to 300 °C at 20 °C/min and held at the final temperature for 5 min. Standard EI mode was used at 70 eV. Masses were detected from 29 to 250 amu. The scan rate was 20 scans/s. System software control was performed through ChromaTOF software from LECO.

2.2. Analysis of Nonvolatile Compounds by Solid Phase Extraction (SPE) and High-Performance Liquid Chromatography with Quadrupole Time-Of-Flight Mass Spectrometry (HPLC-QTOFMS)

Non-volatile compounds were isolated from the juice by solid phase extraction (SPE). Strawberry juices were first centrifuged (6,800xg, 5 min) and the supernatants were passed through C18-SPE cartridges (Discovery DSC18,

500 mg sorbent, 3 mL tubes, Supelco, Bellefonte, PA). The SPE cartridges were conditioned with 3 mL acetonitrile and equilibrated with 3 mL water/acetonitrile (95/5) before loading the juice (7.5 mL) into the cartridge. Afterwards, 2 mL of water/acetonitrile (95/5) were added and elution of the compounds carried out by adding 1 mL of acetonitrile to the cartridge. Samples were then filtered through a 0.2 μm membrane (Millipore Corp., Bedford, MA).

The analysis was performed using an Agilent HP 1200 HPLC system (Agilent Technologies, Palo Alto, CA, USA) interfaced with a quadrupole time-of-flight mass spectrometer (Micromass, Waters, Milford, MA, USA). Samples (5 μL) were analysed by reverse phase HPLC using a Kinetex C18 column (150 x 2.1 mm, 2.6 μm) (Phenomenex, Torrance, CA, USA). The mobile phases were water + 0.1% formic acid (mobile phase A) and acetonitrile + 0.1 % formic acid (mobile phase B). Elution was carried out at a flow rate of 0.3 mL/min using the following linear gradients: from 12 to 22% B over 0-12 min, from 22 to 50% B over 12-22 min, from 50 to 82% B over 22-25 min, and from 82 to 100% B over 25-30 min, with cleaning and conditioning of the column between runs. The total run time was 40 min. The HPLC column effluent was directed into the quadrupole-TOF-MS equipped with an electrospray ionization (ESI) source operating in positive mode. The MS operating conditions were as follows: source temperature: 100 °C, desolvation temperature: 300 °C, desolvation gas: 550 L/h, cone gas: 50 L/h; capillary voltage: 3000 V; cone voltage: 30 V, extraction cone: 4 V; TOF-MS: 50-1500 amu, in continuous mode. System software control was performed with Waters Masslynx software v4.1.

3. Pretreatment of Chemical Data

Data pretreatment was carried out using Genedata Expressionist software (Genedata, Basel, Switzerland). Specifically, the module Refiner MS was used to treat chromatograms, and the module Analyst was used to normalize and reduce the number of variables. Figure 1 summarizes the workflow used for the chemical data pretreatment steps.

3.1. Treatment of Chromatograms Using Refiner MS

Data from chemical analyses (m/z detected along the chromatographic run) were imported into Refiner MS in *.cdf and *.raw format for volatiles and non-volatiles, respectively (Fig. 1). Treatment was different for volatile and non-volatile chromatograms. Default processing parameters were applied unless otherwise specified.

For volatile compounds: noise reduction, by clipping to 0 all data points below threshold intensity of "2000"; nominal mass integration, to sum up centroid data onto their nominal mass values; peak detection (smoothing window 9 scans, curvature-based peak detection, apply consistency threshold of 0.5), and component detection to group all peaks corresponding to the same molecular components (RT tolerance 0.5 min, max deviation 50%, minimum group size 10).

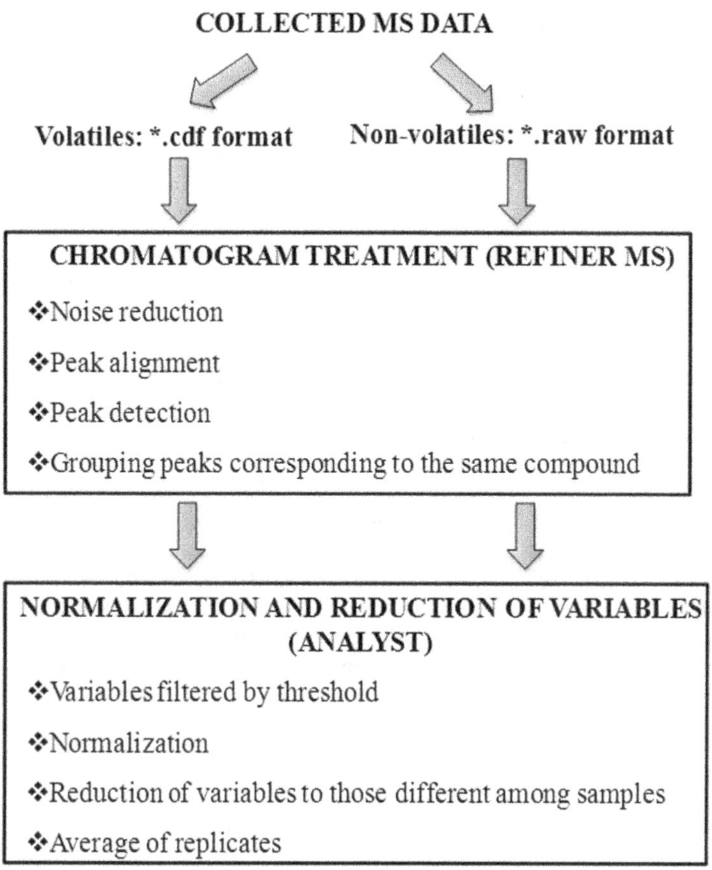

Figure 1. Pretreatment of chemical data.

For the non-volatiles, noise was reduced by a applying the activity chemical noise subtraction (smoothing window 3 scans, structure removal with minimum RT length of 6 scans or minimum m/z length of 3 points); afterwards, retention time alignment was used (RT search interval 1 minute), and peaks were detected with the activity "chromatogram summed peak detection", where peaks are detected on a temporary averaged chromatogram, and the same boundaries are applied to all the chromatograms (smoothing window 0.5 Da). Finally, peaks identified as belonging to the same isotope pattern of a molecule were grouped into peak clusters (m/z tolerance 0.2 points; allow 2 peaks missing with first allowed gap position of 2; restrict cluster size to 0.5).

After treating the chromatograms, data were imported into the module Analyst. Specifically, for the volatiles, the groups of peaks coming from the same compound as the sum of the intensity of the individual peaks, and for the non-volatiles, the isotope clusters.

3.2. Normalization of Variables Using the Module Analyst

Identical treatments were applied to both volatiles and non-volatiles data sets. First, variables were filtered by threshold proportions, so those variables that showed a value higher than the limit in at least 20% of the samples were selected. Secondly, variables were normalized [(intensity of the variable/sum of intensities of all variables in the sample) x 1000]. Furthermore, data were reduced to only those variables that significantly differed among samples ($p<0.01$) by means of ANOVA. Replicates were averaged. Finally, two data sets corresponding to volatiles and non-volatile were obtained.

4. Sensory Data

Descriptive analysis of the strawberry juices was carried out in the Sensory Center of the University of Minnesota, by a panel of 11 judges experienced in descriptive sensory evaluation techniques. About 3 hours before the beginning of each session, juices were thawed. Fifty grams of each sample were placed in coded 100 g soufflé cups with lids. During the training (5 sessions), judges developed a flavor lexicon for the samples, and practiced scaling intensity of the selected attributes in the samples. In total, the lexicon included 36 descriptors including flavor and aroma attributes (both of them were individually evaluated), 3 taste attributes and astringency. All descriptors are shown in Table 2.

Table 2. Descriptors in Strawberry Juices

Aroma/flavor		*Taste*	*Other*
• Fresh strawberry	• Citrus	• Sweetness	• Astringency
• Cooked strawberry	• Vinegar	• Sourness	
• Frozen strawberry	• Seedy	• Bitterness	
• Jam	• Fermented		
• Artifical strawberry	• Tomato		
• Grassy/green	• Oxidized		
• Floral	• Musty		
• Prune	• Hay		

Afterwards, judges participated in ten test sessions in individual booths. They evaluated a different set of five or six samples in each of the first five sessions, and replicated these samples in the final five sessions. Samples were balanced for serving order and carryover effects using a Latin square design. During these sessions, each judge evaluated each sample by rating the intensity of the attributes on 20 point line scales labeled 'none' at the left end and 'intense' at the right end. After carefully inspection of the data, we decided to focus on *cooked* and *fermented* flavor, which corresponded to the flavor reference samples of cooked strawberries, and cooking wine, respectively. The scores obtained for each sample were averaged over judges and replicates, first for *cooked* and then for *fermented* flavor.

5. Multivariate Analysis

For multivariate analyses, chemical data sets with the sensory scores were imported into Simca-P+ statistical software, v. 12.0 (Umetrics, Umea, Sweden).

5.1. Overview of the Chemical Data Sets by Principal Components Analysis (PCA)

Each data set (volatiles, non-volatiles) was first inspected by principal components analysis (PCA). To do so, data were previously mean-centered and scaled to unit variance (1/SDk, where SD is the standard deviation of the variable k). PCA score plots were examined to determine how samples are related to each other.

5.2. Correlation of Chemical and Sensory Data by Partial Least Squares Regression (PLS-R) Analysis

PLS-R was used to relate the chemical composition (matrix X) and sensory scores (matrix Y) by a linear multivariate model. Different models were built by relating individually the volatiles and non-volatiles data with the sensory scores for *cooked* and *fermented* flavor. Additional models were obtained by merging volatiles and non-volatiles data into one data set by carrying out a "mid-level fusion" step prior to its correlation with the sensory scores of *cooked* and *fermented* flavor. "Mid-level fusion" consisted of a preliminary variable selection of each model before merging, and specifically, variables with a VIP (Variable Influence on Projection) value of < 1 were discarded (*11*). Afterwards, a new matrix was built with the selected variables from the volatiles and non-volatiles, and with the sensory scores. When building the model, a weight factor equal to $1/\sqrt{K_{block}}$ (with K_{block} representing the number of variables in a block) was applied to each block, in order to adjust for differences in numerical size of each data set (volatiles, non-volatiles). In addition, for all models, variables were mean-centered and transformed to unit variance. The quality of the model was evaluated through parameters R^2Y and Q^2Y, which correspond to the cumulative fraction of the variation of Y (sensory scores) explained and predicted by the model according to cross validation, respectively.

5.3. Selection and Identification of Markers of Cooked and Fermented Flavor

Variables with the highest positive regression coefficients in prediction model(s) were selected as markers of *cooked* and *fermented* flavor. Markers coming from the volatiles data set were identified by matching their mass spectra with those in the library NIST08. In addition, GC retention indexes were experimentally obtained by reference to a mixture of alkanes (C_6-C_{26}) and comparison with retention indexes sourced from the literature (http://webbook.nist.gov/chemistry/). For markers belonging to the non-volatiles, additional analyses were carried out to obtain the accurate mass (using reserpine as a lockmass) and MS^2 fragmentation patterns. The characteristics were compared with the information available in the online Scripps Center Metlin Database (http://metlin.scripps.edu) and scientific literature about non-volatiles in strawberries.

Results and Discussion

1. Reproducibility of the Methodology and Overview of the Data Sets

In the present study a non-targeted approach, viz., flavoromics, has been used to determine chemical compounds related to *cooked* and *fermented* flavor in strawberry juices. As previously stated, an advantage of using this approach is increased likelihood of discovering new markers of flavor. As volatiles and non-volatiles influence flavor perception, both of them were considered in this study. The methodologies were selected in order to provide as much chemical information as possible, with SBSE-GC-TOFMS and SPE-HPLC-QTOFMS being used for determination of volatiles and non-volatiles, respectively. SBSE shows lower limits of detection for volatile compounds cf. other techniques, such as SPME, while SPE has been extensively used for metabolomics studies (*12*). For both types of analysis, mass spectrometry has been used for its capability to measure compounds present at very low levels and to provide structural information.

After the analysis of chemical compounds, raw MS data were imported into the specialized software Genedata Expressionist. It offers data extraction, normalization, peak alignment and highly sophisticated statistical analysis and visualization tools in one platform (*16*), and has been used in previous studies with a non-targeted approach (*17*, *18*). In our study, chromatograms were treated using the module Refiner MS, concentrating mainly on noise reduction, peak alignment, peak detection and grouping of peaks (mass fragments) corresponding to the same compound. In this way, each group of peaks was a variable in the final data set. Figure 2 presents an example of a chromatogram before pretreatment (a), and a zoomed region of the chromatogram after pretreatment (b) carried out on the non-volatiles data set by applying the data preprocessing steps described in the *Materials and Methods* section (3.1).

As can be seen, chromatograms are displayed in a plane with the x-axis corresponding to mass to charge ratio (m/z) and the y-axis corresponding to retention time. After the treatment (Figure 2b) the noise has largely been eliminated, the peaks have been detected (squares) and those corresponding to the same molecular components grouped together. The result is a table with the intensity of the variables for each sample. Afterwards, by using the module Analyst, variables were normalized and reduced in number. As a result, two data sets corresponding to volatiles (96 variables) and non-volatiles (260 variables) were obtained.

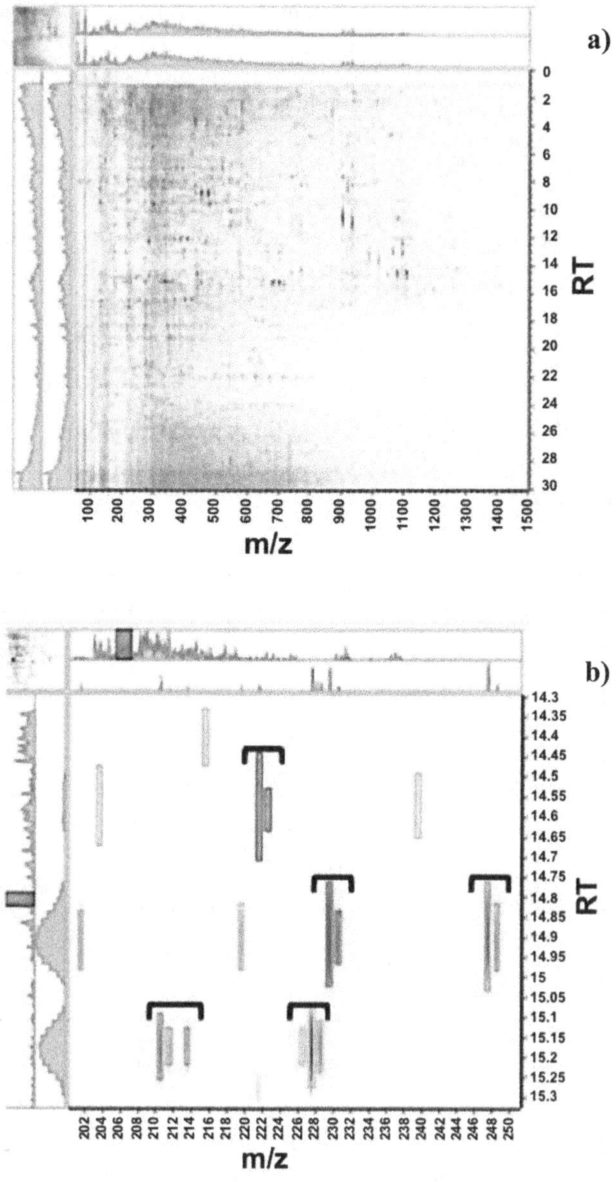

Figure 2. Example of a chromatogram of non-volatile compounds before the pretreatment (a) and a zoomed region of the chromatogram after data pretreatment (b). Rectangles correspond to the peaks detected, and those belonging to the same molecular component are grouped together.

Due to the large number of variables in this type of study, the reproducibility of the analyses was checked by looking at the variability observed for the replicates of the QC sample (Table 3), recalling that this sample was prepared by mixing the 27 samples in equal proportions. To consider that the methodologies are reproducible, according to Gika and collaborators (*15*), the variation of randomly selected variables should exhibit RSD <15%, and the percentage of peaks having RSD <15% should be greater than 60%. Accordingly, as these requirements were indeed fulfilled for both the volatiles and non-volatiles data sets (Table 3), the reproducibility of the methodologies was considered good.

Table 3. Reproducibility of the Chemical Analyses

	% of peaks having RSD <15%	*Variation in randomly selected peaks*		
			Intensity	*RSD (%)*
Volatiles	66	Group_101	174.4±24.1	14
		Group_041	27.19±2.12	8
		Group_010	2.08±0.35	17
		Group_028	1.74±0.16	9
		Group_018	0.93±0.12	13
Non-volatiles	62	Cluster_0168	0.76±0.09	12
		Cluster_0471	0.27±0.02	9
		Cluster_0538	0.43±0.03	7
		Cluster_0747	0.64±0.08	13
		Cluster_1020	0.44±0.05	11

After checking the reproducibility, Principal Components Analysis (PCA) was applied to each chemical data set. The main use of PCA is to represent a multivariate data table as a low-dimensional plane, such that an overview of the data is obtained. Figure 3 shows as an example, the PCA score plot corresponding to the non-volatiles data set.

As can be seen, the QC sample is near the center of the plot, which indicates that the methodologies are not biased to any particular samples. In addition, samples are clustered according to their origin (varieties Wendy, Jewel, or commercial puree) while there are also differences depending on the treatment (heat, storage, freeze-drying).

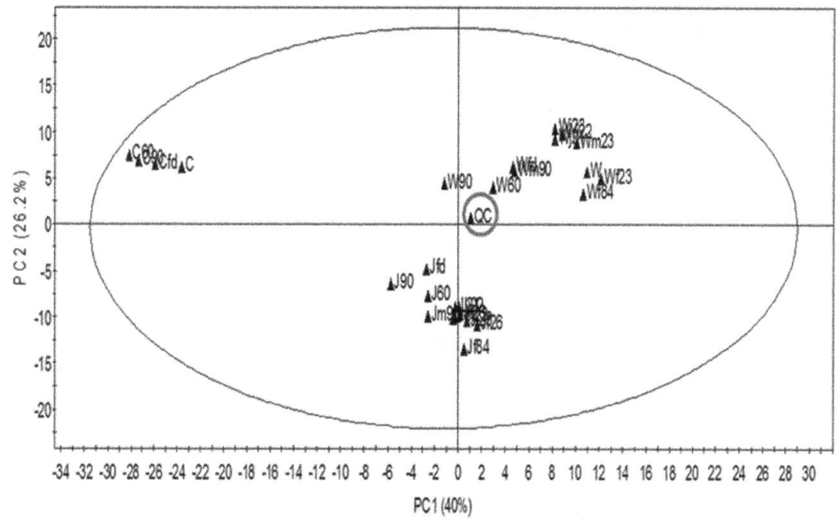

Figure 3. PCA score plot showing the separation of juice samples (including composite QC sample) based on their non-volatile composition.

Table 4. Quality of the PLS-R Models

Model number	*Chemical data set (X)*	*Sensory scores (Y)*	*PCs*	*R^2Y*	*Q^2Y*
1	Volatiles		2[a]	0.84[b]	0.69[c]
2	Non-volatiles	*Cooked*	2	0.71	0.39
3	Volatiles+ non-volatiles		3	0.88	0.76
4	Volatiles		1	0.95	0.94
5	Non-volatiles	*Fermented*	1	0.95	0.93
6	Volatiles + non-volatiles		2	0.97	0.95

[a] Number of significant PLS components; [b] Cumulative fraction of the variation in Y explained by the model; [c] Cumulative fraction of the variation in Y that can be predicted by the model based on cross-validation.

2. Correlation of Chemical Data with *Cooked* and *Fermented* Flavors by Partial Least Squares Regression (PLS-R) Analysis

After reviewing each data set, Partial Least Squares Regression (PLS-R) analysis was used to relate the two data matrices, X (chemical data) and Y (sensory scores), to each other by a linear multivariate model (*19*). Models obtained by PLS-R can be used to predict sensory scores for new samples, as well as to determine chemical compounds responsible for specific attributes (*11*, *20*).

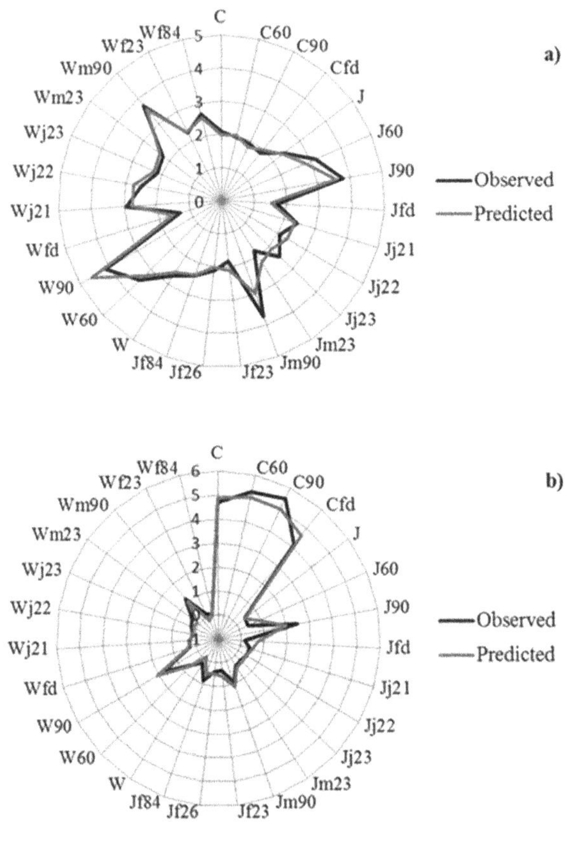

Figure 4. Plots of observed vs. predicted sensory scores in model 3 for cooked flavor (a) and model 6 for fermented flavor (b). See Table 1 for list of sample identification codes.

For each sensory attribute, 3 models were obtained, by relating (a) volatiles data alone, (b) non-volatiles data alone, or (c) a combination of both (obtained as explained in *Materials and Methods* section 5.2). Table 4 presents the main characteristics of the models created for *cooked* (models 1-3) and *fermented* flavors (models 4-6). It is important to note that validation of the models was done using cross-validation, because of the limited number of samples in this study.

As expected, the number of components needed to explain the majority of variation in Y (for either *cooked* or *fermented* flavor) was higher when combining both volatiles and non-volatiles data sets (models 3 and 6) since the complexity of the information modeled was increased. On the other hand, the explanatory and predictive qualities of the models were described by R^2Y and Q^2Y, respectively. When these expressions approach a value of 1, it means that the data contain most of the relevant chemical information related to the variation in the response. As can be seen, these parameters were slightly better when combining both volatiles and non-volatiles data sets (models 3 and 6, for prediction of *cooked* and *fermented* flavor, respectively) suggesting that each of them conveyed complementary information. These results agreed with previous studies that reported improvements in the predictive abilities of the models when different instrumental measurements were combined (*21, 22*). In addition, a model combining both volatile and non-volatile is more interesting for this study, as both can influence flavor perception. Accordingly, models 3 and 6 were selected in order to determine markers of *cooked* and *fermented* flavor, respectively. Figure 4 shows the plot of observed values (sensory scores) vs. the predicted values based on these models, according to cross validation. As can be seen, observed and predicted values were very similar.

3. Selection and Identification of Marker Compounds

Models 3 and 6 were examined to understand which chemical variables correlated in a robust way with *cooked* and *fermented* flavors. Accordingly, the regression coefficient plots obtained for each model were carefully inspected, as these provide information on the magnitude and direction of the relationships between the chemical variables and the sensory scores. Variables having the highest positive regression coefficients in the models were selected as markers. It is important to highlight that for both *cooked* and *fermented* flavor, the markers mainly belonged to the volatiles data set, suggesting that volatiles had a higher impact on *cooked* and *fermented* flavor than did non-volatiles.

3.1. Volatile Markers of Cooked Flavor

Samples showing the highest scores for *cooked* flavor were those juices from either Wendy or Jewel varieties heated to 90 °C, followed by those samples contained in mixtures, and lastly by juices heated to 60 °C (Figure 4a), suggesting that heat was the main treatment affecting the development of *cooked* flavor in our samples. It is not surprising, as heat treatment leads to a significant change in overall strawberry flavor (*14*).

The markers of *cooked* flavor, and their coefficients in model 3, are presented in Table 5. Some of them are important volatile compounds in their own right in strawberries, e.g., mesifuran and the esters, methyl butanoate, hexyl acetate and butyl butanoate, with important green, fruity sweet notes (*23–25*). Others, have not previously been identified in strawberries, e.g., the alcohols α-bisabolol, (*Z,E*)-farnesol and nerol, which are known to present floral and citrus notes. In fact, these alcohols are important volatile compounds in strawberry jams (*26*). Another important marker was γ-undecalactone which, while not present in strawberries, has also been detected in strawberry jams (*26*).

Our results suggest that during heat treatment, the concentrations of these marker compounds increase. The fact that these compounds are important odorants in strawberry jams suggests that they could contribute to the perception of *cooked* flavor, although this should be confirmed by targeted analysis and by compound addition studies (sensory analysis). In addition, a new compound was identified, *E*-geranylacetone, that has not previously been reported in strawberries or derived products. Interestingly, this compound increases during the heating of tomato juice, where it is thought to arise as a result of the degradation of carotenoid pigments (*27*).

Table 5. Identification of Volatile Markers of *Cooked* Flavor

Coeff [a]	*Compound*
0.35	α-Bisabolol
0.33	Methyl butanoate
0.30	Hexyl acetate
0.26	*E*-geranylacetone
0.25	Isomer of (*Z,E*)-Farnesol
0.24	Butyl butanoate
0.22	Mesifuran
0.20	γ-Undecalactone
0.20	Nerol
0.19	Isomer (*Z,E*)-Farnesol
0.15	β-Linalool

[a] Regression coefficients in model 3.

3.2. Volatile Markers of Fermented Flavor

Samples showing the highest sensory scores for *fermented* flavor were those coming from the commercial juices, followed by those heated at 90 °C (Figure 4b). In addition, juices from Jewel fruit stored at 8 °C during 4 days (Jf84), also showed significant fermented flavor.

The chemical compounds that could contribute to *fermented* flavor (or at least be markers thereof) are shown in Table 6, together with their model 6 regression coefficients. As can be seen, they were totally different from those determined in the case of *cooked* flavor. Several markers were esters, namely, butyl isovalerate, ethyl dodecanoate, 2-phenylethyl acetate, ethyl octanoate, amyl acetate and ethyl decanoate. These compounds generally present floral, fruity, sweet notes, and have previously been described in strawberries (*25*). Other markers were α-bisabolol oxide B and epoxy-linalool oxide. The presence of these compounds could be due to degradation of α-bisabolol and epoxylinalool, respectively, both of which have been found in strawberries (*28*, *29*) and in strawberry jam (*26*). Another important marker was 1-octen-3-one, which has not previously been described in strawberries. This compound has been described as an important odorant formed during heating strawberry juices (*14*), raspberries (*30*), and presents intense mushroom-like with vegetative nuances of cabbage and broccoli. Finally, methyl salicylate has been described in strawberries (*28*) and usually presents sweet, wintergreen, and camphoreous notes. On the other hand, benzophenone has previously been found in musk strawberry (*29*), and is characterized by balsam, rose, metallic, powdery and geranium notes.

These results suggest that some chemical compounds contributing to *fermented* flavor could come directly from strawberries, while others could be formed during processing; some of them could contribute off-flavor notes, e.g., 1-octen-3-one.

Table 6. Identification of Volatile Markers of *Cooked* Flavor

Coeff [a]	*Compound*
0.19	Isomer of α-Bisabolol oxide B
0.15	Butyl isovalerate
0.15	Isomer of α-Bisabolol oxide B
0.14	Epoxy-linalool oxide
0.11	Ethyl dodecanoate
0.11	2-phenylethyl acetate
0.11	Ethyl octanoate
0.10	Amyl acetate
0.09	Ethyl decanoate
0.09	Benzophenone
0.09	Methyl salicylate
0.08	1-octen-3-one

[a] Regression coefficients in model 6.

3.3. Nonvolatile Markers of Cooked and Fermented Flavor

As previously stated, the number of non-volatile markers was much lower cf. the number of volatile markers. The first marker for both *cooked* and *fermented* flavor eluted at 8.5 min and showed an accurate mass of 155.0355 [M + H], with an MS^2 ion at m/z 109. According to the Metlin database, this compound was identified as an isomer of dihydroxybenzoic acid (likely, protocatechuic or gentisic acid) and has previously been identified in strawberries (*31*). On the other hand, the second marker was detected only in the case of *cooked* flavor; it eluted at 2.9 min, with a molecular ion of 705 [M^+], and an MS^2 ion at m/z 543. This compound was identified as epiafzelechin(4α→8)pelargonidin 3-O-β-glucopyranoside, a molecule first identified in strawberry by Fossen and collaborators (*32*). Its structure consists of an anthocyanin unit connected via a C-C bond to a flavan-3-ol. However, the possible contribution of these compounds to flavor is not known, as there currently remains a paucity of studies in the literature dealing with the influence of these types of non-volatiles in flavor perception. By using the flavoromics approach, we have identified these new non-volatile markers, whose contribution will be studied in the future by targeted analysis.

4. Conclusions

A relatively new approach called flavoromics has been used to determine chemical markers of the sensory attributes *cooked* and *fermented* flavor in strawberry juices submitted to various process treatments. By building predictive models, the chemical composition of the strawberry juices (both volatiles and non-volatiles) was related to ratings of *cooked* and *fermented* flavor determined by sensory analysis. Some variables were indicated as potential markers, based on having high impact in derived mathematical models. Regarding the markers of *cooked* flavor, most were alcohols that are not usually present in strawberries, but are known to be formed during the heating of fruit juices, together with some important volatile compounds normally found in strawberries. On the other hand, the markers of *fermented* flavor were mainly esters, together with some alcohol oxides and compounds that present unpleasant flavor notes. In addition, two non-volatiles were identified, namely, an isomer of dihydroxybenzoic acid and epiafzelechin(4α→8)pelargonidin 3-O-β-glucopyranoside. The sensory contributions of the markers selected for *cooked* and *fermented* flavor will be further elucidated by targeted analysis and compound addition (sensory) studies.

References

1. Keast, R. S. J.; Dalton, P.; Breslin, P. A. S. Flavor interactions at the sensory level. In *Flavor perception;* Taylor, A. J., Roberts, D. D., Eds.; Blackwell Publishing Ltd.: Ames, Iowa, 2004; pp 228–254.
2. Aishima, T.; Nakai, S. Chemometrics in flavor research. *Food Rev. Int.* **1991**, *7* (1), 33–101.
3. Morales, M. T.; Alonso, M. V.; Rios, J. J.; Aparicio, R. Virgin Olive Oil Aroma: Relationship between Volatile Compounds and Sensory Attributes by Chemometrics. *J. Agric. Food Chem.* **1995**, *43*, 2925–2931.
4. Vilanova, M.; Genisheva, Z.; Masa, A.; Oliveira, JM. Correlation between volatile composition and sensory properties in Spanish albarino wines. *Microchem. J.* **2010**, 240–246.

5. Gonzalez-Alvarez, M.; Gonzalez-Barreiro, C.; Cancho-Grande, B.; Simal-Gandara, J. Relationships between Godello white wine sensory properties and its aromatic fingerprinting obtained by GC-MS. *Food Chem.* **2011**, 890–898.
6. Vilanova, M.; Campo, E.; Escudero, A.; Grana, M.; Masa, A.; Cacho, J. Volatile composition and sensory properties of *Vitis vinifera* red cultivars from North West Spain: correlation between sensory and instrumental analysis. *Anal. Chim. Acta* **2012**, *720*, 104–111.
7. Baldwin, E. A.; Scott, J. W.; Einstein, M. A.; Malundo, T. M. M.; Carr, B. T.; Shewfelt, R. L.; Tandon, K. S. Relationship between sensory and instrumental analysis for tomato flavor. *J. Am. Soc. Hortic. Sci.* **1998**, *123* (5), 906–915.
8. Ruiz, J.; Garcia, C.; Diaz, M. C.; Cava, R.; Tejeda, J. F.; Ventanas, J. Dry cured Iberian ham non-volatile components as affected by the length of the curing process. *Food Res. Int.* **1999**, *32* (9), 643–651.
9. Chira, K.; Pacella, N.; Jourdes, M.; Teissedre, PL. Chemical and sensory evaluation of Bordeaux wines (Cabernet-Sauvignon and Merlot) and correlation with wine age. *Food Chem.* **2011**, *126*, 1971–1977.
10. Reineccius, G. A. Flavoromics - the next frontier? Abstracts of Papers, 235th ACS National Meeting, New Orleans, LA, United States; American Chemical Society: Washington, DC, 2008; p AGFD-061.
11. Charve, J. I. M. Prediction of Mandarin Juice Flavor: A Flavoromic Approach. Ph.D. Thesis, University of Minnesota, Saint Paul, MN, 2011.
12. Dettmer, K.; Aronov, P. A.; Hammock, B. D. Mass spectrometry-based metabolomics. *Mass Spectrom. Rev.* **2007**, *1*, 51–78.
13. Shamaila, M.; Powrie, W. D.; Skura, B. J. Sensory evaluation of strawberry fruit stored under modified atmosphere packaging (MAP) by quantitative descriptive analysis. *J. Food Sci.* **1992**, *5*, 1168–1184.
14. Schieberle, P. Heat-induced changes in the most odour-active volatiles of strawberries. In *Trends in Flavour Research;* Marrse, H., Van der Heij, D. G., Eds.; Elsevier Sciences: New York, 1994; pp 345–351.
15. Gika, H. G.; Theodoridis, G. A.; Wingate, J. E; Wilson, I. D. Within-day reproducibility of an HPLC-MS-based method for metabonomic analysis: Application to human urine. *J. Proteome Res.* **2007**, *8*, 3291–3303.
16. Theodoris, G. A.; Gika, H. G.; Want, E. J.; Wilson, I. D. Liquid chromatography-mass spectrometry based global metabolite profiling: a review. *Anal. Chim. Acta* **2012**, *711*, 7–16.
17. Levin, A. M.; de Vries, R. P.; Conesa, A.; de Bekker, C.; Talon, M.; Menke, H.; van Peij, N. N. M; Wosten, H. A. B. Spatial differentiation in the vegetative mycelium of *Aspergillus niger*. *Eukaryotic Cell* **2007**, 2311–2322.
18. Kin, N. W.; Crawford, D. M.; Liu, J.; Behrens, T. W.; Kearney, J. F. DNA microarray gene expression profile of marginal zone versus follicular B cells and idiotype positive marginal zone B cells before and after immunization with *Streptococcus pneumoniae*. *J. Immunol.* **2008**, *180*, 6663–6674.

19. Eriksson, L.; Johansson, E.; Kettaneh-Wold, N.; Trygg, J.; Wikstrom, C.; Wold, S. *Multi- and Megavariate Data Analysis. Part I. Basic Principles and Applications*, 2nd ed.; Umetrics: Umea, Sweden, 2006; pp 21−38.
20. Parker, M.; Pollnitz, A. P.; Cozzolino, D.; Fracis, I. L.; Herderich, M. J. Identification and Quantification of a Marker Compound for "Pepper" Aroma and Flavor in Shiraz Grape Berries by Combination of Chemometrics and Gas Chromatography-Mass Spectrometry. *J. Agric. Food Chem.* **2007**, *55*, 5948–5955.
21. Brás, L. P.; Bernardino, S. A.; Lopes, J. A.; Menezes, J. C. Multiblock PLS as an approach to compare and combine NIR and MIR spectra in calibrations of soybean flour. *Chemom. Intell. Lab. Syst.* **2005**, *1*, 91–99.
22. Cozzolino, D.; Smyth, H. E.; Cynkar, W.; Janik, L.; Dambergs, R. G.; Gishen, M. Use of direct headspace-mass spectrometry coupled with chemometrics to predict aroma properties in australian Riesling wine. *Anal. Chim. Acta* **2008**, *1*, 2–7.
23. Forney, C. F.; Kalt, W.; Jordan, M. A. The composition of strawberry aroma is influenced by cultivar, maturity, and storage. *HortScience* **2000**, *6*, 1022–1026.
24. Ozcan, G.; Barringer, S. Effect of enzymes on strawberry volatiles during storage, at different ripeness level, in different cultivars, and during eating. *J. Food Sci.* **2011**, *2*, 324–333.
25. Zabetakis, I.; Holden, M. A. Strawberry Flavour: Analysis and Biosynthesis. *J. Sci. Food Agric.* **1997**, *74*, 421–434.
26. Barron, D.; Etievant, P. X. The volatile constituents of strawberry jam. *Z. Lebansm.-Unters Forsch.* **1990**, *191*, 279–285.
27. Crouzet, J.; Chairote. G.; Rodriguez, F. Volatile components modifications during heat treatment of fruit juices. In *Instrumental Analysis of Food V2: Recent Progress*; Charalambous, G., Inglett, G., Eds.; Academic Press, Inc: Orlando, FL, 1983; pp 119−135.
28. Aubert, C.; Baumann, S.; Arguel, H. Optimization of the analysis of flavor volatile compounds by liquid-liquid microextraction (LLE). Application to the aroma analysis of melons, peaches, grapes, strawberries and tomatoes. *J. Agric. Food Chem.* **2005**, *53*, 8881–8895.
29. Petka, J.; Leitner, E.; Parameswaran, B. Musk strawberries: the flavour of a formerly famous fruit reassessed. *Flavour Fragrance J.* **2012**, *27*, 273–279.
30. Roberts, D. D.; Acree, T. E. Effects of Heating and Cream Addition on Fresh Raspberry Aroma Using a Retronasal Aroma Simulator and Gas Chromatography Olfactometry. *J. Agric. Food Chem.* **1996**, *44*, 3919–3925.
31. Russel, W. R.; Labat, A.; Scobbie, L.; Duncan, G. J.; Duthie, G. G. Phenolic acid content of fruits commonly consumed and locally produced in Scotland. *Food Chem.* **2009**, *115*, 100–104.
32. Fossen, T.; Rayyan, S.; Andersen, O. M. Dimeric anthocyanins from strawberry (*Fragaria ananassa*) consisting of pelargonidin 3-glucoside covalently linked to four flavan-3-ols. *Phytochemistry* **2004**, *65*, 1421–1428.

Chapter 22

Determining Causal Relationships between Physical Structure and Sensory Function in Food Systems via Partial Least Squares (PLS) Path Modelling

Sean A. Smith,[*,1] Brian Guthrie,[1] Adam Steinbach,[1] Tim Lindgren,[1] and Stephane Debon[2]

[1]Cargill Global Food Research, 2301 Crosby Road, Wayzata, Minnesota 55193, United States
[2]Cargill Global Food Research, Havenstraat 84, 1800 Vilvoorde, Belgium
[*]E-mail: sean_smith@cargill.com.

The concept of structure:function modeling of food systems is demonstrated by using the Partial Least Squares (PLS) path modelling technique to quantify the time-dependent relationships between various physical and sensory measurements made on a set of food system prototypes intended to approximate a commercially available target product. An iterative process was used to deduce which instrumental parameters (derived from the physical measurements) best served as determinants of structure that dominated sensory perception at five points along a hypothesized breakdown path from the time at which the food system was introduced to the mouth, through mastication, and so on until after it was swallowed. Once these relationships were verified, the resultant PLSpath model could be used to form simple regression relationships between the relevant instrumental parameters and the three active ingredients in the prototype recipe. Based on these results, it is anticipated that such models will enable an accelerated process for qualifying alternative ingredients, that were not used to build the model, in order to drive sensory properties of novel food systems.

Introduction

Novel food system design is becoming increasingly important to manufacturers who must seek out alternatives formulations of popular products to replace expensive (e.g. dairy) and/or controversial (e.g. palm oil) ingredients with cheaper and preferably lower-calorie alternatives. Usually, new product development occurs through a trial-and-error approach that relies largely on anecdotal evidence to make progress. However, at best, statistical experimental design methods that are used to deduce the relationships between ingredient composition/processing and sensory performance can study only a few product variables at a time, because this approach quickly becomes tedious and time-consuming as the number of alternative ingredients to the traditional product recipe is increased. It would be more efficient to instead understand how the physical properties of the traditional product affect sensory perception independent of ingredient composition (Figure 1). If the mode of operation of the product is understood, then ingredient alternatives to the traditional recipe can be sought out based on how well their analytical properties (e.g. rheology, texture analysis, tribology) approximate those of the traditional ingredients.

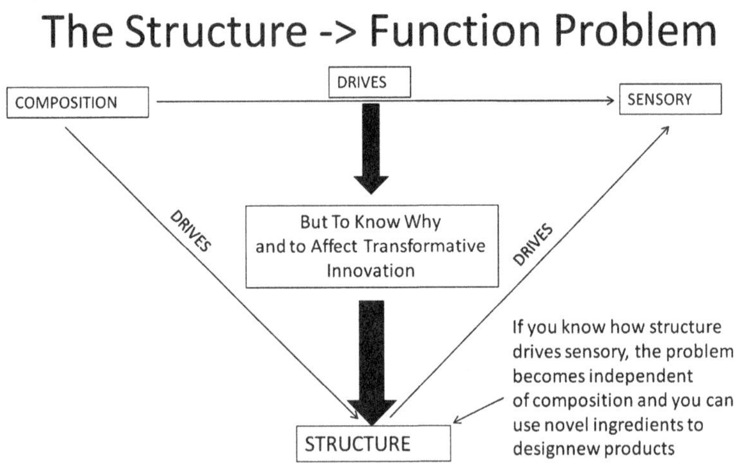

Figure 1. Visual representation of the structure:function model paradigm.

Published research has shown that analytical mouthfeel attributes of food systems as determined by trained taste panels are linked to consumer satisfaction, and these are also sensitive to formulation differences (*1*). While an analytical relationship has been demonstrated between sensory perception of bolus "thickness" and shear stress in certain food systems (*2*), in general, mouthfeel attributes of many foods have been found to be insufficiently described by bulk rheology properties alone (*3, 4*). It is also known that tribology measurements done in combination with rheology can provide a more comprehensive characterization of food structure in the mouth (*5*), but the interrelationship of these properties is likely to be unique for a given food system application. Furthermore, the sensory experience of a food is determined not only by its properties at the moment it's placed in the mouth but by how it follows a time-dependent breakdown path as it is chewed and then swallowed. Thus, the complete characterization of a food system requires physical measurements such as rheology, tribology, and solvation/mixing rate as a function of mastication-induced bolus breakdown along with corresponding sensory measurements. This results in dozens of possible time-dependent regression relationships between the physical and sensory parameters that must be deduced and then articulated as a single, cohesive model. Fortunately, there is a method designed to accomplish this task.

PLSpath modeling grew out of Structural Equation Modelling (SEM), which originated decades ago as a way to quantify and validate complex relationships between multiple blocks of data. SEM has been applied mainly to business (e.g. marketing) and social science data (*6*, *7*). As such, SEM practitioners usually have access to datasets composed of thousands of samples with only a handful of predictor variables. The analytical scientist, however, usually encounters the opposite conditions: samples are expensive and/or time-consuming to prepare, but physical and chemical measurements consisting of dozens or even hundreds of (usually highly correlated) possible predictor variables are relatively easy to make, resulting in "short" but "wide" datasets. Data analysis in this situation requires latent variable methods such as Partial Least Squares (PLS) (*8*, *9*). PLSpath modeling relies on PLS to reduce the dimensionality of the dataset to a few significant underlying relationships that can be linked together as a path. In the embodiment described herein, this means forming a multi-step sequence of regression relationships that describe the time-dependent sensory behavior of a model food system as a function of its measurable physical properties. This model food system is presented here in generic terms in order to preserve the proprietary nature of the discoveries made as a result of this work. The intent of this chapter is to describe the analytical approach as opposed to reporting specific application results.

PLSpath Model Construction

Path modeling seeks to estimate and validate a mathematical (linear) model of a complex process. The theoretical path model is defined as a set of interconnected cause-and-effect relationships among hypothesized concepts, known as Latent Structures (LS's), that can be estimated via some combination of real-world measurements called Manifest Variables (MV's). The parameters of the theoretical model are determined by first estimating a so-called Measurement or Outer Model that relates all of the MV's according to the constraints of the theoretical model in order to estimate the LS's; then, a Path or Inner Model is formed by regressing those estimated LS's against each other. Several reviews on the adaptation of PLS to the path modeling concept are available (*10*, *11*). The actual algorithm used to build the model described herein was taken from (*12*, *13*).

In practice, the hypothetical path that relates a set of n data blocks to each other via a series of "causal" (ie. regression) relationships is codified as a binary matrix, c, of size [n,n], in which the [i,j] entry in this matrix is 1 if data blocks i and j are connected in the path:

$$c_{nxn} = \begin{bmatrix} 1/0 & \cdots & 1/0 \\ \vdots & \ddots & \vdots \\ 1/0 & \cdots & 1/0 \end{bmatrix} \quad (1)$$

PLS then forms the Outer Model by calculating a set of weights, w_i, for every data block, X_i, that taken together maximize the total explained covariance among ALL the X_i->X_j relations defined in the path hypothesis:

$$w_i = \sum_{j=1, j \neq i}^{N} c_{i,j} X_i^T X_j \,/ \left\| X_i^T X_j \right\| \quad (2)$$

These weights are used, as in ordinary PLS, to calculate a set of scores vectors, t_i, that define each of the hypothesized LS's (or latent variables) in path modeling terminology:

$$t_i = X_i w_i \quad (3)$$

These scores are then regressed against each other via multiple linear regression, again obeying the constraints of the hypothesized path model, to form the Inner Model. Then, confidence limits for the parameterized Inner Model regression coefficients are then estimated by bootstrapping (*14*) in order to gauge the validity of the hypothesized path.

In this work, the conceptual bolus structure breakdown model is outlined in the Introduction above is estimated by forming LS's that describe 1.the physical structure of the bolus at different points along its breakdown path and 2. the sensory perceptions at those points. According to path modeling terminology, these LS's were formed reflectively after standardizing all the MV's to zero mean and unit standard deviation, such that each MV within a given LS represented a unique physical (or sensorical) measurement that correlates to each other and to the same underlying phenomenon. Use of such 'reflective indicators' is convenient because it provides several statistical means of assessing model quality (see the discussion regarding Confirmatory Factor Analysis in section Results and Discussion, below).

Overview of the PLSpath model building steps:

1. Analyze correlations within the sensory data to establish the time-dependent food breakdown path
2. Assess the relationship of the Structural data to the Sensory data

    1. Using PLS1 to derive an initial guess of the outer model
    2. Using CFA to arrive at a final subset of Structure Measurements that define the latent variables in the inner model

3. Assess the relationship of Composition to Structure

The PLSpath algorithm was coded in Matlab, version R2013a, and all preliminary data analysis and iterative model-building steps were done in Matlab after importing the raw Instrumental parameters and Sensory scores from Microsoft Excel2007.

Methods and Materials

The food system prototypes consisted of a set of 16 samples made from systematic variations of 3 active ingredients. These were analyzed by a trained sensory panel for 10 standardized sensory attributes. In additions, the prototypes were measured by a suite of analytical measurements that consisted of 3 multivariate rheology methods, tribology friction curves, dynamic mixing curves, and tribology and one rheology measurement done on expectorated, Orally-Processed (OP) samples. Various parameters were extracted from each of these multivariate curves, resulting in a final set of 109 instrumental parameters. Figure 2 shows how all of these various data blocks relate within the structure->function paradigm described in the Introduction.

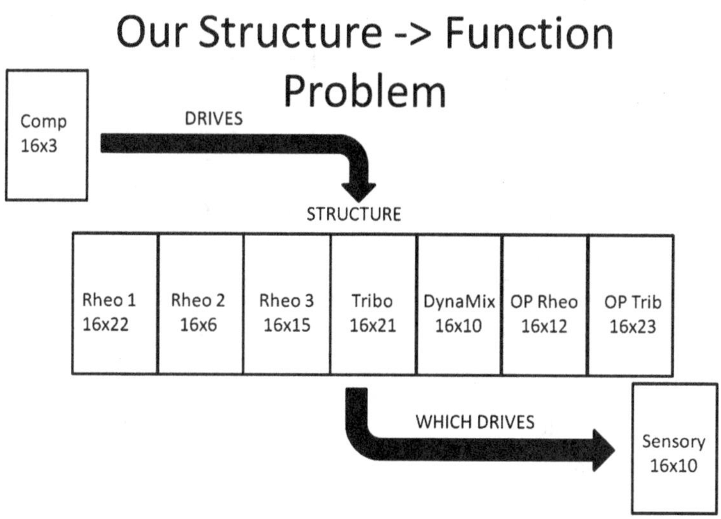

Figure 2. Visually relating the various data collected in terms of the structure:function model paradigm established in Figure 1.

Sensory Measurements

Eight trained descriptive panelists received a minimum of 65 h of training on the descriptive analysis of food flavor and mouthfeel/texture attributes using the Spectrum™ technique to develop a lexicon for ten sensory attributes specific to the type of food system studied herein (*15*). This lexicon consisted of definitions and scale normalizations for the panelists to place each attribute on a quantitative 0- 15 scale as they tasted the prototypes according to a standardized sample procedure. This procedure dictated a 1 hour fast before any measurements took place. The mouth was rinsed with mineral water before approximately 5 grams of sample was placed on a spoon and placed in the mouth. The sample was then masticated by mashing it between the tongue and palatte 20 times with up and down motions, prior to expectoration. A statistical summary of the sensory panel results is given in Table 1. Note from the table that there's relatively little variance in the initial bulk and afterfeel attributes, whereas the panelists were most sensitive to the surface and breakdown attributes. These statistical metrics have implications for the strengths of the regression relationships shown in the path model below.

Table 1. Summary Statistics for the Ten Sensory Attributes

Attribute	*Mean*	*St. Dev.*
initial bulk	6.6	1.2
breakdown_1	11.2	1.7
initial surface	6.7	2.2
breakdown 2	10.6	1.9
late bulk	6.1	1.3
late surface	7.4	2.2
swallow 1	10.2	1.6
afterfeel 1	9.3	1.1
swallow 2	6.2	1.2
afterfeel 2	0.7	0.2

Instrumental Measurements

All rheology and tribology measurements were made using a control stress Anton Paar Rheometer (Physica MCR 301 series) with 50mm parallel plate geometry at a gap of 1mm. Rheology included: 1.amplitude sweeps (at constant frequency of 10rad/s), 2.flow curves (at shear rates from 1e-6 – 1000 sec^{-1}), and 3.frequency sweeps (from 0.1 to 100 rad/sec) on the as-is samples; in addition, flow curves were measured for the orally-processed samples after they were masticated and exporated following the same standardized procedure as that used by sensory panel. Samples were removed from a 5°C refrigerator and then equilibrated to 20°C and shaken vigorously for 10-20 seconds immediately prior to sampling. All measurements were made in duplicate in serial, randomized order (that is, the entire series of samples was run in a random order to generate one replicate of each, then the entire series was run again in random order to generate the duplicates).

For the tribology measurement, a steel ball was used on the instrument's torque drive, but the friction contacts were replaced with a pliable elastomer to better mimic the surface of the mouth (*16*). Friction curves were measured while ramping the torque drive speed stepwise from 1 to 2240min^{-1} (logarithmically) while maintaining a constant normal force and temperature. The normal force tolerance is set at ± 0.02N. Tribology was also done on the orally-processed sample following the same procedure as was done for the flow curve measurements.

A custom dynamic mixing cell was used to measure the time required for the diluted samples to reach a constant viscosity. The instrument consisted of a sample cup affixed to an auger screw mixing blade that was driven by the toque drive of the Anton Paar rheometer. To perform the measurement, the auger screw was set to turn in the deposited samples at 100rpm prior to depositing an amount of water calculated to equal 20% of the volume of the initial sample volume. The rheometer then recorded the solution viscosity until a steady-state was reached.

Results and Discussion

The main goal of the PLSpath model is to explain the sensory properties of the food system in terms of a sequence of physically observable breakdown structures, which we infer from the suite of instrumental measurements made on the 'as-is' and orally-processed samples.

To begin determining which of the >100 derived parameters were the most strongly related to the various sensory attributes, individual PLS models were constructed to predict each of the sensory attributes from the appropriate collection of parameters (for example, this means that the measurements made from the orally-processed samples were not included in the PLS predictions of the initial sensory attributes).

As an example of this process, Figure 3 below shows the results of a two-component PLS model fit, chosen using leave-one-out cross validation, in terms of how important each of the relevant instrumental parameters was to predicting breakdown_1 sensory attribute. Parameters that ended up in the path model are labelled. The Variable Importance to the Projection (VIP) metric essentially sums the relative contribution of each parameter (via its weights) to all the latent variable components that formed the regression vector of the PLS model (*17*). Parameters with a VIP score greater than 1 were retained for inclusion in the final PLSpath model. After each individual PLS model was constructed, a global set of the most important parameters was formed by referencing across these ten PLS models and finding those parameters which had the highest rate of occurrence and strongest influence with respect to predicting the sensory properties.

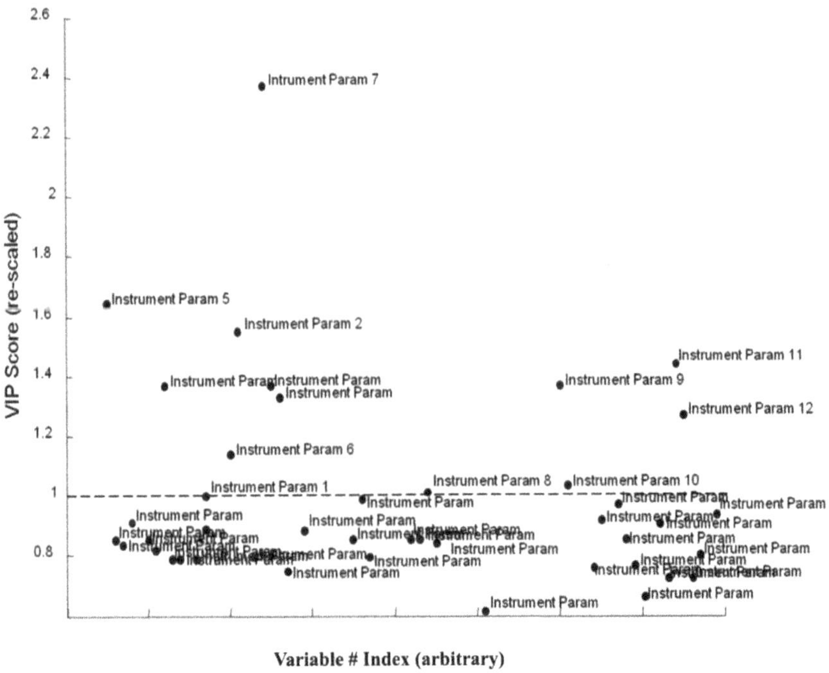

Figure 3. Variable Importance to the Projection Scores for the 2-component PLS model relating rate of dispersion to the suite of instrumental measurements. Scores >1 indicate a significant contribution for that variable to the prediction. Instrumental parameters that contributed to the path model are labelled with a number from 1-12.

As a result of the ten individual model fits, the most prevalent predictor variables were grouped into data blocks from which path model latent variables could be estimated according to how well correlated they were to each other (this iterative process is described more specifically below). These assignments can be seen on the left hand side of Figure 4 below, along with the correlation coefficients of each parameter to its assigned latent variable.

The sensory attributes, too, were grouped into latent sensory variables according to their inter-correlations and their time-dependent definitions from the lexicon; for example, the initial bulk and surface attributes were defined to occur before bolus breakdown, and late bulk and surface attributes were defined to occur after bolus breakdown. These sensory attribute inter-correlations can be seen in Table 2. As a result of these values and the attribute definitions, the two breakdown attributes were grouped into one latent variable called "breakdown", the late surface and late bulk characteristics attributes were grouped into one latent variable called "late sensory", the two swallow attributes were grouped into one latent variable called "swallowing", and the two afterfeel attributes were grouped into one latent variable called "residual".

Table 2. Matrix of Inter-Correlations among Sensory Attributes

<--TIME--	initial bulk	initial surface	breakdown_1	breakdown 2	late bulk	late surface	swallow1	swallow 2	afterfeel 1	afterfeel 2
	-----	-----	-----	-----	TIME-->	-----	-----	-----	-----	-----
initial bulk	1.00	0.01	0.08	0.19	0.56	0.02	0.02	0.00	0.11	0.25
initial surface		1.00	0.83	0.82	0.46	0.97	0.82	0.59	0.44	0.43
breakdown_1			1.00	**0.88**	0.55	**0.85**	**0.75**	0.48	0.40	0.35
breakdown_2				1.00	**0.74**	**0.85**	**0.66**	0.44	0.46	0.47
late bulk					1.00	0.54	0.37	0.15	0.50	0.53
late surface						1.00	0.80	0.54	0.49	0.47
swallow_1							1.00	0.80	0.48	0.55
swallow_2								1.00	0.24	0.38
afterfeel_1									1.00	0.57
afterfeel_2										1.00

Having grouped the ten sensory attributes into six sensory data blocks and the 19 "best" instrumental parameters (as defined by their VIP scores in the series of PLS1 model) into five structure data blocks, Table 3 shows the binary path matrix, c, that was used to define the PLSpath model that relates all of these together. The PLSpath model algorithm forms latent variables (equation 3) from these 11 blocks using the constraints defined in this matrix according to equation 2.

Table 3. The Binary Path Model Matrix c, As Defined by Equation 1

Init Bulk	Init Surf	Break	Late Sense	Swallow	Residual	Init Struct	Early Struct	Mid Struct	Late Struct	Res Struct	
0	0	0	0	0	0	1	0	0	0	0	Init Bulk
0	0	0	0	0	0	0	1	1	0	0	Init Surf
0	0	0	0	0	0	0	1	1	0	0	Break
0	0	1	0	0	0	0	0	1	1	0	Late Sense
0	0	1	0	0	0	0	0	0	1	1	Swallow
0	0	0	0	0	0	0	0	0	0	1	Residuals
0	0	0	0	0	0	0	0	0	0	0	InitStruct
0	0	0	0	0	0	1	0	0	0	0	EarlyStruct
0	0	0	0	0	0	0	1	0	0	0	MidStruct
0	0	0	0	0	0	0	1	0	0	0	LateStruct
0	0	0	0	0	0	0	0	1	0	0	ResStruct

Table 4 summarizes a Confirmatory Factor Analysis (CFA) of the calculated latent variables (7). The Reliability metric in the table intends to measure how well each parameter reflects the latent variable it is assigned within; it's defined as:

$$Reliability = \sum l_i^2 \Big/ \left[\sum l_i^2 + \sum(1 - l_i^2)\right] \quad (4)$$

Where the l_i are the loadings of each parameter on its latent variable. Since all the parameters were standardized to mean zero and unit variance, these loadings vary between -1 to 1 and are equal to the correlation coefficient of the parameter to the latent variable. Reliability metrics of >0.9 for each latent variable indicate good internal consistency for all constructs (that is, no constructs contain contributions from spurious measurement parameters), whereas the composite average parameter variance explained (AVE) metric >=0.87 for each latent variable indicates that the constructs do a good job of fitting their respective parameters (*18*). Finally, the fact that the cross-correlations in this table are all less than the AVE's indicates that none of the latent constructs are redundant or misidentified in the final path model.

Of course, arriving at this final CFA analysis was the result of iterating through several sequences of defining the number of latent variables to include in the model as well as defining which measurement parameters to include in each of the latent variables. In addition, the global Goodness-of-Fit metric was monitored at each of these iterations to see whether or not inclusion of particular measurements and/or particular changes in the path matrix (Table 3) had a significant effect. With respect to PLSpath modeling, the Goodness-of-Fit is calculated by:

$$GoF = \sqrt{mult.R^2} \times \sqrt{Ave.Communality} \quad (5)$$

Where *mult* R^2 is the average r-squared correlation amongst all connected latent variables in the inner part of the PLSpath model, and *Ave. Communality* is the average r-squared correlation of all manifest variable to their corresponding latent variables (*13*). These iterations began from a parsimonious starting point: each latent structure variable began as only the one manifest variable corresponding to the top PLS1 VIP score for its most relevant sensory attribute; from here, the other measurement parameters were added to these LS variables until either the CFA assumptions were violated or the Goodness-of-Fit metric did not improve.

The PLSpath model is shown in Figure 4. This diagram shows how the latent variables defined as combinations of the measured parameters relate to each other in terms of their hypothesized causal relationships. The various hypothesized latent breakdown structures are represented in ovals situated along the left-hand side of the figure, and the order of the bolus breakdown sequence proceeds from top to bottom. These LS's are labeled according to their time dependence within the proposed breakdown path. For example, Init Struct denotes the most prominent and highly correlated set of instrumental measurement parameters with regards to how well they predict the Initial Bulk attribute that, in turn, intends to capture the sensory reponse immediately upon placing the food sample in the mouth. The corresponding sensory components of the breakdown sequence are represented as ovals aligned from top to bottom on the right-hand side of the figure. However, because the initial bulk and initial surface constructs were actually measured as independent variables, they are shown in rectangular boxes along with the other measured variables (MV's), according to path modelling formalism. Each endogenous construct (i.e. an LS that is defined in part, as having been "caused" by at least one other LS) has its explained variance stated in parentheses. Note that the initial structural LS is labeled "exo" to indicate that it is exogenous; that is, it does not depend on any of the other structural LS's. The arrows indicating the various causal relationships are superimposed with their path coefficients (standardized regression coefficients) and underscored in parentheses with the p-value of the regression as determined by the bootstrapping procedure (using n=100 re-samples). Each endogenous LS also has its regression fit r-square shown in parentheses under the structure name. All p-values are <=0.05 except for that of the link between the Early Struct LS and Initial Surface LS, which is 0.06. However, given the limited size of the sample set, and the fact that the prototype samples were 'expensive' in terms of conducting the sensory panel and the suite of instrumental measurements on them, it was assumed that, if additional samples were composed that spanned the difference in composition between the original sample set, the relationship between the early structure LS and initial slipperiness would be confirmed.

Table 4. Average Variance Extracted (AVE), Composite Reliability, and Interconstruct Correlations for the Latent Variables in the Model. the Reliability Scores above 0.9 Indicate Good Internal Consistency for Each Construct; the Fact That Every Construct'S AVE Is Higher than Its Cross-Loadings with the Constructs Demonstrates Adequate Convergent Validity for the Model Components

	Reliability (#items)	Init Bulk	Init Surf	Break	Late Sense	Swallow	Resid.	Init Struct	Early Struct	Mid Struct	Late Struct	Res Struct
Init Bulk	1 (1)	1										
Init Surf	1 (1)	0.11	1									
Breakdown	0.98 (2)	-0.37	-0.92	0.99								
Late Sense	0.93 (2)	0.51	0.88	-0.93	0.96							
Swallowing	0.97 (2)	-0.07	-0.86	0.79	-0.71	0.98						
Residuals	0.93 (2)	-0.43	-0.7	0.71	-0.81	0.69	0.96					
InitStruct	0.94 (4)	0.6	-0.01	-0.22	0.22	0.16	0.18	0.94				
EarlyStruct	0.92 (4)	0.2	0.74	-0.79	0.73	-0.51	-0.34	0.27	0.9			
MidStruct	0.98 (4)	0.24	0.76	-0.73	0.82	-0.69	-0.87	-0.24	0.54	0.97		
LateStruct	0.98 (4)	0.24	0.59	-0.62	0.64	-0.23	-0.23	0.21	0.82	0.36	0.98	
ResStruct	0.89 (3)	-0.05	-0.58	0.49	-0.52	0.74	0.79	0.49	-0.25	-0.82	0.03	0.87

The PLSpath model diagram also shows the predicted variance for each endogenous LS (in parentheses in the ovals), as determined by the jackknife procedure described. The predicted variances are acceptably high for all the sensory constructs, indicating that the breakdown sequence proposed by the model can be measured instrumentally and can be used to predict the sensory dynamics of unknown samples. The predicted variance of the initial bulk attribute is fairly low, but this is merely because all the prototype samples were composed such that their initial bulk attribute matched that of a commercially-available, full-calorie target product; this means that any difference in initial bulk among the samples is due only to the inability to perfectly match this characteristic with the three active ingredients that were varied in the model food system. Thus, it was expected that the predicted variance of initial bulk would be low. Furthermore, the predicted variances of most of the latent breakdown structure constructs was fairly low (though significant, as indicated by the p-values being <0.05 for each regression along the breakdown path), but this is to be expected, since each of LS's represents a new state of the masticated bolus. It should be loosely related to the previous state of the bolus, and thus have a statistically significant regression relationship to its breakdown structure precedent, but it should also contain significant new information that isn't sufficiently described by the other latent structures.

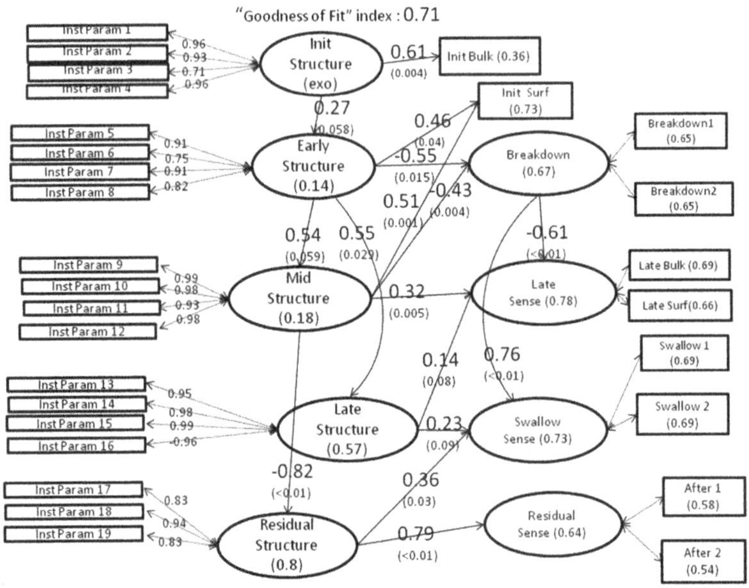

Figure 4. The PLSpath model relating physically measured breakdown structures to sensory perceptions.

Table 5 compares the R-squared model fits of the series of PLS1 models performed on each sensory attribute (using all relevant instrumental parameters as the X-block) the model fits of the PLSpath model. Due to constraints imposed by the various path relationships and the use of less X-block parameters (that is, MV's) in each PLSpath regression, it is expected that some predictability will be lost relative to the "brute force" PLS1 models. This is not surprising, since the relationships of the predictors to the sensory attributes in the individual PLS models are not constrained in any way, whereas in the path model certain parameters are forced into an indirect relationship with one or more sensory attributes – and the sensory attributes themselves are grouped into latent structures according to their inter-correlations and breakdown definitions - as a result of their role in the hypothesized structural breakdown sequence. However, the goal of the path model is not optimal prediction of every sensory attribute; it is to provide a means for gauging the validity of a global model that encompasses (and helps to explain mechanistically!) the inter-relationships between all of the sensory attributes.

Table 5. Comparison r-Squared Metrics for the Sensory Attributes Based on Leave-One-Out Cross Validation of the Individual PLS1 Model Fits (R-Sq CV) to Their Corresponding PLSpath Model Fits. The Number of Latent Variables (LVs) and Path Regression Coefficients Used to Model Each Attribute Are Also Given

Attribute	*#LV's*	*R-Sq CV*	*Path R-Sq*	*Path Coeffs*
initial bulk	1	0.39	0.36	1
breakdown_1	2	0.71	0.65	2
initial surface	2	0.73	0.73	2
breakdown 2	2	0.83	0.65	2
late bulk	2	0.81	0.69	2
late surface	2	0.79	0.66	2
swallow 1	3	0.89	0.69	2
afterfeel 1	1	0.92	0.58	2
swallow 2	2	0.90	0.69	2
afterfeel 2	1	0.26	0.54	2
AVERAGE		**0.72**	**0.62**	

Examination of Individual Structure:Function Relationships

The Inner Model allows us to examine each hypothesized (PhysioChemical) Structure-> (Sensory) Function relationship via a series of simple bi-variate plots. Trends based on composition can also be seen within such plots by labeling the data points (even though these trends are formally quantified later on by regressing the latent variable model scores onto the compositions – see the subsection below entitled Examination of Composition:Structure Relationships). There are many regression relationships within the path model, so this process generates a lot of plots, but these can be quite valuable in assessing exactly how composition drives structure, which in turn drives function (sensory). Note that, in the plots to follow in this subsection it is the latent variable scores that are being visualized, and not the measured data themselves. Thus, the correlations between various latent variables in the path model will usually be somewhat stronger than the correlations between the individual manifest variables that make up the path model constituents. However, the Confirmatory Factor Analysis in Table 4 demonstrates that the manifest variables (ie. the experimental data) are all highly correlated to their corresponding latent variables, so it can be assumed that the manifest variables combined within each of the latent variables all have similar relationships as illustrated in the regression plots.

As an example of this model investigation process, Figure 5 shows the relationship between the Early and Mid Struct latent variable scores and the Breakdown latent variable scores. From the size of the standardized coefficients in the regression equation, it can be seen that the Early and Mid Struct latent variable contribute fairly equally to the prediction of Breakdown in this food system. It can be seen, too, that Ingredient #1 is the main determiner of Breakdown rate in this food system.

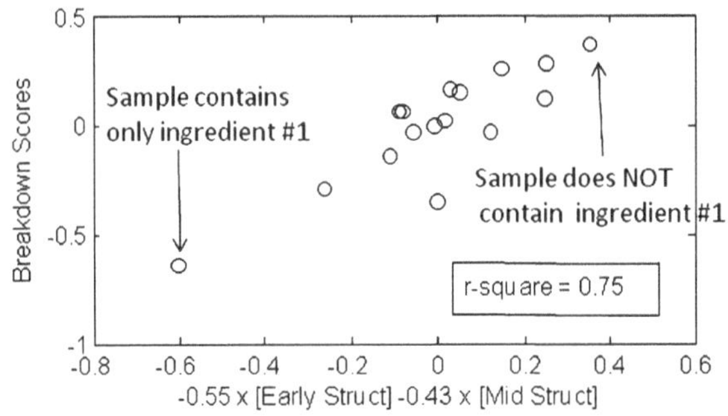

Figure 5. Regression relationship between the Early and Mid Structure and Breakdown rate.

Figures 6 and 7 further elaborate this particular structure-function relationship by showing how the two latent structures each correlate to Breakdown. Close examination of these two plots shows that Early Struct has a somewhat stronger dependence on Ingredient 1 than Mid Struct, which depends more on Ingredient 3. As implied by the magnitude of their regression coefficients from Figure 5, each has essentially the same correlation to Breakdown (r values -0.79 and -0.73, respectively). However, while Figure 6 shows that Init Struct scores (which capture rheological features of the samples in the linear visco-elastic region) follow a continuous linear trend as a fuction of Ingredient 1, Figure 7 reveals that Ingredient 3 segregates the samples into two distinct groups according to the set of measurement parameters (that capture non-linear rheological deformations as well as certain friction properties of the samples) that had been united in the so-called Mid Struct latent variable. Furthermore, the Breakdown scores within each subgroup are still anti-correlated to the amount of Ingredient 1 in each sample. That is, the sample with the most Ingredient 1 of each subgroup in Figure 7 lies on the lower right-hand extreme of that subgroup's extent on the graph. Also, the range of Ingredient 1 inclusion in each subgroup is the same; thus, it can be seen that Ingredient 3 not only exerts its own effect on Breakdown and mid-mouth-processing structure, it compresses the effect that Ingredient 1 has on these latent variables.

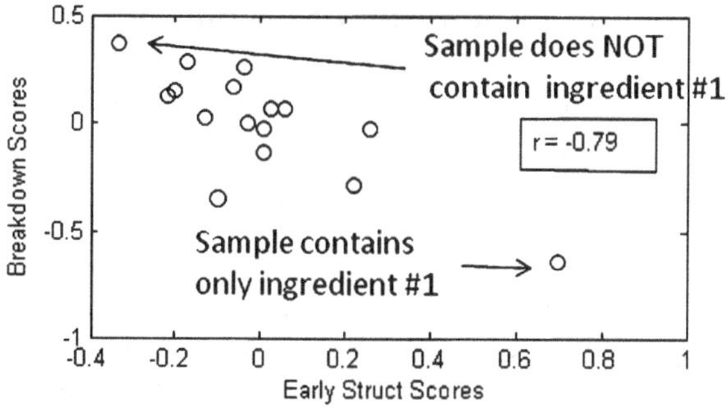

Figure 6. Correlation of Breakdown rate to Early Struct.

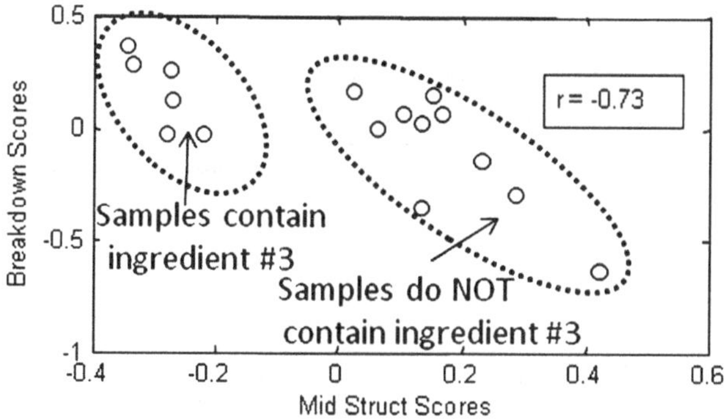

Figure 7. Correlation of Breakdown rate to Mid Struct.

In a similar way, the other "causal" connections that are indicated by the single-headed arrows in the PLSpath diagram of Figure 4 can be dissected in terms of a series of bivariate plots in order to decipher how the ingredients act to alter various time-dependent bolus structural features that occur in the mouth and in turn how these structure relate to sensory perception. Although these relationships are explicitly formed as functions of latent variables in the inner path model, the influence of the manifest variables (that is, the actual measurements) on those latent variables can be deduced by looking at the loading correlation coefficients that are superimposed on the double-headed arrows in the PLSpath diagram. The benefit of primarily examining latent variables, though, is that it allows for a more concise overview of the time-dependent inter-relationships among the dozens of individual parameters that compose the overall model.

Examination of Composition:Structure Relationships

A concise overview of the relationship of sample composition to the various components of the PLSpath model is shown by the correlation matrix in Table 6. Ingredient 2 seems to counteract the effects of Ingredient 1 on the initial and late sensory attributes – much like Ingredient 3. However, it doesn't affect the swallowing or residual characteristics to the extent that Ingredient 3 does. It is seen that initial and late bulk and surface characteristics are directly related to the amount of ingredient 1, while the breakdown rates are inversely related to ingredient 1. This demonstrates that while ingredient 1 acts as a bulking agent, it doesn't break down in the mouth, and can result in a stubborn bolus that slides through the mouth without leaving a residue on the tongue. Conversely, ingredient 3 is proportional to breakdown rate and swallowing & residual characteristics, indicating that it tends to separate from the bolus and deposit on the tongue. Ingredient 2 seems to counteract the effects of Ingredient 1 on the initial and late sensory attributes – much like Ingredient 3. However, it doesn't affect the swallow or residual characteristics to the extent that Ingredient 3 does.

Table 6. Matrix of Correlation Coefficients (r-Values) of Sample Composition to Sensory and Physical Structure Latent Variables

	Ingredient 1	*Ingredient 2*	*Ingredient 3*
Init Bulk	0.41	0.09	-0.14
Init Surface	0.66	-0.29	-0.54
Breakdown	-0.75	0.29	0.51
Late Sense	0.77	-0.20	-0.57
Swallow	-0.33	0.14	0.66
Residuals	-0.31	-0.19	0.79
InitStruct	0.45	-0.20	0.39
EarlyStruct	0.83	-0.61	-0.24
MidStruct	0.41	0.09	-0.87
LateStruct	0.92	-0.68	0.03
ResStruct	0.02	-0.29	0.94

This analysis of the correlation matrix enables a qualitative, mechanistic interpration of the effects of the active ingredients on the structural and sensory performance. A quantitative model can be built, too, via multiple linear regression of the latent variable scores onto the composition matrix. In addition to allowing for "dialing in" an optimal recipe, visualization of these results allows for refining mechanistic interpretations in a way that is similar to the discussion around

Figures 5-7 above. Note that, in the following plots, the Ingredient concentrations have been standardized by subtracting their mean and dividing the resultant by their standard deviation.

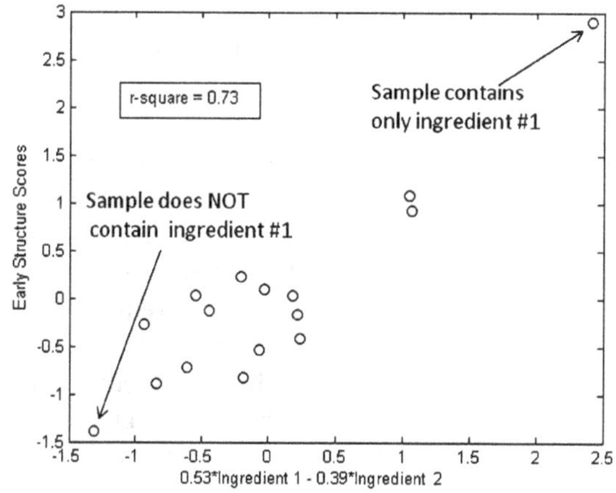

Figure 8. Regression of Early Struct onto sample composition.

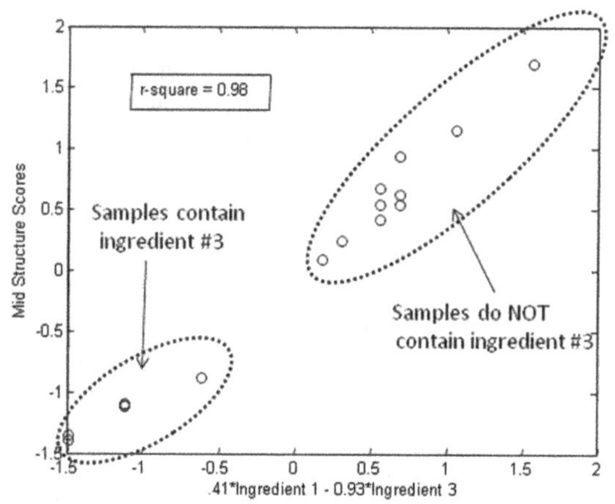

Figure 9. Regression of the Mid Struct latent variable on sample composition.

The multiple regression results to follow will focus on the Early and Mid Struct latent variables, as was done in Figures 5-7. Figure 8 shows, as expected from that previous qualitative assessment, that the Early Struct latent variable is well-explained as a function of Ingredient 1 with some help from Ingredient 2. Because the variables in the regression have been standardized, the fraction of variance explained by each can be deduced by multiplying its regression coefficient by the absolute value of its univariate correlation coefficient. So Ingredient 1 explains [0.83 x 0.53] = ~49% of the variance in Early Struct scores while Ingredient 2 explains [0.61 x 0.39] = ~23%; the sum of these fractions, 0.49 + 0.23, equals the total r-squared variance explained of ~0.73 or 73%.

In the same way, Figure 9 shows that the Mid Structure latent variable can be explained by the amounts of Ingredient 1, which explains [0.41 x 0.41] = ~15% of the variation in the Mid Structure scores, and ingredient 3, which explains [0.87 x 0.93] = ~83% of the variation in Mid Structure scores.

This regression analysis can be repeated for all of the hypothesized latent variables in the PLSpath model, allowing for assessment (and quantitative prediction) of the importance of each active ingredient to the components of the hypothesized breakdown path. While this does allow for "dialing in" or optimizing the performance of the system of ingredients that were used to build the model, the main utility of this modelling step is to further refine the understanding of the mechanisms by which the ingredients interact. This is because ultimately these mechanisms will be used to guide the choice of other (cheaper, lower calorie, etc.) ingredient systems via their ability to reproduce certain functional properties (ie. rheology, tribology, mixing, etc.) that are revealed through the various components of the PLSpath model to be important in affecting sensory attributes.

Conclusions

The Partial Least Squares (PLS) path modelling technique was used to quantify the time-dependent relationships between various physical measurements and sensory attributes made on a set of food system prototypes intended to approximate a commercially available target product. An iterative process based on assessment of the VIP scores from individual PLS1 models of each sensory attribute against all the physical measurements was used to deduce which instrumental parameters (derived from the physical measurements) served as the best surrogates of discrete structural elements that dominated sensory perception at five points along the bolus breakdown path from the time at which the food system was introduced to the mouth, through mastication, and so on until after it was swallowed. These sensory points were assigned as Initial Bulk characteristics, Initial Surface characteristics, Breakdown behavior, Late Bulk characteristics, Swallow consistency, and post-swallowing Afterfeel in the mouth, based upon the inter-correlations within a sensory attribute data matrix that consisted of ten individual attribute evaluations by a trained taste panel.

Examination of the individual regression relationships that made up the breakdown sequence depicted in the resultant path model revealed how the oral processing of prototypes involves time-dependent changes in bulk and mouth surface properties of the bolus. Taken together, these relationships composed a structure:function analysis of the food system. The use of PLSpath latent variables, as opposed to conventional PLS1 regression latent variables, allowed for a more concise overview of the time-dependent inter-relationships among the dozens of individual parameters that composed the overall PLSpath model. It also allowed for testing specific sequences of time-dependent relationships between the physical and sensory measurements until a final PLSpath model was arrived at that enabled a mechanistic interpretation of the samples' physical and sensory behavior in the mouth. This model was not quite as predictive as a series of PLS1 regressions on the individual sensory attributes, but it was more explainable in mechanistic terms. Specifically, these latent variables revealed that the structural properties of the samples were a result of the interplay of ingredient 1, that was resistant to breakdown via mastication, and ingredient 3, which separated from the bolus during mastication and deposited on the surface of the mouth. The ratio of ingredient 1 to ingredient 2 had the strongest effect on the bulk physical and sensory measurements while the ratio of Ingredient 1 to ingredient 3 had the strongest effect on the rate of sample breakdown and mouthcoating during mastication. Incorporating these learnings into a unified mechanistic model of this particular food system's behavior allowed for novel prototypes to be constructed using cheaper, reduced-calorie ingredients that closely matched the sensory properties of a commercially available target product.

References

1. de Wijk, R. A.; Prinz, J. F. Fatty versus creamy sensations for custard desserts, white sauces, and mayonnaises. *Food. Qual. Preference* **2007**, *18*, 641–650.
2. Kokini, J. L.; Kadane, J. B.; Cussler, E. L. *J. Texture Stud.* **1977**, *8*, 195–218.
3. Malone, M. E.; Appelqvist, I. A. M.; Norton, I. T. Oral behavior of food hydrocolloids and emulsion I.T. Oral behavior of food hydrocolloids and emulsions: Part 1, Lubrication and deposition considerations. *Food Hydrocolloids* **2003**, *17*, 763–773.
4. Akhtar, M.; Murray, B.; Dickinson, E. Perception of creaminess of model oil-in-water dairy emulsions: Influence of the shear-thinning nature of a viscosity-controlling hydrocolloid. *Food Hydrocolloids* **2006**, *20*, 839–847.
5. Stokes, J. R. 'Oral' Rheology and 'Oral' Tribology. In *Food Oral Processing: Fundamentals of Eating and Sensory Perception*; Chen, J., Engelen, L., Eds.; Wiley-Blackwell: West Sussex, U.K., 2012; pp 227–288.
6. Fornell, C.; Johnson, M. D.; Anderson, E. W.; Cha, J.; Bryant, B. E. The American Customer Satisfaction Inde B.E. The American Customer Satisfaction Index: Nature, purpose, and findings. *J. Mark.* **1996**, *60* (4), 7–18.
7. Agarwal, R.; Karahanna, E. Time flies when you're having fun: Cognitive absorption and beliefs about information technology usage. *Manage. Inf. Syst. Q.* **2000**, *24* (4), 665–694.
8. Gerlach, R. W.; Kowalski, B. R.; Wold, H. O. A. Partial least-squares path modeling with latent variables. *Anal. Chim. Acta.* **1979**, *112* (4), 417–421.
9. Pagès, J.; Tenenhaus, M. Multiple factor analysis combined with PLS path modelling. Application to the analysis of relationships between physicochemical variables, sensory profiles and hedonic judgements. *Chemometr. Intell. Lab. Syst.* **2001**, *58* (2), 261–273.

10. Chin, W. The Partial Least Squares Approach to Structural Equation Modeling. In *Modern Methods for Business Research*; Marcoulides, G. A., Ed.; Lawrence Erlbaum Associates: Mahwah, NJ, 1998.
11. Vinzi, V. E.; Trinchera, L.; Amato, S. PLS Path Modeling: From Foundations to Recent Developments and Open Issues for Model Assessment and Improvement. In *Handbook of Partial Least Squares, Concepts, Methods and Applications*; Vinzi, V. E., Chin, W. W., Henseler, J., Wang, H., Eds.; Springer-Verlag: Berlin, 2010; pp.47–82.
12. Lofstedt, T.; Hanafi, M.; Mazerorlles, G.; Trygg, J. OnPLS path modeling. *Chemometr. Intell. Lab.* **2012**, *118*, 139–149.
13. Tenenhaus, M.; Hanafi, M. A bridge between PLS path modeling and multi-block data analysis. In *Handbook of Partial Least Squares, Concepts, Methods and Applications*; Vinzi, V. E., Chin, W. W., Henseler, J., Wang, H., Eds.; Springer-Verlag: Berlin, 2010; pp 99–124.
14. Fox, J. *Applied Regression Analysis and Generalized Linear Models*, 2nd ed.; Sage Publications, Inc.: Los Angeles, 2008; pp 587–606.
15. Meilgaard, M. C.; Carr, B. T.; Civille, G. V. *Sensory Evaluation Techniques*, 3rd ed.; CRC Press: Boca Raton, FL, 1999; pp 173–230.
16. Steinbach, A.; Guthrie, G.; Smith, S.; Lindgren, T.; Debon, S. Normal force-controlled tribological measurement of soft drinks and lubrication additives. *J. Food Meas. Charact.* **2014**, *8* (2), 142–148.
17. Chong, I. G.; Jun, C. H. Performance of some variable selection methods when multicollinearity is present. *Chemometr. Intell. Lab.* **2005**, *78* (1), 103–112.
18. Hair, J.; Black, W.; Babin, B.; Anderson, R. *Multivariate Data Analysis*, 7th ed.; Prentice-Hall, Inc.: Upper Saddle River, NJ, 2010.

Chapter 23

Odor-Structure Relationship Studies of Indan, Tetralin, and Isochroman Musks

Barry K. Lavine* **and Collin White**

107 Physical Science, Department of Chemistry, Oklahoma State University, Stillwater, Oklahoma 74078
***E-mail: bklab@chem.okstate.edu.**

One hundred and eighty-seven tetralin-, indan-, and isochroman-like compounds were combed from the published literature to investigate the relationship between chemical structure and musk odor quality. Utilizing Breneman's transferable atom equivalent (TAE) and property encoded surface translator (PEST) descriptors and comprehensive descriptors for structural and statistical analysis (CODESSA), each compound in this database was represented by 1369 descriptors. A genetic algorithm (GA) for pattern recognition analysis was used to identify a subset of the 1369 molecular descriptors that could differentiate musks from nonmusks in a plot of the two largest principal components of the data. The 20 molecular structural descriptors selected by the pattern recognition GA for variable selection contained information about the shape, electronic properties, and intermolecular interactions of these compounds. Due to the risk of model interpretation when performing variable selection, model cross-validation was performed using an external validation set of 19 compounds. Discriminants (in the form of a principal component plot of the 20 molecular descriptors and the 168 compounds comprising the training set) were successfully validated using the external validation set of 19 compounds.

Introduction

Musks are one of the most universally appreciated and widely investigated odor groups because of their characteristic odor quality which has been described as warm, animal, and natural (*1*, *2*). Almost all perfumes and toiletries sold in North America and Western Europe contain some musk (*3*). Musk compounds impart sensuality and are a crucial constituent of perfumes. For some perfumes, musk is the principal odor, whereas the role of musk in other perfumes is to deepen and enhance other odors.

Musks are unique as they possess a characteristic odor found within a large variety of different structural manifolds (*4*). Musk odor is rarely confused with other odors. Among the different structural classes of musks, there are subtle odor differences. Naturally occurring musk (*5*) was originally obtained from the glands of the musk deer (Moschus moschiferus) in Central Asia. This dark red to black-brown grainy secretion is used by the male musk deer to mark territorial boundaries and to attract females. Ruzicka (*6*) succeeded in determining the chemical structure of this secretion. Since the 1980s when Albert Baur (*7*) discovered that the odor of natural musks can be imitated by nitrated derivatives of benzene, the fragrance industry has turned to synthetic musks. The chemical structure of synthetic musks can be divided into three broad classes: macrocyclic musks, aromatic nitro musks and polycyclic musks. Currently, polycyclic musks are the largest and most important group of synthetic musks as they include some of the most powerful musks known, e.g., phantolide and galaxolide.

Many theories have attempted to explain the relationship between structure and odor quality (*8–12*). However, these theories cannot explain why so many different classes of compounds are able to elicit musk odor. Analysis of odor-structure relationships (OSR) using computer generated molecular descriptors and pattern recognition techniques can provide a practical approach to the study of musks. The heart of this approach is finding a set of molecular descriptors from which discriminating relationships can be found to differentiate musks from nonmusks. Using this approach, Brugger (*13*) investigated the relationship between molecular structure and odor quality for a set of 60 musks and 240 nonmusk compounds. A discriminant developed from 13 fragment-based molecular descriptors for the training set of 300 compounds was able to correctly predict nine unknown compounds as musk odorants. Ham (*14*) correctly classified 71 monocyclic nitrobenzenes as musks or nonmusks using substructure driven descriptors and nonparametric pattern recognition methods. Narvaez (*15*) investigated the relationship between molecular structure and musk odor quality for a set of 148 bicyclic and tricyclic benzenoid compounds (67 musks and 81 nonmusks). A discriminant developed from a set of 14 molecular (topological, electronic, and geometric) descriptors correctly classified every compound in the training set, and 14 of 15 compounds in the validation set. Using computer automated structure evaluation (CASE) methodology, Klopman and Ptscelintsev (*16*) analyzed 152 nonnitro aromatic musks and their odorless structural analogues and identified nine substructure driven descriptors correlated to musk odor quality as well as seven dearomatizing fragments. Charastrette (*17*) and Cherqauoi (*18*) employed back propagation neural networks and linear free energy relationship parameters as descriptors for well characterized aromatic substituents to discriminate tetralin and indan musks from their nonmusk analogues. Employing Breneman's Transferrable Atom Equivalent (TAE)-based reconstruction (RECON) methodology (*19–21*) and spatially resolved shape/property hybrid electron density-based property encoded surface translation (PEST) descriptors, Lavine (*22*) was able to model the OSR of aromatic nitro musks, which was previously not well understood because of the complex substitution pattern, highly impure and informational rich complex odors, and the polyfunctional character of the nitro group. Using six TAE derived descriptors, a discriminant was developed that correctly classified every compound in the training set. The discriminant was successfully validated using a set of 17 nitro-aromatic compounds. The six descriptors used for discriminant development contained information about molecular interactions important in nitro-aromatic musks.

Utilizing Breneman's RECON and PEST methodology and Katritzky's (*23*) comprehensive descriptors for structural and statistical analysis (CODESSA), 168 tetralins, indans, and isochromans (87 musks and 81 nonmusks) were investigated to develop a better understanding of the relationship between molecular structure and odor quality for the three different structural manifolds comprising this set of polycyclic musks. Previous studies undertaken in our laboratory using topological and fragment-based descriptors to develop an odor structure relationship for these compounds could not differentiate tetralin and indan musks from their nomusk analogues. In this study, each compound was represented by 1369 descriptors. A genetic algorithm (GA) for pattern recognition analysis was used to identify a subset of the 1369 molecular descriptors that could differentiate musks from nonmusks in a plot of the two largest principal components of the data (*24–27*). The 20 molecular structural descriptors selected by the pattern recognition GA for feature selection contained information about the size, shape, electronic properties, and intermolecular interactions of these compounds. Discriminants developed in this study using these 20 molecular descriptors were validated using a separate external validation set of 19 compounds (11 nonmusks, four indan and tetralin musks, and four isochroman musks).

Experimental

Dataset

An olfactory database of 187 compounds was compiled from the published literature (*28–35*) for investigating the relationship between chemical structure and musk odor quality. For this study, nonmusks were chosen to be as similar in structure to the musks as possible (see Figure 1). This not only provided an additional challenge of differentiating among similar structures according to their odor quality but also increased our understanding of how small structural changes can influence musk odor quality. Of the 187 compounds comprising the database, 95 are musks and 92 are nonmusks. The musks are strong, medium, or weak odor intensity whereas the nonmusks are odorless or have an odor other than musk, e.g., amber or woody. The training set consisted of 168 compounds and the validation set consisted of 19 compounds. Validation set compounds were selected by random lot. A list of the 168 training set compounds is given in Table 1 and a list of the 19 compounds comprising the validation set is given in Table 2.

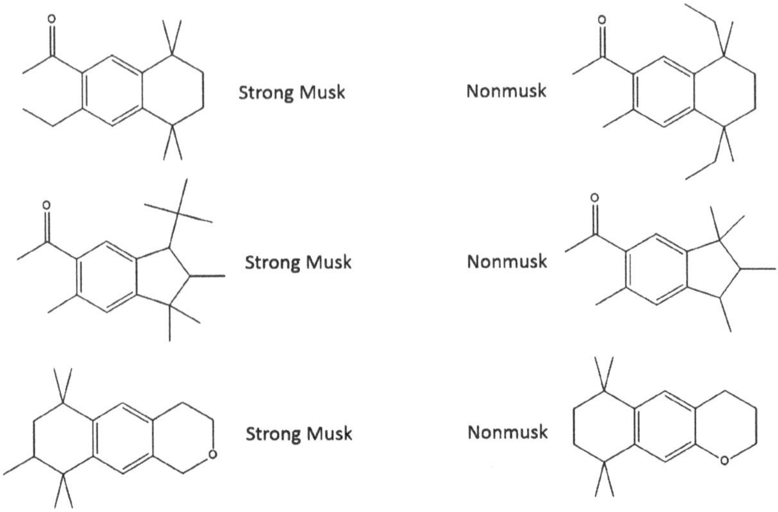

Figure 1. Examples of musk and nonmusks in the training set.

Molecular Descriptor Generation

In terms of musks, the key features are molecular shape, electronic properties of the molecule and intermolecular interactions. To characterize and quantitate these features, it was necessary to develop a formalism that permitted integration of molecular modeling and conformational analysis into the calculation of molecular descriptors which were then used for the development of the OSR. The approach taken in this study was as follows. First, a 3-dimensional molecular model was generated for each compound by a force-field molecular mechanics model building routine utilizing the CHARMm force field (Molecular Simulations). Second, TAE descriptors of three general types were generated: molecular surface property histogram descriptors, surface property wavelet coefficient descriptors, and surface property hybrid descriptors. For some compounds, it was necessary to take into account specific conformational effects by developing new atom types in the TAE descriptor generation routines. Furthermore, the shape signature ray tracing approach used to generate PEST descriptors for molecular shape analysis was standardized to allow for better correlations between shape and odor quality. Third, traditional topological, geometric, and fragment-based descriptors were generated for each molecule using CODESSA (CompuDrug International, Sedona AZ). In this study, descriptors based on experimental data were not used because of the difficulty to obtain this type of data from the literature for a large number of compounds. Furthermore, any OSR that contains experimental data could not be used as a virtual screening tool to identify new musks since these compounds would have to be synthesized in order to obtain the necessary data.

Table 1. Training Set Compounds

Compound Name	*Odor Quality*
6-(1,1-dimethylethyl)-4-ethyl-2,3-dihydro-1,1-dimethyl-1H-indene	OLES[1]
7-acetyl-3,4-dihydro-1,1,4,4,5-pentamethyl-2(1H)-naphthalenone	OLES
8-acetyl-3,4-dihydro-1,1,4,4,5-pentamethyl-2(1H)-naphthalenone	OLES
5-acetyl-3,4-dihydro-1,1,4,4,6-pentamethyl-2(1H)-naphthalenone	OLES
7-acetyl-3,4-dihydro-1,1,4,4,6-pentamethyl-2(1H)-naphthalenone	OLES
8-acetyl-3,4-dihydro-1,1,4,4,6-pentamethyl-2(1H)-naphthalenone	OLES
5-acetyl-3,4-dihydro-1,1,4,4,7-pentamethyl-2(1H)-naphthalenone	OLES
6-acetyl-3,4-dihydro-1,1,4,4,7-pentamethyl-2(1H)-naphthalenone	OLES
5-acetyl-3,4-dihydro-1,1,4,4,8-pentamethyl-2(1H)-naphthalenone	OLES
7-acetyl-3,4-dihydro-1,1,4,4,8-pentamethyl-2(1H)-naphthalenone	OLES
8-acetyl-3,4-dihydro-1,1,4,4,7-pentamethyl-2(1H)-naphthalenone	OLES
1-(2,3-dihydro-1,1-dimethyl-1H-inden-4-yl)-ethanone	OLES
1-(2,3-dihydro-3,3-dimethyl-1H-inden-4-yl)-ethanone	OLES
1-(2,3-dihydro-1,3,3,6-tetramethyl-1-propyl-1H-inden-5-yl)-ethanone	OLES
1-(5,8-dihydro-5,5,7,8,8-pentamethyl-1-naphthalenyl)-ethanone	OLES
1-[3-(1,1-dimethylethyl)-5,6,7,8-tetrahydro-1-naphthalenyl]-ethanone	OLES
1-(2,3-dihydro-1,1-dimethyl-1H-inden-5-yl)-ethanone	OLES

Continued on next page.

Table 1. (Continued). Training Set Compounds

Compound Name	*Odor Quality*
1-(2,3-dihydro-3,3-dimethyl-1H-inden-5-yl)-ethanone	OLES
1-(2,3-dihydro-1,2,3,3,6-pentamethyl-1H-inden-5-yl)-ethanone	OLES
5,6,7,8-tetrahydro-3-methoxy-5,5,8,8-tetramethyl-2-naphthalenecarboxaldehyde	OLES
1-(5,6,7,8-tetrahydro-3-methoxy-5,5,8,8-tetramethyl-5-naphthalenyl)-ethanone	OLES
1-(5,6,7,8-tetrahydro-1-methoxy-3,5,5,8,8-pentamethyl-2-naphthalenyl)-ethanone	OLES
5,6,7,8-tetrahydro-4-methoxy-5,5,8,8-tetramethyl-2-naphthalenecarboxaldehyde	OLES
3-acetyl-5,6,7,8-tetrahydro-5,5,8,8-tetramethyl-2-naphthalenecarbonitrile	OLES
5,6,7,8-tetrahydro-3,5,5,8,8-pentamethyl-2-(3-methylbutoxymethyl)-naphthalene	OLES
1-(5,8-diethyl-5,6,7,8-tetrahydro-3,5,8-trimethyl-2-naphthalenyl)-ethanone	OLES
1-(3-ethyl-5,6,7,8-tetrahydro-8,8-dimethyl-2-naphthalenyl)-ethanone	OLES
1-[2,3-dihydro-1,3,3,6-tetramethyl-1-(2-methylpropyl)-1H-inden-5-yl]-ethanone	OLES
1-[5,6,7,8-tetrahydro-5,5,8,8-tetramethyl-3-(1-methylethyl)-2-naphthalenyl]-propanone	OLES
1-(5,8-dihydro-5,5,6,8,8-pentamethyl-2-naphthalenyl)-ethanone	OLES
1-(5,8-dihydro-5,5,6,8,8-pentamethyl-1-naphthalenyl)-ethanone	OLES
1-[5,6,7,8-tetrahydro-3,8,8-trimethyl-5-(1-methylethyl)-2-naphthalenyl)-ethanone	OLES
1-[5,6,7,8-tetrahydro-3,5,5-trimethyl-8-(1-methylethyl)-2-naphthalenyl)-ethanone	OLES
4-(5,6,7,8-tetrahydro-5,5,8,8-tetramethyl-2-naphthalenyl)-3-buten-2-one	OLES

Compound Name	***Odor Quality***
1,2,3,4-tetrahydro-5-methoxy-1,1,4,7,7-pentaethylnaphthalene	OLES
5,6,7,8-tetrahydro-1-methoxy-4,5,5,8,8-pentaethylnaphthalene	OLES
5,6,7,8-tetrahydro-1,3-dimethoxy-5,5,8,8-tetraethylnaphthalene	OLES
3-amino-5,6,7,8-tetrahydro-5,5,8,8-tetramethyl-2-naphthalenecarboxylic acid methyl ester	OLES
5,6,7,8-tetrahydro-5,5,8,8-tetramethyl-2-naphthalenol	OLES
5,6,7,8-tetrahydro-2-methoxy-5,5,8,8-tetramethylnaphthalene	OLES
2-ethoxy-5,6,7,8-tetrahydro-5,5,8,8-tetramethylnaphthalene	OLES
5,6,7,8-tetrahydro-3-methylbutoxy-5,5,8,8-tetramethylnaphthalene	OLES
5,6,7,8-tetrahydro-3,5,5,8,8-pentamethyl-2-naphthalenol	OLES
5,6,7,8-tetrahydro-2-methoxy-3,5,5,8,8-pentamethylnaphthalene	OLES
5,6,7,8-tetrahydro-2,3-dimethoxy-5,5,8,8-tetramethylnaphthalene	OLES
3-ethyl-5,6,7,8-tetrahydro-2-methoxy-5,5,8,8-tetramethylnaphthalene	OLES
2-methoxy-1-(5,6,7,8-tetrahydro-3-methoxy-5,5,8,8-tetramethyl-2-naphthalenyl)-ethanone	OLES
3-acetyl-5,6,7,8-tetrahydro-5,5,8,8-tetramethyl-2-naphthalenecarboxylic acid methyl ester	NONM[2]
1-[5,6,7,8-tetrahydro-3-(methoxymethyl)-5,5,8,8-tetramethyl-2-naphthalenyl]-ethanone	NONM
1-(5,6,7,8-tetrahydro-3-hydroxy -5,5,8,8-tetramethyl-2-naphthalenyl)-ethanone	NONM
5,6,7,8-tetrahydro-5,5,8,8-tetramethyl-2,3-naphthalenedicarboxylic acid diethyl ester	NONM
1-(3,5,8-triethyl-5,6,7,8-tetrahydro-5,8-dimethyl-2-naphthalenyl)-ethanone	NONM

Continued on next page.

Table 1. (Continued). Training Set Compounds

Compound Name	*Odor Quality*
1-{5,8-diethyl-5,6,7,8-tetrahydro-5,8-dimethyl-3-(1-methylethyl)-2-naphthalenyl]-ethanone	NONM
1-(5,6,7,8-tetrahydro-3,8,8-trimethyl-2-naphthalenyl)-ethanone	NONM
1-(3-fluoro-5,6,7,8-tetrahydro-5,5,8,8-tetramethyl-2-naphthalenyl)-ethanone	NONM
1-(3-chloro-5,6,7,8-tetrahydro-5,5,8,8-tetramethyl-2-naphthalenyl)-ethanone	NONM
1-(3-bromo-5,6,7,8-tetrahydro-5,5,8,8-tetramethyl-2-naphthalenyl)-ethanone	NONM
1-(3-iodo-5,6,7,8-tetrahydro-5,5,8,8-tetramethyl-2-naphthalenyl)-ethanone	NONM
1-[2,3-dihydro-1,1,6-trimethyl-3-(1-methylethyl)-1H-inden-5-yl]-ethanone	NONM
1-[6-ethyl-2,3-dihydro-1,1-dimethyl-3-(1-methylethyl)-1H-inden-5-yl]-ethanone	NONM
1-[2,6-diethyl-2,3-dihydro-1,1-dimethyl-3-(1-methylethyl)-1H-inden-5-yl]-ethanone	NONM
1-(2,3-dihydro-1,1,2,3,3-pentamethyl-1H-inden-5-yl)-1-propanol	MMUS[3]
1-[2,3-dihydro-1,1,3,6-tetramethyl-3-(1-methylethyl)-1H-inden-5-yl]-ethanone	MMUS
1-[3-(1,1-dimethylethyl)-2,3-dihydro-1,1,2,6-tetramethyl-1H-inden-5-yl]-ethanone	MSTR[4]
1-(2,3-dihydro-1,1,2,3,3,6-hexamethyl-1H-inden-5-yl)-ethanone	MSTR
1-[2,3-dihydro-1,1,2,2,6-pentamethyl-3-(1-methylethyl)-1H-inden-5-yl]-ethanone	MUSK[5]
1-[6-ethyl-3-dihydro-1,1,2,2-tetramethyl-3-(1-methylethyl)-1H-inden-5-yl]-ethanone	MUSK
1-[2,3-dihydro-1,1,2,6-tetramethyl-3-(1-methylethyl)-1H-inden-5-yl]-1-propanone	MUSK
1-ethyl-2,3-dihydro-1,3,3,6-tetramethyl-1H-indene-5-carboxyaldehyde	MSTR

Compound Name	*Odor Quality*
2,3-dihydro-1,1,2,3,3,6-hexamethyl-1H-indene-5-carboxaldehyde	MSTR
2,3-dihydro-1,1,3,3,6-pentamethyl-1H-indene-5-carboxyaldehyde	MSTR
5,6,7,8-tetrahydro-3,5,5,6,8,8-hexamethyl-2-naphthalenylcarboxaldehyde	MSTR
Trans-5,6,7,8-tetrahydro-3,5,5,6,7,8,8-heptamethyl-2-naphthalenyl-carboxaldehyde	MSTR
Cis-5,6,7,8-tetrahydro-3,5,5,6,7,8,8-heptamethyl-2-naphthalenyl-carboxaldehyde	MSTR
1-(2,3-dihydro-1,1,2,3,3,6-hexamethyl-1H-inden-5-yl)-ethanone ("Phantolid")	MSTR
1-(2,3-dihydro-1,1,2,3,6-pentamethyl-1H-inden-5-yl)-ethanone	MWEA[6]
1-(2-ethyl-2,3-dihydro-1,1,3,3,6-pentamethyl-1H-inden-5-yl)-ethanone	MMUS
1-(6-ethyl-2,3-dihydro-1,1,3,3-tetramethyl-1H-inden-5-yl)-ethanone	MMUS
1-(6-ethyl-2,3-dihydro-1,1,2,3,3-tetramethyl-1H-inden-5-yl)-ethanone	MMUS
1-(2,3-dihydro-1,1,3,3,5,6-hexamethyl-1H-inden-4-yl)-ethanone	MMUS
1-(2,3-dihydro-1,1,2,3,3,-pentamethyl-1H-inden-5-yl)-ethanone	MMUS
5,6,7,8-tetrahydro-5,5,8,8-tetramethyl-3-(1-methylethyl)-2-naphthalenecarboxaldehyde	MMUS
1-[2,3-dihydro-1,1,2,3,3-pentamethyl-6-(1-methylethyl)-1H-inden-5-yl]-ethanone	MUSK
1-[2,3-dihydro-1,1,3,3-pentamethyl-6-(1-methylethyl)-1H-inden-5-yl]-ethanone	MUSK
1-[2,3-dihydro-1,1,2,6-tetramethyl-3-(trimethylsilyl)-1H-inden-5-yl]-ethanone	MSTR
2,3-dihydro-1,1,2,6-tetramethyl-3-(1-methylethyl)-1H-indene-5-carboxaldehyde	MSTR
1-[2-ethyl-2,3-dihydro-1,1,6-trimethyl-3-(1-methylethyl)-1H-inden-5-yl]-ethanone	MSTR

Continued on next page.

Table 1. (Continued). Training Set Compounds

Compound Name	*Odor Quality*
1-[2,3-dihydro-1,1,2,6-tetramethyl-3-(1-methylethyl)-1H-inden-5-yl]-ethanone	MSTR
(5,6,7,8-tetrahydro-1,3,5,5,8,8-hexamethyl-2-naphthalenylcarboxaldehyde	MSTR
3-ethyl-5,6,7,8-tetrahydro-1-methoxy-5,5,8,8-tetramethyl-2-naphthalenecarboxaldehyde	MMED[7]
5,6,7,8-tetrahydro-1-methoxy-3,3,5,8,8-pentamethyl-2-naphthalenecarboxaldehyde	MSTR
5,6,7,8-tetrahydro-1-hydroxy-3,3,5,8,8-pentamethyl-2-naphthalenecarboxaldehyde	MSTR
5,6,7,8-tetrahydro-5,5,8,8-tetramethyl-2,3-naphthalenedicarboxaldehyde	MMED
5',6',7',8'-tetrahydro-5',5',8',8'-tetramethyl-3'-(methylthio)-2'-acetonaphthone	MUSK
5',6',7',8'-tetrahydro-5',5',6',8',8'-pentamethyl-3'-(methylthio)-2'-acetonaphthone	MUSK
5',6',7',8'-tetrahydro-5',5',6',7',8',8'-pentamethyl-3'-(methylthio)-2'-acetonaphthone	MUSK
3'-(ethylthio)-5',6',7',8'-tetrahydro-5',5',8',8'-tetramethyl-2'-acetonaphthone	MUSK
1-(3-ethyl-5,6,7,8-tetrahydro-5,5,8,8-tetramethyl-2-naphthalenyl)-propanone	MWEA
1-(3-ethyl-5,6,7,8-tetrahydro-5,5,8,8-tetramethyl-2-naphthalenyl)-ethanone ("Versalide")	MSTR
1-(5,6,7,8-tetrahydro-3,5,5,6,8,8-hexamethyl-2-naphthalenyl)-ethanone ("Tonalid")	MSTR
1-(5,6,7,8-tetrahydro-2-naphthalenyl)-ethanone	MSTR
5,6,7,8-tetrahydro-3,5,5,8,8-pentamethyl-2-naphthalenecarboxaldehyde	MSTR
3-ethyl-5,6,7,8-tetrahydro-5,5,8,8-tetramethyl-2-naphthalenecarboxaldehyde	MSTR
5,6,7,8-tetrahydro-1,3,5,5,7,8,8-heptamethyl-2-naphthalenecarboxaldehyde	MSTR

Compound Name	***Odor Quality***
Trans-1-(5,6,7,8-tetrahydro-3,5,5,6,7,8,8-heptamethyl-2-naphthalenyl)-ethanone	MSTR
1-(5,8-dihydro-5,5,8,8-tetramethyl-2-naphthalenyl)-ethanone	MMUS
1-(5,6,7,8-tetrahydro-5,5,8,8-tetramethyl-3-(1-methylethyl)-2-naphthalenyl)-ethanone	MWEA
1-(5-ethyl-5,6,7,8-tetrahydro-3,5,8,8-tetramethyl-2-naphthalenyl)-ethanone	MWEA
1-(5,6,7,8-tetrahydro-5,5,6,8,8-pentamethyl-2-naphthalenyl)-ethanone	MMUS
6-ethyl-2,6,8,8-tetramethyl-cyclopenta[*g*]-2-benzopyran	MMUS
1,3,4,6,7,8-hexahydro-6-(1-methylethyl)-4,7,8,8-tetramethyl-cyclopenta[*g*]-2-benzopyran	MMUS
1,3,4,6,7,8-hexahydro-8-(1-methylethyl)-4,6,6,7-tetramethyl-cyclopenta[*g*]-2-benzopyran	MMUS
3,4-dihydro-4-methyl-5,7-bis-(1-methylethyl)-1H-2-benzopyran	MMUS
2,3-dihydro-β-(1-methylethyl)-1H-Indene-5-ethanol	MMUS
4-ethyl-1,3,4,6,7,8-hexahydro-6,6,7,8,8-pentamethyl-cyclopenta[*g*]-2-benzopyran	MMXT[8]
4,6-diethyl-1,3,4,6,7,8-hexahydro-6,8,8-trimethyl-cyclopenta[*g*]-2-benzopyran	MMXT
3,4,6,7,8,9-hexahydro-6,6,8,9,9-pentamethyl-1H-naphtho[2,3-C]pyran	MSTR
3,4,6,7,8,9-hexahydro-6,6,7,9,9-pentamethyl-1H-naphtho[2,3-C]pyran	MSTR
1,2,3,4,6,7,8,9-octahydro-6,6,9,9-tetramethyl-benz[*g*]isoquinoline	MWEA
1,2,3,4,6,7,8,9-octahydro-2,6,6,9,9-pentamethyl-benz[*g*]isoquinoline	MWEA
1-(3-ethyl-2,3-dihydro-1,1,3-trimethyl-1H-inden-4-yl)-ethanone	MWEA
1-(5-ethyl-5,6,7,8-tetrahydro-3,5,8-trimethyl-2-naphthalenyl)-ethanone	MWEA

Continued on next page.

Table 1. (Continued). Training Set Compounds

Compound Name	*Odor Quality*
1-(1-ethyl-2,3-dihydro-1,3,3,5,6-pentamethyl-1H-inden-4-yl)-ethanone	MWEA
1-(3-ethyl-2,3-dihydro-1,1,3,5,6-pentamethyl-1H-inden-4-yl)-ethanone	MWEA
1-(1-ethyl-2,3-dihydro-1,3,3-trimethyl-1H-inden-4-yl)-ethanone	MWEA
1-(3-ethyl-2,3-dihydro-1,1,3,6-tetramethyl-1H-inden-5-yl)-ethanone	MWEA
1-(2,3-dihydro-1,1,3,3-tetramethyl-1H-inden-5-yl)-ethanone	MWEA
1-[3-(1,1-dimethylethyl)-5,6,7,8-tetrahydro-5,5-dimethyl-1-naphthalenyl]-ethanone	MWEA
1-(5,6,7,8-tetrahydro-4,5,5,8,8-pentamethyl-2-naphthalenyl)-ethanone	MWEA
1-(5,6,7,8-tetrahydro-4,5,5,8,8-pentamethyl)-2-naphthalenecarboxaldehyde	MWEA
1-(5,6,7,8-tetrahydro-1,3,4,5,5,8,8-heptamethyl)-2-naphthalenecarboxaldehyde	MWEA
1-(5,6,7,8-tetrahydro-1,4,5,5,8,8-hexamethyl-2-naphthalenyl)-carboxaldehyde	MWEA
1-(5,8-dihydro-3,5,5,8,8-pentamethyl-2-naphthalenyl)-ethanone	MWEA
1,2,3,4,6,7,8,9-octahydro-4,6,6,9,9-pentamethyl-benz[*g*]isoquinoline	MMED
6,9-diethyl-3,4,6,7,8,9-hexahydro-4,6,9-trimethyl-1H-naphtho [2,3-C]pyran	MWEA
1,3,4,6,7,8-hexahydro-4,6,6,7,8,8-hexamethyl-cyclopenta[*g*]-2-benzopyran	MSTR
3,4,6,7,8,9-hexahydro-6,6,9,9-tetramethyl-1H-naphtho[2,3-C]-pyran	MSTR
4-ethyl-3,4,6,7,8,9-hexahydro-6,6,9,9-tetramethyl-1H-naphtho[2,3-C]pyran	MSTR
5,6,7,8-tetrahydro-1,3,5,5,6,8,8-heptamethyl-2-naphthalenecarboxaldehyde	MSTR

Compound Name	***Odor Quality***
3,4,6,7,8,9-hexahydro-6,6,9,9,10-pentamethyl-1H-naphtho[2,3-C]pyran	MMED
3,4,6,7,8,9-hexahydro-6,6,7,8,9,9-hexamethyl-1H-naphtho[2,3-C]pyran	MMED
1,3,4,6,7,8-hexahydro-6,6,8,8-tetramethyl-cyclopenta[*g*]-2-benzopyran	MWEA
1,3,4,6,7,8-hexahydro-4,6,6,8,8-pentamethyl-cyclopenta[*g*]-2-benzopyran	MWEA
1,3,4,6,7,8-hexahydro-6,6,7,8,8-pentamethyl-cyclopenta[*g*]-2-benzopyran	MWEA
3,4,6,7,8,9-hexahydro-4,6,6,7,9,9-hexamethyl-1H-naphtho[2,3-C]-pyran	MMED
1,3,4,6,7,8-hexahydro-4,4,8,8-tetrahydro-cyclopenta[*g*]-2-benzopyran	OLES
6,8-diethyl-1,3,4,7-tetrahydro-6,8-dimethyl-cyclopenta[*g*]-2-benzopyran	OLES
Dodecahydro-6,6,9,9-tetrahydro-1H-naphtho[2,3-C]pyran	OLES
3,4,6,7,8,9-hexahydro-1,6,6,9,9-pentamethyl-1H-naphtho[2,3-C]pyran	OLES
3,4,6,7,8,9-hexahydro-3,6,6,9,9-pentamethyl-1H-naphtho[2,3-C]pyran	OLES
1,3,5,6,7,8-hexahydro-5,5,8,8-tetramethyl-naphtho[2,3-C]thiophene	OLES
1,3,5,6,7,8-hexahydro-5,5,8,8-tetramethyl-naphtho[2,3-C]thiophene-2-oxide	OLES
6,7,8,9-tetrahydro-6,6,9,9-tetramethyl-benzo[*g*]-phthalazine	OLES
2,3,7,8-tetraahydro-4,4,6,6,8,8-hexamethyl-cyclopenta[*g*/-1-benzopyran	OLES
6,9-diethyl-3,4,7,8-tetrahydro-6,9-dimethyl-1H-naphtho[2,3-C]pyran	OLES
3,4,6,7,8,9-hexahydro-6,6,9,9-tetramethyl-2H-naphtho[2,3-B]pyran	NONM
2,3,4,7-tetrahydro-4,6,6,7,8,8-hexamethyl-cyclopenta[*g*]-1-benzopyran	NONM

Continued on next page.

Table 1. (Continued). Training Set Compounds

Compound Name	*Odor Quality*
1-(1,2,3,4,5,6,7,8-octahydro-2,3,8,8-tetramethyl-2-naphthalenyl)-ethanone	AMBER[9]
1-(1,2,3,4,5,6,7,8-octahydro-3,8,8-trimethyl-2-naphthalenyl)-ethanone	WOODY[10]
1-(1,2,3,5,6,7,8,8a-octahydro-4,8,8-trimethyl-2-naphthalenyl)-ethanone	WOODY
2,3-dihydro-β,1,1,2,3,3-hexamethyl-1H-indene-5-ethanol	WOODY
1-(1,2,3,4,5,6,7,8-octahydro-4,8,8-trimethyl-2-naphthalenyl)-ethanone	WOODY
1-[5,6,7,8-tetrahydro-5,5-dimethyl-3-(1-methylethyl)-2-naphthalenyl)-ethanone	WOODY
1-(1,2,3,4,6,7,8,8a-octahydro-4,8,8-trimethyl-2-naphthalenyl)-ethanone	WOODY
1-(3-ethyl-5,6,7,8-tetrahydro-5,5,6,8,8-pentamethyl-2-naphthalenyl)-ethanone	MSTR
5,6,7,8-tetrahydro-5,5,8,8-tetramethyl-2,3-naphthalenedicarboxylic acid dimethyl ester	NONM
5,6,7,8-tetrahydro-5,5,8,8-tetramethyl-2,3-naphthalenedimethanol	MWEA
1-(1-ethyl-2,3-dihydro-1,6-dimethyl-1H-inden-5-yl)-ethanone	MWEA
1-(5,6,7,8-tetrahydro-5,5,8,8-tetramethyl-2-naphthalenyl)-ethanone	NONM
1-(2,3-dihydro-1,1,2,3,3-pentamethyl-1H-inden-5-yl)-1-propanone	MMUS
1-(1,2,6,7,8,8a-hexahydro-3,6,6,8a-tetramethyl-4-acenaphthalenyl)-ethanone	MSTR
1,2,6,7,8,8a-hexahydro-3,6,6,8a-tetramethyl-4-acenaphthalenecarboxaldehyde	MSTR

Compound Name	*Odor Quality*
(4,5,5a,6,7,8-hexahydro-1,4,4,6-tetramethyl-2-acenaphthalenyl)-carboxaldehyde	MWEA
1-(2,3,6,7,8,9-hexahydro-1,1-dimethyl-1H-benz[e]inden-4-yl)-ethanone	MSTR

[1] OLES: Odorless. [2] NONM: Nonmusk. [3] MMUS: Moderately Strong Musk. [4] MSTR: Strong Musk. [5] MUSK: Musk of Unspecified Odor Intensity. [6] MWEA: Weak Musk. [7] MMED: Medium Strength Musk. [8] MMXT: Mixture of Musk with another Odor. [9] AMBER: Amber odor. [10] WOODY: Woody odor.

Table 2. Validation Set Compounds

Compound Name	*Odor Quality*
1-(5,6,7,8-tetrahydro-1,3,5,5,8,8-hexamethyl-2-naphthalenyl)-ethanone	OLES
1-(5,6,7,8-tetrahydro-1,3,5,5,6,8,8-heptamethyl-2-naphthalenyl)-ethanone	OLES
2-methyl-1-(5,6,7,8-tetrahydro-5,5,8,8-tetramethyl-2-naphthalenyl)-1-propanone	OLES
1,3,3-trimethyl-1-propyl-1H-indene	NONM
5,6,7,8-tetrahydro-2-methoxymethyl-3,5,5,8,8-pentamethylnaphthalene	OLES
5,6,7,8-tetrahydro-1,3-dimethoxy-5,5,8,8-tetramethylnaphthalene	OLES
1-(5,8-dihydro-5,5,7,8,8-pentamethyl-2-naphthalenyl)-ethanone	OLES
1,3,5,6,7,8-hexahydro-5,5,8,8-tetramethyl-naphtho[2,3-C]furan	MSTR
5,6,7,8-tetrahydro-5,5,8,8-tetramethyl-2,3-naphthalenedicarboxylic acid	NONM
6-acetyl-3,4-dihydro-1,1,4,4,8-pentamethyl-2(1H)-naphthalenone	OLES
1-[6-ethyl-2,3-dihydro-1,1,2,2-tetramethyl-3-(1-methylethyl)-1H-inden-5-yl]-ethanone	MUSK
6-acetyl-3,4-dihydro-1,1,4,4,5-pentamethyl-2(1H)-naphthalenone	OLES
3,4-dihydro-4-methyl-6,8-bis-(1-methylethyl)-1H-2-benzopyran	MSTR
2,3-dihydro-1,1,3,3,6-pentamethyl-1H-Indene-5-carboxaldehyde	MSTR
3,4,6,7,8,9-hexahydro-4,4,6,6,9,9-hexamethyl-1H-naphtho[2,3-C]pyran	MMED
3,4,6,7,8,9-hexahydro-4,6,6,9,9-pentamethyl-1H-naphtho[2,3-C]pyran	MSTR
1-(5,6,7,8-tetrahydro-3-methoxy-5,5,8,8-tetramethyl-2-naphthalenyl)-ethanone	OLES

Compound Name	***Odor Quality***
1-(5,6,7,8-tetrahydro-3,5,8,8-tetramethyl-2-naphthalenyl)-ethanone	MMED
1-(2,3-dihydro-1,1,2,3,3,5,6-heptamethyl-1H-inden-4-yl)-ethanone	MWEA

Molecular Descriptor Selection

The basic premise underlying our approach to OSR development highlighted in this study is that all data analysis methods will work well when the data analysis problem is simple. By identifying the appropriate molecular descriptors, a "hard" problem can be reduced to a "simple" one. Therefore, our goal is variable selection. To ensure identification of all relevant molecular descriptors, it is best that a multivariate approach to variable selection be employed. The approach must take into account the existence of redundancies in the data.

In this study a GA for pattern recognition analysis was used to identify molecular descriptors from which discriminating relationships can be found. The pattern recognition GA selects descriptors that increase clustering while simultaneously searching for descriptors that optimize the separation of the classes in a plot of the two or three largest principal components of the data. The principal component analysis routine embedded in the fitness function of the pattern recognition GA acts as an information filter, significantly reducing the size of the search space, since it restricts the search to descriptors whose principal component plots show clustering on the basis of the class label of the compound. In addition, the algorithm focuses on those compounds and/or odorant classes that are difficult to classify as it is training by boosting the class and compound weights. Compounds that consistently classify correctly are not as heavily weighted as compounds that are more difficult to classify. Over time, the algorithm learns its optimal parameters in a manner similar to a neural network.

The fitness function of the pattern recognition GA emulates human pattern recognition through machine learning to score the principal component plots and thereby identify descriptors that optimize the separation of the odorant classes in a plot of the two or three largest principal components of the data. To facilitate the tracking and scoring of the principal component plots, class and compound weights, which are an integral part of the fitness function, are computed (see Equations 1 and 2) where CW(c) is the weight of class c (with c varying from 1 to the total number of classes in the dataset), and $SW_c(s)$ is the weight of compound s in class c. The class weights sum to 100, and the compound weights for the molecules comprising a particular class sum to a value equal to the class weight of the class in question.

$$\mathrm{CW(c)} = 100\frac{\mathrm{CW(c)}}{\sum_{\mathrm{c}}\mathrm{CW(c)}} \tag{1}$$

$$\mathrm{SW}(s) = \mathrm{CW(c)}\frac{\mathrm{SW}(s)}{\sum_{s\in\mathrm{c}}\mathrm{SW}(s)} \tag{2}$$

Each principal component plot generated for each subset of descriptors after it has been extracted from its chromosome is scored using the K-nearest neighbor classification algorithm (*36*). For each compound in the training set, Euclidean distances are computed between it and every other compound (which are represented as data points) in the principal component plot. These distances are arranged from smallest to largest. A poll is then taken of the compound's K_c nearest neighbors. For the most rigorous classification of the data, K_c equals the number of compounds in the class to which the compound is a member. (K_c often has a different value for each class due to differences in class size.) The number of K_c nearest neighbors with the same class label as the data point (i.e., compound) in question, the so-called sample hit count, SHC(s), is computed ($0 \leq \mathrm{SHC(s)} \leq K_c$) for each compound. It is then a simple matter to score a principal component plot (see Equation 3). First, the contribution to the overall fitness by each compound in class 1 is computed, with the scores of the compounds comprising the class summed to yield the contribution by this class to the overall fitness. This same calculation is repeated for the other classes with the scores from each class summed to yield the overall fitness, *F (d)*.

$$F(d) = \sum_{c} \sum_{s \in c} \frac{1}{K_c} \times \mathrm{SHC(s)} \times \mathrm{SW(s)} \qquad (3)$$

To better understand the scoring of the principal component plots, consider the following binary classification problem. Class 1 has 10 compounds, and class 2 has 20 compounds. Each class has been assigned equal weights. For uniformly distributed compound weights, each compound from class 1 has a weight of 5 and each compound from class 2 has a weight of 2.5, since each class has a weight of 50 and the sum of the sample weights in each class must be equal to the class weight. If a compound in class 1 has, as its nearest neighbors, seven class 1 compounds in a principal component plot developed from a particular subset of molecular descriptors, then $\mathrm{SHC}(c)/K_c = 7/10$, and the contribution of this compound to the fitness score equals 0.7×5 or 3.5 for this set of molecular descriptors. Multiplying SHC/K_c by $\mathrm{SW}(s)$ for each compound and then summing up the corresponding product terms (see Equation 3) for the 30 compounds in the dataset yields the fitness score for this particular set of descriptors.

The fitness function of the pattern recognition GA is able to focus on specific compounds and classes that are difficult to classify by boosting their weights over successive generations. In order to boost the weights, it is necessary to compute both the sample-hit rate (SHR), which is the mean value of SHC/K_c over all descriptor subsets produced in a particular generation (see equation 4), and the class-hit rate (CHR), which is the mean sample hit rate of all samples in a class (see equation 5).

$$\mathrm{SHR(s)} = \frac{1}{\phi} \sum_{i=1}^{\phi} \frac{\mathrm{SHC}_i(s)}{K_c} \qquad (4)$$

$$\mathrm{CHR_g(c)} = AVG(\mathrm{SHR_g(s)} : \forall_{s \in c}) \qquad (5)$$

φ in equation 5 is the number of chromosomes in the population, and AVG in equation 5 refers to the average or mean value. During each generation, class and sample weights are adjusted by a perceptron (see Equations 6 and 7) with the momentum, P, set by the user and with g + 1 being the current generation and g being the previous generation. Classes with a lower class hit rate are boosted more heavily than those classes that score well.

$$CW_{g+1}(s) = CW_g(s) + P(1 - CHR_g(s)) \tag{6}$$

$$SW_{g+1}(s) = SW_g(s) + P(1 - SHR_g(s)) \tag{7}$$

Boosting is crucial to the successful operation of the pattern recognition GA as it modifies the fitness landscape by adjusting the values of both the class and compound weights. This helps to mitigate the problem of premature convergence. Hence, the fitness function of the pattern recognition GA changes as the population is evolving towards a solution.

Results and Discussion

The first step in the study was to apply principal component analysis to the training set data. Principal component analysis is a method for transforming the original descriptors into new uncorrelated variables called principal components. Each principal component is a linear combination of the original molecular descriptors. Using this procedure is analogous to finding a new coordinate system that is better at conveying the information present in the data than axes defined by the original molecular descriptors. Often, only two or three principal components are necessary to explain all of the information present in a dataset if there are a large number of interrelated (i.e., correlated) descriptors. Using principal component analysis, dimensionality reduction, classification, and identification of clusters in high dimensional data are possible.

Figure 2 shows the results of a principal component mapping for the 168 compounds and 1369 molecular descriptors comprising the training set (see Table 1). The principal component plot explains 28.5% of the total cumulative variance of the data. Each compound in the training set is represented by a point in the plot. The 1's are nonmusks, the 4's are indan and tetralin musks, and the 7's are isochroman musks. The overlap of the musks and nonmusks in the principal component plot of the 1369 molecular descriptors is evident and is not surprising in view of the similarity of their chemical structure.

The pattern recognition GA was used to identify molecular descriptors correlated to musk odor quality for the tetralin, indan, and isochroman musks. Key descriptors were identified by sampling key feature subsets and scoring their principal component plots, while tracking those classes and compounds that were difficult to classify. The pattern recognition GA identified 20 molecular descriptors whose principal component plot (see Figure 3) showed clustering of the compounds in the training set on the basis of odor.

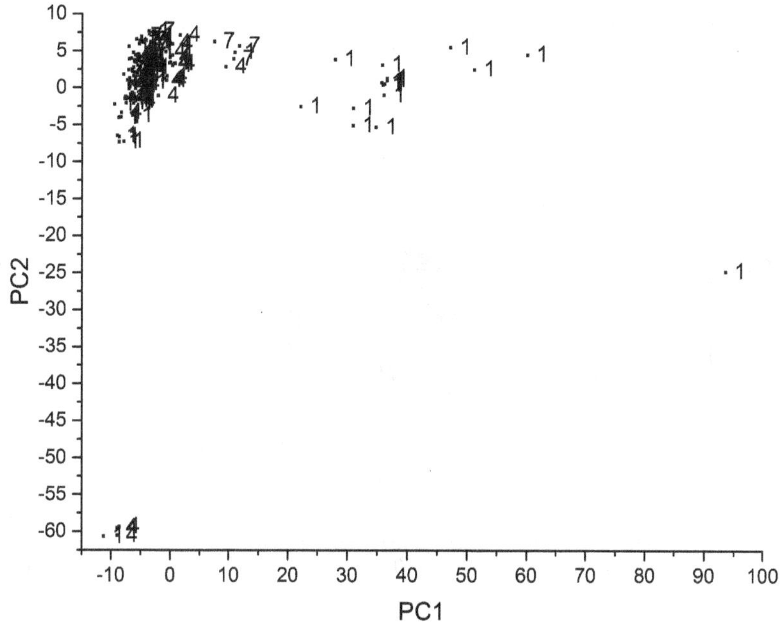

Figure 2. Principal component plot of the 168 training set compounds and the 1369 molecular descriptors. 1 = nonmusks, 4 = indan and tetralin musks, and 7 = isochroman musks.

To identify informative descriptors, it was necessary to configure the pattern recognition GA in the asymmetric classification mode (*37–39*). The tetralin and indan musks represented as 4's in the principal component plot of the high dimensional measurement space spanned by the 20 descriptors occupy a small but well-defined region and the isochroman musks (represented as 7's) occupy another small but well-defined region in the same principal component plot, whereas the nonmusks are randomly distributed in the plot. The two distinct clusters of compounds in the principal component plot strongly suggest that isochroman musks possess a unique odor that differs from tetralin and indan musks. Interestingly enough, a trained perfumer can differentiate the unique odor associated with each of the eight structural classes of musks.

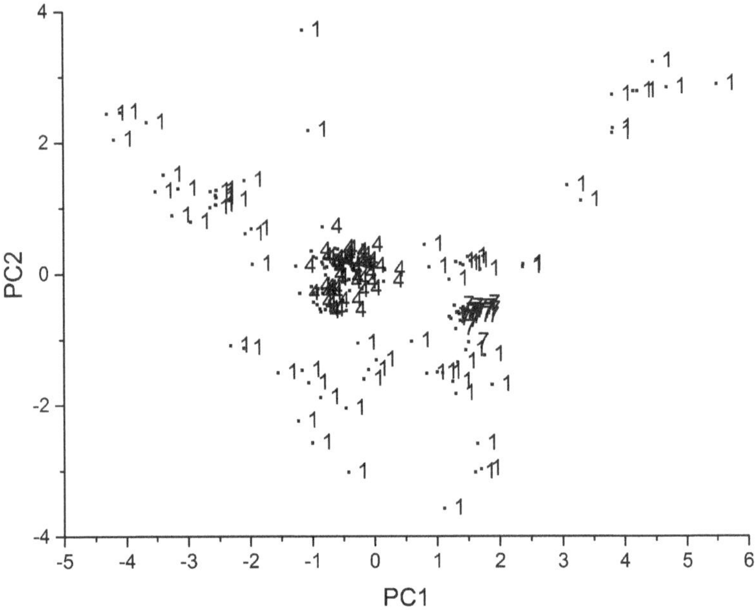

Figure 3. Principal component plot of the 168 training set compounds and the 20 molecular descriptors identified by the pattern recognition GA. 1 = nonmusks, 4 = indans and tetralins, and 7 = isochromans.

The predictive ability of the 20 molecular descriptors identified by the pattern recognition GA was assessed using a validation set consisting of 19 compounds (see Table 2). 11 of these compounds were nonmusks, two were indan musks, two were tetralin musks and four were isochroman musks. Of the 11 nonmusks, nine were odorless and two had an odor other than musk. The 19 musks and nonmusks were mapped onto the principal component plot defined by the 168 compounds and the 20 molecular descriptors. Figure 4 shows the projection of the 19 validation set compounds onto the principal component plot developed from the training set samples and the 20 descriptors identified by the pattern recognition GA. All validation set samples were correctly classified, i.e., they lie in a region of the map with compounds that have the same class label. For this mapping, the validation set samples are designated as N (nonmusk), M for tetralin and indan musks, and S for isochroman musks. Tetralin and indan musks (M) in the validation set lie in a region of the map with tetralin and indan musks from the training set and isochroman musks from the validation set (S) lie in a region of the map containing isochroman musks from the training set.

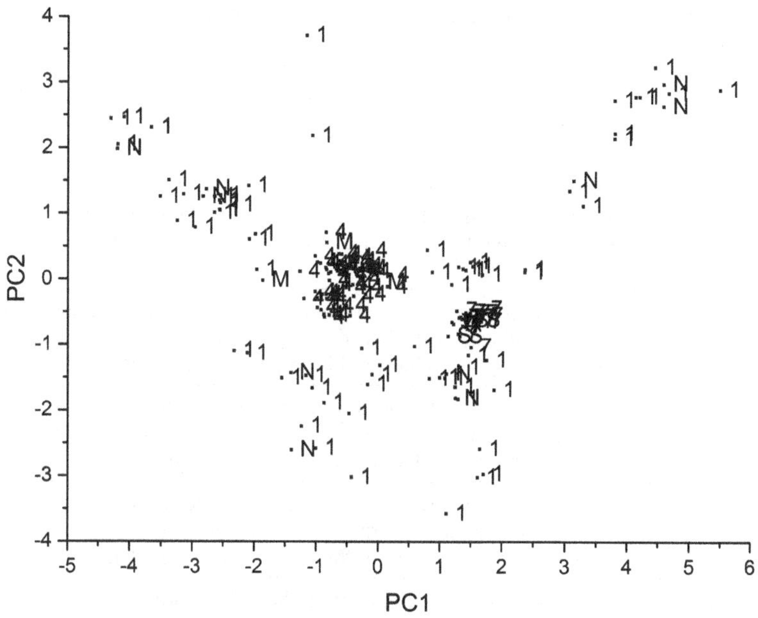

Figure 4. Projection of the 19 validation set samples onto the principal component plot defined by the 168 compounds in the training set and the 20 descriptors identified by the pattern recognition GA. 1 = nonmusk (training set), 4 = indan and tetralin musks (training set), 7 = isochroman musks (training set), N = nonmusk (validation set), M = indan or tetralin musk (validation set), and S = isochroman musk (validation set).

The 20 molecular descriptors identified by the pattern recognition GA are listed in Table 3. Nine of the 20 descriptors contain information about molecular shape. Seven of the nine shape descriptors are RECON and PEST descriptors whereas the other two shape descriptors are CODESSA descriptors. The two shape CODESSA descriptors, Moment Product AB and YZ shadow, favor planar disk shaped molecules. The seven RECON and PEST descriptors selected by the pattern recognition GA capture more of the interior volume (i.e., local shape) as opposed to conformational information. BNP.B40 contains information about shape and polarity, whereas both PIP.B20 and PIP.B21 contains information about molecular shape and hydrogen bonding. FUK.B02 is similar to the two Politzer ionization potential (PIP) shape descriptors as it has information about the local ionization potential of the molecule. ANGLE.B66 is a shape descriptor that also favors planar disk shape molecules. DKN.B10 is derived from ray traces of the rate of change of the K kinetic energy density on the molecular surface. The information conveyed by DKN.B10 suggests that it is crucial for areas with a low rate of change in the K kinetic energy to be relatively far apart in the molecule. DGN.B32, which is a PEST descriptor derived from ray traces of the G kinetic energy normal to and away from the surface of the molecule, suggests that dispersion interactions are important.

Table 3. Descriptors Selected by Pattern Recognition GA

DKN.B10	Shape descriptor developed from the rate of change of the k kinetic energy. The one and zero indicates that both short ray lengths and low values of the k kinetic energy density normal to and away from the surface of the molecules are important.
DGN.B32	Shape descriptor developed from the rate of change of the G kinetic energy density. The "3" indicates that intermediate rays are important, and the "2" indicates that low values of this property are important.
PIP.B20	Shape descriptor developed from the Politzer ionization potential. The "2" indicates that short ray lengths are important and the "0" indicates that low values of this property are important.
PIP.B21	Shape descriptor developed from the Politzer ionization potential. The "2" indicates that short ray lengths are important and the "1" indicates that low values of this property are important.
BNP.B40	Shape descriptor developed from the bare nuclear potential. The "4" indicates that rays of intermediate length are important and the "0" indicates that low values of this property are important.
FUK.B02	Shape descriptor developed from the Fukui Radical Reactivity Indices, which are related to the Politzer ionization potential. The "0" indicates that short rays are important and the "2" indicates that low values of the Fukui indices are important.
ANGLE.B66	Pure shape descriptor
Moment Product AB	Shape descriptor developed from the product of the first and second principal moments of inertia
YZ Shadow	Shape descriptor developed from the rectangle encompassing the area defined by the projection of the molecule onto its YZ plane
# S Atoms	Fragment descriptor denoting the number of sulfur atoms in the molecule
DKN.W13	Detail coefficient wavelet descriptor ("W") emphasizing the importance of the rate of fall off of the electron density
DKN.STDN	Describes the rate of change of the k kinetic energy density normal to and away from the surface of the molecule
DGN.W17	Detail coefficient wavelet descriptor ("W") describing the rate of change of the G kinetic energy density normal to and away from the surface.
DGN.W24	Detail coefficient wavelet descriptor ("W") describing the rate of change of the G kinetic energy density normal to and away from the surface.
G.W18	Detail coefficient wavelet descriptor ("W") emphasizing the importance of the electronic kinetic energy density G.
K.W19	Detail coefficient wavelet descriptor ("W") describing the k kinetic energy reconstruction normal to and away from the surface of the molecule.

Continued on next page.

Table 3. (Continued). Descriptors Selected by Pattern Recognition GA

EP.AVGP	Wavelet descriptor ("W") describing the electrostatic potential on the surface of the molecule
PIP.STD	Wavelet descriptor ("W") for the Politzer Ionization Potential
FUK.W17	Detail coefficient wavelet descriptor ("W") for Fukui Radical Reactivity Indices
FUK.W28	Detail coefficient wavelet descriptor ("W") for Fukui Radical Reactivity Indices

Molecules with sulfur atoms have interactions that are different from those of the other molecules in the dataset and this is denoted by the fragment-based descriptor # S Atoms. The remaining 10 molecular descriptors selected by the pattern recognition GA are TAE surface property wavelet descriptors. DKN.W13 and DKN.STDN convey information about hydrophobicity and polarizability of the molecule. DGN.W18 and DGN.W24 are correlated to weak bonding interactions and probably describe some facet of the interaction between the musk and the receptor. Both G.W18 and K.W19 describe hydrogen bonding interactions. EP.AVG is correlated to the solvation energy of the molecule, whereas PIP.STD conveys information about the local ionization potential of the molecule. FUK.W17 and FUK.W28 describe the high value range of the Fukui Radical Reactivity Index, which is related to the PIP index.

According to current theories of olfaction (*12*), perception and recognition of odor is mediated by G-protein coupled transmembrane protein receptors located on the surface of cilia cells at the olfactory epithelium. An examination of the molecular descriptors identified by the pattern recognition GA indicates that volatility and transport are not important factors for differentiating musks from nonmusks. This result, which conflicts with the theory that diffusion rates through the mucous membrane is a major determinant of odor, can be explained by the fact that nonmusks selected for this study are similar in structure to the musks. The molecular descriptors selected by the pattern recognition GA indicate that receptor binding and activation appear to be the factors for discrimination of musks from nonmusks in the training set. This conclusion is further supported by the asymmetric data structure obtained for the training set compounds in this study.

Conclusions

The results of this study demonstrate the advantages of using the genetic algorithm for variable selection and odor classification. Utilizing a machine learning perspective, subtle patterns within descriptor fields have been identified to develop odor structure relationships. The advantages of using transferrable atom based descriptors to develop structure property relationships have also been demonstrated in this study.

References

1. Jennings-White, C. *Perfum. Flavor.* **1985**, *9*, 46–58.
2. Turin, L. *The Secret of Scent*; Harper Perennial: New York, 2006
3. Pybus, D. H.; Sell, C. S. *The Chemistry of Fragrances*; RSC: Cambridge, U.K., 1999
4. Beets, M. J. *Structure-Activity Relationships in Human Chemoreception*; Applied Science Publishers: London, 1978.
5. Diete, C. *Manufacture of Perfumery*; Henry Carey & Baird Company: Philadelphia, PA, 1982.
6. Ruzicka, L. *Helv. Chim. Acta.* **1926**, *715*, 1008.
7. Bauer, A. *Ber. Dtsch. Chem. Ges.*, *466*, 2832–2843.
8. Wright, R. *J. Theor. Biol.* **1977**, *64*, 473–502.
9. Beets, M. *Molecular Structure and Organoleptic Quality*; MacMillan: New York, 1957.
10. Theimer, E.; Davies, J. *J. Agric. Food Chem.* **1967**, *15*, 6–14.
11. Amoore, J. *Molecular Basis of Odor*; Charles C. Thomas: Springfield, IL, 1970.
12. Axel, R. *Sci. Am.* **1995**, 154–159.
13. Brugger, W.; Jurs, P. *J. Agric. Food Chem.* **1977**, *25*, 1158–1164.
14. Ham, C.; Jurs, P. *Chem. Senses* **1985**, *10*, 491–505.
15. Narvaez, J.; Lavine, B.; Jurs, P. *Chem. Senses* **1986**, *11*, 145–156.
16. Klopman, G.; Ptscelintsev, D. *J. Agric. Food Chem.* **1992**, *40*, 2244–2251.
17. Charastrette, M.; Zakarya, D.; Peyraud, J. *Eur. J. Med. Chem.* **1994**, *29*, 343–348.
18. Cherqaoui, D.; Esseffar, M.; Villemin, D.; Cense, J.; Chastrette, M.; Zakarya, D. *New J. Chem.* **1998**, *22*, 839–843.
19. Breneman, C.; Thompson, T.; Rhem, M.; Dung, M. *Comput. Chem.* **1995**, *19*, 161–172.
20. Breneman, C.; Sundling, C.; Sukumar, N.; Shen, L.; Katt, W.; Embrechts, M. *J. Comput.-Aided Mol. Des.* **2003**, *17*, 231–240.
21. Whitehead, C. E.; Breneman, C. M.; Sukumar, N.; Ryan, M. D. *J. Comput. Chem.* **1997**, *18*, 182–197.
22. Lavine, B.; Davidson, C.; Breneman, C.; Katt, W. *J. Chem. Inf. Comput. Sci.* **2003**, *43*, 1890–1905.
23. Katritzky, A. R.; Perumal, S.; Petrukhin, R.; Kleinpeter, E. *J. Chem. Inf. Comput. Sci.* **2001**, *41*, 569–578.
24. Lavine, B. K.; Davidson, C. E.; Rayens, W. T. *Comb. Chem. High Throughput Screening* **2004**, *7*, 115–131.
25. Lavine, B. K.; Davidson, C. E.; Breneman, C. M.; Katt, W. In *Chemoinformatics – Concepts, Methods, and Tools for Drug Discovery*; Bajorath, J., Ed.; Humana Press: New York, 2004; Vol. 275, pp 399–426.
26. Lavine, B. K.; Nuguru, K; Mirjankar, N. *J. Chemom.* **2011**, *25*, 116–129.
27. Karasinski, J.; Andreescu, S.; Sadik, O. A.; Lavine, B. K.; Vora, M. N. *Anal. Chem.* **2005**, *77*, 7941–7949.
28. Wood, T. F. *Chemistry of the Aromatic Musks*; Givaudanian: Clifton, NJ, 1970; pp 1–37.

29. Tenahsi, R. In *Gustation and Olfaction*; Ohloff, G., Thomas, A. F., Eds.; Academic Press: London, 1971
30. Beets, M. G. J. In *Structure Activity Relationships*; Cavalitto, C. J., Ed.; Pergamum Press: New York, 1973.
31. Beets, M. G. J. In *Contribution of Chemistry to Food Supplies*; Morton, I., Rhodes, D. N., Eds.; Butterworth: London, 1974; pp 99–152.
32. *Fragrance Chemistry, Science of the Sense of Smell*; Theimer, E. T., Ed.; Academic Press: New York, 1982
33. Fehr, C.; Galindo, J; Haubrichs, R.; Perret, R. *Helv. Chim. Acta.* **1989**, *72*, 1537–1553.
34. Bersuker, I. B.; Dimoglo, A. S.; Gorbachov, M. Y; Vlad, P. F.; Pesaro, M. *New J. Chem.* **1991**, *15*, 307–320.
35. Ohloff, G. *Scent and Fragrances: The Fascination of Odors and their Chemical Perspectives*; Springer Verlag: New York, 1994.
36. James, M. *Classification*; John Wiley & Sons: New York, 1992.
37. Dunn, W. J.; Wold, S. *J. Med. Chem.* **1980**, *23*, 595–599.
38. Dunn, W. J.; Wold, S. In *Chemometric Methods in Molecular Design*; van de Waterbeemd, H., Ed.; VCH: New York, 1995.
39. Rose, V. S.; Wood, J.; MacFie, H. J. H. *QSAR* **1992**, *11*, 492–504.

Editor's Biographies

Brian Guthrie

Brian Guthrie currently performs research to understand the chemical and physical origins of human sensations during the oral processing of food. He has worked in the food and ingredients industries with responsibilities spanning from knowledge building, utilizing fundamental science, to formulation and product development. Brian has also worked extensively in food sensory science, from studies on the peripheral cellular events of olfaction and gustatory signal transduction to developing the understanding of consumer preference and choice. He has also been involved exploring the use of sensor technology in food and ingredient applications.

Jonathan Beauchamp

Jonathan Beauchamp is a physicist (MSci, University College London, U.K., 2002) with expertise in the detection of volatile organic compounds (VOCs) – including odorants – primarily using proton-transfer-reaction mass spectrometry (PTR-MS). After completing his PhD in environmental physics (University of Innsbruck, Austria, 2005), Jonathan worked for a small private company on R&D of instrumentation for VOC detection, specifically in the medical and bio-pharmaceutical sciences, principally breath gas analysis. In 2008 Jonathan relocated to Germany, joining the Department of Sensory Analytics at Fraunhofer IVV in Freising, where he is currently a research associate and deputy head of department. His research interests and activities focus on olfactory detection and perception of aromas, as well as the temporal development and release of volatiles from food, both desired and undesired. Jonathan also maintains an active role in the breath gas analysis community and is a member of the editorial board of *Journal of Breath Research.*

Andrea Buettner

Dr. Buettner studied Food Chemistry and is Professor of Aroma Research. Her work has demonstrated the importance of the combined effects of the food matrix, physiology and behavior on flavor release and perception. She has identified new aroma compounds and elucidated structure-odor relationships. Some of her experimental methods include chemical trace analysis, quantification via isotope dilution analysis, monitoring the physiology during eating, sensory psychophysics, and determination of odor-active compounds in vivo. At the University of Erlangen and Fraunhofer IVV, Freising, Dr. Buettner has broadened her research interests to include the field of odorants in the physiological context, with monitoring of uptake, distribution and biotransformation of odorants, as well as further physiological impact in humans.

Barry K. Levine

Dr. Barry K. Lavine is a Professor of Chemistry at Oklahoma State University in Stillwater, OK. He has published more than 100 papers in chemometrics and is on the editorial board of several journals including the Journal of Chemometrics and the Microchemical Journal. He is the Assistant Editor of Chemometrics for Analytical Letters. Lavine s research interests encompass odor structure relationships and many aspects of the applications of computers in chemical analysis including spectral library matching, pattern recognition, and multivariate curve resolution. His laboratory is a leader in the field of evolutionary computations, and the application of pattern recognition methods to the forensic examination of automotive paints using infrared and Raman imaging techniques.

Indexes

Author Index

Subject Index

A

B

C

E

F

N

O

P

S

T

U

W